AF411419

Advanced Energy Harvesting Technologies

Advanced Energy Harvesting Technologies

Editor

Dibin Zhu

MDPI • Basel • Beijing • Wuhan • Barcelona • Belgrade • Manchester • Tokyo • Cluj • Tianjin

Editor
Dibin Zhu
Shanghai Jiao Tong University
China

Editorial Office
MDPI
St. Alban-Anlage 66
4052 Basel, Switzerland

This is a reprint of articles from the Special Issue published online in the open access journal *Energies* (ISSN 1996-1073) (available at: https://www.mdpi.com/journal/energies/special_issues/energy_harvesting_technologies).

For citation purposes, cite each article independently as indicated on the article page online and as indicated below:

LastName, A.A.; LastName, B.B.; LastName, C.C. Article Title. *Journal Name* **Year**, *Volume Number*, Page Range.

ISBN 978-3-0365-3829-7 (Hbk)
ISBN 978-3-0365-3830-3 (PDF)

Contents

About the Editor

Dibin Zhu is an Associate Professor at Shanghai Jiao Tong University, China and an Honorary Associate Professor at the University of Exeter, UK. He received a BEng in Information and Control Engineering from Shanghai Jiao Tong University, and an MSc in RF Communication Systems and a PhD in Electrical and Electronic Engineering, both from the University of Southampton, in 2004, 2006 and 2009, respectively. He then worked as a Research Fellow, and later as a Senior Research Fellow, at Southampton. He became an academic from 2015 and was a Lecturer at Coventry University and Senior Lecturer at the University of Exeter. In January 2022, he joined Shanghai Jiao Tong University in China as an Associate Professor. His research interests include energy harvesting, wireless power transfer, and self-powered systems as well as advanced materials for energy-harvesting applications. He has been General Chair or co-Chair in several international conferences, including IEEE PowerMEMS 2021. He has also served as a reviewer for various prestigious international journals and conferences, as well as funding bodies. He has been a Guest Editor for several Special Issues in energy harvesting. He is currently an Associate Editor of IEEE Open Journal of Ultrasonics, Ferroelectrics, and Frequency Control. He has over 80 publications, with a total citation rate of over 2400 and an H-factor of 23.

Editorial

Advance Energy Harvesting Technologies

Dibin Zhu

School of Electronic Information and Electrical Engineering, Shanghai Jiao Tong University, Shanghai 200240, China; d.zhu@sjtu.edu.cn

Citation: Zhu, D. Advance Energy Harvesting Technologies. *Energies* **2022**, *15*, 2366. https://doi.org/10.3390/en15072366

Received: 22 March 2022
Accepted: 23 March 2022
Published: 24 March 2022

Publisher's Note: MDPI stays neutral with regard to jurisdictional claims in published maps and institutional affiliations.

Energy harvesting is the conversion of unused or wasted energy in the ambient environment into useful electrical energy. It can be used to power small electronic systems such as wireless sensors and is beginning to enable the widespread and maintenance-free deployment of Internet of Things (IoT) technology. This Special Issue is a collection of the latest developments in both fundamental research and system level integration.

This Special Issue features two review papers covering two of the hottest research topics in the area of energy harvesting: 3D printed energy harvesting and triboelectric nanogenerators (TENGs).

The fast-evolving 3D printing technology has enabled researchers to manufacture energy harvesting devices in a simple and expedient way. Gawron et al. [1] provide a comprehensive review of the development of hybrid and non-hybrid 3D printed electromagnetic vibration energy harvesters. Various harvesting approaches, their utilized geometry, functional principle, power output and the applied printing processes are discussed. This Special Issue analyzes the advantages and challenges of 3D printed harvesters and identifies research gaps in the field.

On the other hand, TENGs are recognized as one of the most promising solutions for future energy harvesting. In the review by Nazar et al. [2], TENGs targeting various environmental energy resources are systematically summarized, including various TENGs for ocean and other applications as well as the hybridization of TENGs with other energy harvesters to improve performance. The advantages and disadvantages of various TENG structures are explored. A high-level overview in this Special Issue explores the connection of TENGs with structural health monitoring, artificial intelligence and highlights the future developments of this technology. This is a must-read paper for those who are interested in TENGs.

In addition to these two review papers, this Special Issue also includes ten research papers covering a wide range of energy harvesting techniques, including electromagnetic and piezoelectric wideband vibration, wind, current-carrying conductors, thermoelectric and solar energy harvesting, etc. Not only are the foundations of these novel energy harvesting techniques investigated, but the numerical models, power conditioning circuitry and real-world applications of these novel energy harvesting techniques are also presented.

Bradai et al. [3] study a novel energy-autonomous wireless sensor system powered by a wideband electromagnetic energy harvester via passive energy management. The self-powered sensor system can detect machine failure and deliver alerts without the use of an external energy supply. The solution can also detect failure without the use of additional sensors by utilizing the Analog Digital Converters (ADCs) of the Wireless Sensor Nodes (WSNs) themselves, making it more compact with lower energy consumption.

A dual-resonance vibration electromagnetic energy harvester (EMEH) for wideband operation is proposed by Feng et al. [4]. Compared with the conventional dual resonance harvester, the proposed system realizes an enhanced "band-pass" harvesting characteristic with a significant improvement in the average harvested power. Furthermore, two resonant frequencies are decoupled in the proposed system, which leads to a more straightforward design. Experimentally, the proposed dual resonance EMEH has demonstrated a higher

Normalised Power Density over a wider frequency range twice that of the state-of-the-art wideband EMEHs.

Bäumker et al. [5] introduce a novel animal-tracking system powered solely by thermal energy harvesting. The proposed tracker harvests electrical power using the temperature difference between the animal's fur and the environment. The steps to enhance power generation are presented and validated in a field test using a system which fulfils common tracking tasks, including GPS, activity and temperature measurements, and wireless data transmission via LoRaWAN. Furthermore, the ultra-low-power design has extremely low overall sleep power consumption and can operate when temperature differences are minimal.

A novel approach is presented by Khan et al. [6] to harness energy from a small-scale wind turbine to support communication primitives in electric vehicles, enabling a variety of applications in the Internet of Vehicles (IoV). The harvested power is processed through regulation circuitry to achieve the desired power supply for the end loads, such as the battery and supercapacitor. The orientation for optimum conversion efficiency is proposed through an ANSYS-based aerodynamics analysis and verified in experiments.

Russo et al. present the dynamic experimental identification of an inductive energy harvester for the conversion of vibration energy into electric power [7]. The proposed energy harvesting technique is based on an asymmetrical magnetic suspension and addresses structural monitoring applications on vehicles. The design of the interfaces for the electrically, magnetically, and structurally coupled systems forming the harvester are described using dynamic modelling and simulation. The experimental results are also compared with the harvester's dynamic response, calculated via numerical simulations; good correspondence was obtained.

Liu et al. [8] validate the potential of unequal-length section-varied geometries in developing an orthoplanar spring-based piezoelectric vibration energy harvester (PVEH). A basic quad-trapezoidal-leg orthoplanar spring (QTOPS) is theoretically analyzed, and the structurally effective stress and eigenfrequency are formulated to determine the main dimensions. An improved QTOPS with additional intermediations is also constructed and simulated. It is experimentally verified that the proposed approach is more suitable to construct a high-performance PVEH than the orthoplanar spring with equal-length or rectangular legs.

Sun et al. [9] introduce a new method of electricity generation using a Wiegand sensor. The design and verification of a self-oscillating boost converter circuit are presented in this paper. A DC voltage obtained by rectifying and smoothing the pulse voltage generated from the Wiegand sensor is boosted by the proposed circuit. A quantitative analysis of the power generated by the Wiegand sensor reveals a suitable voltage-current range for application in self-powered devices and battery-less modules.

In the work by Jiang et al. [10], the acoustic resonance of a circular pipe and an equivalent diameter square pipe are simulated with the realizable k-ε DDES model. The influences of the intersecting line of the pipe on the acoustic resonance are discussed. Furthermore, the influence of the branch angle on acoustic resonance is also studied. The simulation results identified methods to improve output power of the mean flow wind harvesting and suppress acoustic resonance in industrial pipelines.

Dai et al. [11] review the existing applications of PV facilities in road structures and facilities, including two systematic large-scale applications in China where the PV facilities are applied in spare spaces without traffic loads. The existing practices of PV facilities in pavement structures with vehicle loads are also presented and discussed. In addition, a novel integration of amorphous silicon (a-Si) PV cells and glass-fiber-reinforced polymer (GFRP) profiles is proposed. The existing mechanical experiments on the a-Si PV cells and their integrations with GFRP profiles are introduced and their mechanical performance is evaluated experimentally.

The study by Han et al. [12] focuses on metallic submicron-sized silicon particles used to improve the absorption properties of absorber plates for solar still applications. Various concentrations of silicon are mixed with black paint and applied to the aluminum plate. Indoor and outdoor testing are conducted using an improvised apparatus for concentration optimization. Although this research is not directly linked to traditional energy harvesting technologies, it does provide a unique insight into an important application of solar energy.

As the Guest Editor of this Special Issue, I hope that readers will find the papers intriguing and that this collection can stimulate the research community to further advance this important research area in the years to come. Finally, I would like to thank all the authors, reviewers and editorial staff who have contributed to this Special Issue.

Funding: This research received no external funding.

Conflicts of Interest: The authors declare no conflict of interest.

References

1. Gawron, P.; Wendt, T.; Stiglmeier, L.; Hangst, N.; Himmelsbach, U. A Review on Kinetic Energy Harvesting with Focus on 3D Printed Electromagnetic Vibration Harvesters. *Energies* **2021**, *14*, 6961. [CrossRef]
2. Matin Nazar, A.; Idala Egbe, K.; Abdollahi, A.; Hariri-Ardebili, M. Triboelectric Nanogenerators for Energy Harvesting in Ocean: A Review on Application and Hybridization. *Energies* **2021**, *14*, 5600. [CrossRef]
3. Bradai, S.; Bouattour, G.; El Houssaini, D.; Kanoun, O. Vibration Converter with Passive Energy Management for Battery-Less Wireless Sensor Nodes in Predictive Maintenance. *Energies* **2022**, *15*, 1982. [CrossRef]
4. Feng, Z.; Peng, H.; Chen, Y. A Dual Resonance Electromagnetic Vibration Energy Harvester for Wide Harvested Frequency Range with Enhanced Output Power. *Energies* **2021**, *14*, 7675. [CrossRef]
5. Bäumker, E.; Conrad, L.; Comella, L.; Woias, P. A Fully Featured Thermal Energy Harvesting Tracker for Wildlife. *Energies* **2021**, *14*, 6363. [CrossRef]
6. Khan, Z.; Sherazi, H.; Ali, M.; Imran, M.; Rehman, I.; Chakrabarti, P. Designing a Wind Energy Harvester for Connected Vehicles in Green Cities. *Energies* **2021**, *14*, 5408. [CrossRef]
7. Russo, C.; Lo Monaco, M.; Fraccarollo, F.; Somà, A. Experimental and Numerical Characterization of a Gravitational Electromagnetic Energy Harvester. *Energies* **2021**, *14*, 4622. [CrossRef]
8. Liu, Y.; Mo, S.; Shang, S.; Wang, H.; Wang, P.; Yang, K. Quad-Trapezoidal-Leg Orthoplanar Spring with Piezoelectric Plate for Enhancing the Performances of Vibration Energy Harvester. *Energies* **2020**, *13*, 5919. [CrossRef]
9. Sun, X.; Iijima, H.; Saggini, S.; Takemura, Y. Self-Oscillating Boost Converter of Wiegand Pulse Voltage for Self-Powered Modules. *Energies* **2021**, *14*, 5373. [CrossRef]
10. Jiang, L.; Zhang, H.; Duan, Q.; Liu, X. Numerical Simulation of Acoustic Resonance Enhancement for Mean Flow Wind Energy Harvester as Well as Suppression for Pipeline. *Energies* **2021**, *14*, 1725. [CrossRef]
11. Dai, Y.; Yin, Y.; Lu, Y. Strategies to Facilitate Photovoltaic Applications in Road Structures for Energy Harvesting. *Energies* **2021**, *14*, 7097. [CrossRef]
12. Han, S.; Ghafoor, U.; Saeed, T.; Elahi, H.; Masud, U.; Kumar, L.; Selvaraj, J.; Ahmad, M. Silicon Particles/Black Paint Coating for Performance Enhancement of Solar Absorbers. *Energies* **2021**, *14*, 7140. [CrossRef]

Review

A Review on Kinetic Energy Harvesting with Focus on 3D Printed Electromagnetic Vibration Harvesters

Philipp Gawron *, Thomas M. Wendt, Lukas Stiglmeier, Nikolai Hangst and Urban B. Himmelsbach

Department of Business and Industrial Engineering, Offenburg University of Applied Sciences, Klosterstraße 18, 77723 Gengenbach, Germany; thomas.wendt@hs-offenburg.de (T.M.W.); lukas.stiglmeier@hs-offenburg.de (L.S.); nikolai.hangst@hs-offenburg.de (N.H.); urban.himmelsbach@hs-offenburg.de (U.B.H.)
* Correspondence: philipp.gawron@hs-offenburg.de

Abstract: The increasing amount of Internet of Things (IoT) devices and wearables require a reliable energy source. Energy harvesting can power these devices without changing batteries. Three-dimensional printing allows us to manufacture tailored harvesting devices in an easy and fast way. This paper presents the development of hybrid and non-hybrid 3D printed electromagnetic vibration energy harvesters. Various harvesting approaches, their utilised geometry, functional principle, power output and the applied printing processes are shown. The gathered harvesters are analysed, challenges examined and research gaps in the field identified. The advantages and challenges of 3D printing harvesters are discussed. Reported applications and strategies to improve the performance of printed harvesting devices are presented.

Keywords: energy harvesting; 3D printed; vibration harvester; electromagnetic; hybrid

Citation: Gawron, P.; Wendt, T.M.; Stiglmeier, L.; Hangst, N.; Himmelsbach, U.B. A Review on Kinetic Energy Harvesting with Focus on 3D Printed Electromagnetic Vibration Harvesters. *Energies* **2021**, *14*, 6961. https://doi.org/10.3390/en14216961

Academic Editor: Dibin Zhu

Received: 22 September 2021
Accepted: 18 October 2021
Published: 22 October 2021

Publisher's Note: MDPI stays neutral with regard to jurisdictional claims in published maps and institutional affiliations.

1. Introduction

Energy harvesting is the utilisation of ambient energy in order to power electronics such as wireless sensor nodes (WSN) or wearables without the need of batteries [1–3]. This allows to operate the node over a much longer time period compared to battery-powered devices along with lower maintenance efforts. Furthermore, the low-maintenance requirements allow to operate these WSNs in environments with limited or no accessibility [4,5]. Energy harvesting can be categorised into four types of energy sources: solar/light, mechanical motion/vibration, thermoelectricity and electromagnetic radiation. Each type of energy source has different concepts to convert the ambient energy into electrical energy [1]. Testing different energy harvesters often requires specific power conditioning depending on the type of harvester. In [6], a reconfigurable power management was presented, allowing fast adaptation of the power management.

Solar energy is converted into electrical energy via solar panels [4,5]. Compared to outdoor applications the indoor available energy is 10 to 100 times lower [7]. The light's spectral composition and the applied photovoltaic cell significantly impact the generated power, even if the same illuminance level is given [8]. Thermoelectric generators (TEG) utilise the Seebeck-effect to convert thermal into electrical energy [9]. The TEG's hot side is mounted on a heat source and the cold side on a heat sink. A temperature gradient between both sides creates an electric voltage [1,10]. RF harvesting utilises electromagnetic (EM) radiation in the near- or far-field [11]. Near-field sources apply EM induction or resonance methods for power transfer. For example, a sensor powered via near field communication (NFC) by a smart phone was presented in [12]. However, near-field harvesting might limit scaling for WSNs distributed over a wide area [11]. Far-field sources (range up to few kilometres) are for instance cellular base stations, TV stations or WiFi access points [1]. The available energy in the far field depends on the harvesting device's location and surroundings. Kinetic harvesters convert motions or vibrations into electrical energy via electromagnetic [13–15], electrostatic [16–20] or piezoelectric [21–23] transduction methods [1]. Electromagnetic (EM) harvesters consist of magnets and coils. Based on Faraday's law, a varying magnetic flux induces an electric voltage across the coil [2]. The induced voltage depends on the number of the coil's turns as well as the size of coil and magnet. Thus, achieving small harvesters with a high power output is tough [19]. Applications with high amplitudes and low frequencies are suitable for EM harvester [24]. Piezoelectric harvesters utilise the piezoelectric effect to transduce mechanical into electrical energy when stress is applied on the geometry [2]. Electrostatic harvester exploit the ambient motion (for instance vibrations) to vary the distance of charged capacitor plates, resulting in a changing capacitance and thus energy transfer [2]. Ambient mechanical energy occurs in different forms such as rotation or vibration. Therefore, various harvester designs and techniques for these forms exist. Three-dimensional printed EM harvesters utilising rotation are presented in [25–30]. Three-dimensional printed EM vibration harvesters are reviewed in this paper. Furthermore, a recent review on various techniques for mechanical energy harvesting was provided in [31].

To achieve a higher energy output, different techniques or sources may be combined in hybrid harvesters [1]. Figure 1 shows the energy sources and their combinations. Combining multiple sources achieves multiple opportunities for energy harvesting (EH).

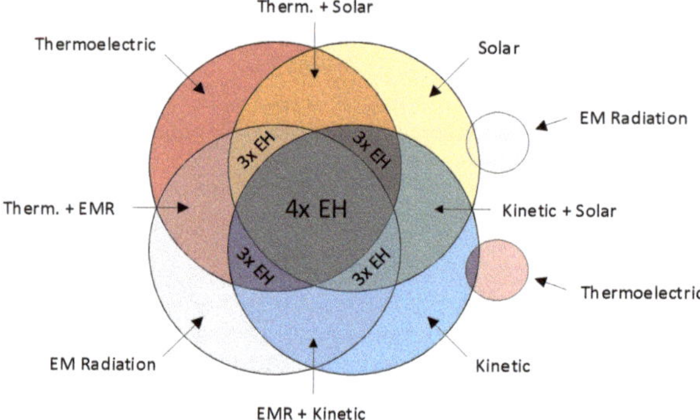

Figure 1. Energy sources and combinations.

2. 3D Printing

Additive manufacturing (AM) offers many benefits compared to traditional technologies such as moulding. The development is faster due to the seamless transfer from 3D Computer-Aided Design (CAD) to AM. No moulds are required, meaning different variants can be produced without additional costs and fabricating complex shapes requires no additional process steps [32]. Furthermore, AM has the potential of waste-, lead time- and cost-reduction. Assembled parts with relative motion can be printed together when utilising soluble support material and appropriate clearance between the parts [33]. Studies addressing the drawback of mass production capability along with opportunities such as mass customisation have been reported [34,35]. Printed harvesters either utilise the AM-process to print the housing of the harvester or to print functional parts such as springs or magnets of the device. Three-dimensional printing is also capable of printing multiple materials [36] at once as well as functional materials [37,38]. Three-dimensional printing or AM-processes apply the layer-by-layer-principle to create three-dimensional objects from 3D CAD-models [33]. There are seven AM-categories, each containing different technologies [32,33]. In this section, the most common printing techniques applied for energy harvesting devices are described: Binder jetting, fused filament fabrication and inkjet. Table 1 summarises the advantages and disadvantages of these three printing technologies.

2.1. Fused Filament Fabrication

Fused filament fabrication (FFF, also known as FDM) is categorised as material extrusion and utilises plastic filament which is melted in the hot end and then extruded through a nozzle onto the printing bed [33,37]. Figure 2a displays the process. As presented in the study common materials are PLA, ABS or PC but also advanced materials such as PVDF or PEEK. There is also the opportunity to utilise functional materials such as conductive filament to print electronic circuits [39].

Table 1. Advantages and disadvantages of Fused filament fabrication (FFF), Inkjet, Binder Jetting (BJT).

Technology	Advantages	Disadvantages
FFF	• Multi-material printing [33] • Functional materials available [39] • Printing assembled, movable parts (soluble support material) [33]	• Building speed [32] • Accuracy [32] • Surface roughness (depending on layer height) [33] • Material density [32]
Inkjet	• Multi-material-capabilities [33] • High layer-resolution (10–30 µm) [33] • High speed [32] • High scalability [32]	• High absorption of electromagnetic waves for photoactive compounds [33] • Limited choice of materials [32] • Lower accuracy for large parts [32]
BJT	• Large build volumes (up to 4 m in length) [32] • No support material required [32] • Combination of powder and binder-additives [32] • Fast printing process [32] • Various materials available (metal, polymer, ceramic, composite) [32] • Appropriate for low to medium batch production [32]	• Post processing required (infiltration) [32] • Lower accuracy than inkjet [32] • Poor surface finish [32]

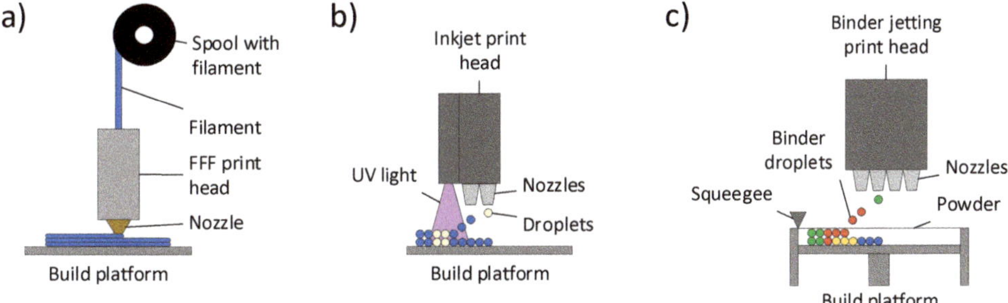

Figure 2. Schematic printing processes, (**a**) FFF with one extruder, (**b**) inkjet with photopolymers (blue = model material, beige = support material), (**c**) binder jetting with multiple colours.

2.2. Inkjet

Inkjet is a material jetting-process that utilises print heads with arrays of nozzles to eject material (for instance photopolymers, metals or ceramics) onto the print bed [32,33,37]. As showed in Figure 2b UV light is applied to cure the printed photopolymers and support-material. In addition, high layer-resolutions and multi-material-capabilities are mentioned for this process [33].

2.3. Binder Jetting

Binder Jetting (BJT), as described in [32], applies binder droplets to form the geometry for each powder-layer as well as to bond it to prior layers. Materials mentioned for BJT are, for instance, plaster, metal or polymers. As stated in the monograph, the BJT print heads consist of multiple, parallel nozzles offering a fast and scalable process. Figure 2c displays the process. According to the authors, a post processing is necessary, which involves removing the part from the bed as well as remaining powder from the part. Infiltration or sintering may be used to improve the part's properties, depending on the material [32].

3. Review Methodology

3.1. Objective of the Review

The goal of this review is to provide a compact overview of the current printed vibration harvesting approaches, their principle of operation, the applied printing technique, the utilised materials, achieved power output, the degree of exploitation of 3D printing technology as well as open challenges. Details on the most common printing techniques were shortly explained in the previous section. A dedicated review on 3D printed vibration harvesters was provided in [41] in 2016. The contained harvesters were picked up in this review and extended with presented research from 2016 on.

3.2. Categories of Harvesters

The review is divided into non-hybrid and hybrid harvesters. Additionally, the harvesters are categorised based on their shape, resulting in tube-shaped, spherical, cantilevers, pendulum and others.

3.3. Formal Rules

As a blueprint for the review, [42] was applied. The search period was from 2016 to 2021. The latest search was carried out in September 2021. The reviewed literature was gathered via IEEE Xplore, SCOPUS and by checking the references of reviewed papers for relevant literature. The keywords applied were 3D printing, 3D printed, energy harvester, energy harvesting, electromagnetic, vibration.

3.4. Criteria for Exclusion

The EM harvester's central parts have to be 3D printed, except the coil and magnet. Meaning the printed parts should have an active function other than just offering a simple housing. For example, a harvester consisting of a printed spring with non printed coil and magnet is considered a "printed harvester" in this work. On the other hand, if the same harvester would utilise a non printed spring (e.g., wound steel wire) with just a printed housing, it is no longer considered a printed harvester. Printed parts offering a guiding-function as well as acting as housing (for instance a tube guiding a magnet ball) are considered active as well. However, few remarkable approaches found during the review process are presented in a dedicated section instead of being excluded. These papers might provide useful ideas or inspiration for new EM vibration harvesters.

4. 3D Printed Non-Hybrid Vibration Harvester

This section presents non-hybrid harvesters categorised by their shape.

4.1. Tube-Shaped

In [43], a vibration harvester with magnetic levitation was proposed. The device consisted of a cylindrical housing made of PLA, four magnets and a coil (1000 turns). Figure 3 shows the schematic and the printed harvester. Two magnets (NdFeB N35) were fixed the cylinders ends. The remaining two magnets (NdFeB N45) acted as inertial mass. They were glued together with a spacer between and faced each other with their north pole. The coil was wound around the cylinder. Under vibration, the inertial mass's movement induced a voltage in the coil. Guide rails inside the cylinder decreased the fraction between magnets and cylinder. The reported power output at resonance frequency of 10 Hz with 0.5 g was 1.4 mW with a 500 Ω load. A giant magnetoimpedance (GMI) accelerometer was implemented with the harvester. In order to achieve a self-powered sensor, the authors aimed to expand the harvester's bandwidth for a better overall energy conversion.

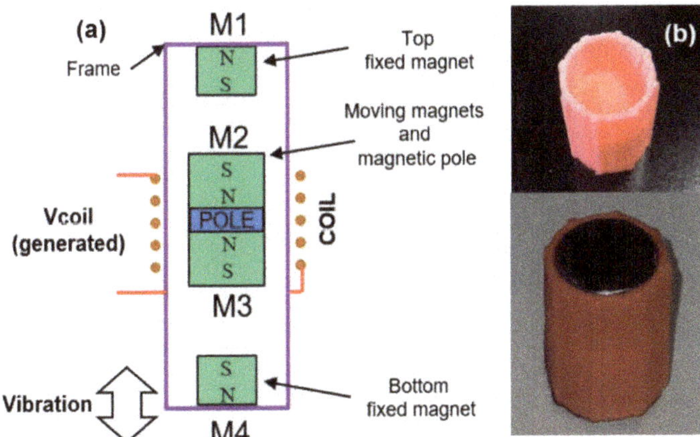

Figure 3. (**a**) Schematic of a straight tube harvester with magnetic springs [43], (**b**) printed harvester made of PLA with and without magnets [43].

In [44], a vibrational harvester was printed via inkjet (3D Systems ProJet 3510, materials: Visijet M3 Crystal and S300). The harvester was made of a printed body with a centred pad for a seismic mass. The pad was attached via springs printed in-plane. Multiple variants of spring geometries and thickness were tested. One or two magnets (1.5 mm diameter, 1 mm height) were placed on the central pad. A 7.36 mH SMD 812 coil was positioned 2 mm above the magnets as inductor. Figure 4 shows the harvester with one variant of the spring geometry as well as it's dimensions. The spring-geometry had a significant impact on the achievable voltage. 23.7 µW at a resonant frequency of 100 Hz were shown as the highest power output with a circular shaped spring. The harvester's reported total volume was less than 1 cm^3.

In [45], a linear tube-shaped, magnetic repulsion harvester was presented. Since the repulsive force between magnets is not linear such systems showed a non-linear frequency response. The authors exploited this to broaden the utilisable frequency range. The tube's inner surface showed the characteristic ridges from 3D printing. These increased the friction between tube and magnet-stack. In consequence, the authors turned the next prototype and did not print it. They also applied a ferrofluid to the magnet to reduce the friction. This improved the coupling between magnets and coils and thus increased the harvester's performance by 15%. Drilled magnets were applied as end caps to act as magnetic spring and improve the air flow affected by the ferrofluid. However, the authors pointed out that there were evaporation issues with the ferrofluid and that a sealed container might solve this. Five coils with 350 or 1400 turns in total were mounted on the tube. With a 75 Ω load, they calculated around 60 µW power output.

In [46], a 3D printed wrist-wearable harvester for human frequencies ≤5 Hz was reported. A hollow cycloid tube, printed with PLA, contained a 10 mm NdFeB N35 magnet ball. Two serially connected coils (600 turns each) were wrapped around the tube. A cycloid shape offered a shorter travel time of the ball from one end to the other, thus increasing the induced voltage into the coils. The authors also investigated straight and circular structures with the cycloid structure providing a 1.3 or 1.45 times higher open circuit voltage, respectively. The device was tested with a custom built swinging arm under 5 Hz excitation frequency and 2.5 g acceleration. The average power output was 8.8 mW with a 104.7 Ω load at 5 Hz. Vibrating hand motion of 5 s duration a wristwatch could be powered for 34 min.

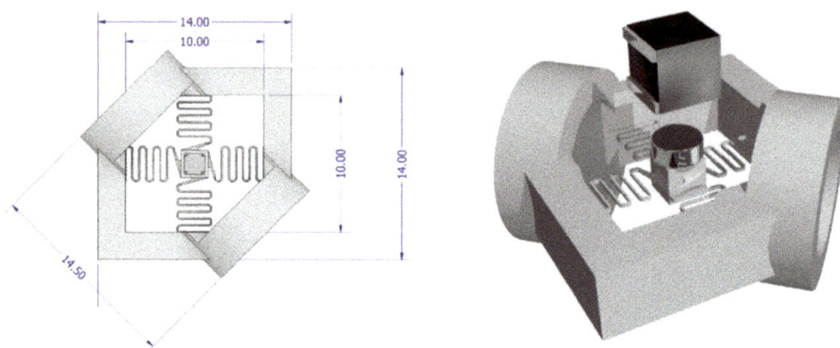

Figure 4. (**left**) Top view of inkjet printed harvester with dimensions in mm [44], (**right**) harvester with magnet and coil [44].

In [47], tube harvesters with 2 degrees of freedom (2-DOF) were investigated. The authors combined springs as linear and magnets as non-linear oscillators in four configurations to realise a 2-DOF harvester. The structure consisted of an outer tube containing an inner tube. The inner tube had a magnet (10 mm diameter and 10 mm thickness) inside. The magnet was either supported with springs or magnets. The inner tube itself was supported either with springs or magnets, resulting in four possible combinations. The tubes were printed via UV curable resin and had a coil (170 turns) wound around. All magnets were made of NdFeB. The authors pointed out that differences in their results between simulation and experiment might come due to intermittent sliding contact between magnet and tube because of the tube's deviation from vertical direction or from assembly. Configuration C with magnetic springs for the outer tube and linear springs for the magnet inside the inner tube was reported with a power output of 5.9 mW at 0.5 g. Configuration A with two linear springs showed 6.1 mW at 0.5 g.

In [48], a straight tube harvester was proposed. The authors applied a cylindrical N52 magnet and utilised the whole tube's surface for the coils. The harvester's dimensions were 14 mm in diameter and 50 mm in height (standard AA cell battery). An total of 6 mm in height was reserved for a future power management. The 0.5 mm thick tube was printed with resin and had 3 mm holes in the top and bottom for air flow. Different numbers of coils were investigated with four coils showing the best results. The average power output with four coils was 63.9 mW at 5 Hz hand shaking which equalled a power density of 9.42 mW/cm^3). A total of 4.2 mW was achieved when the harvester was worn at the ankle at 1 Hz walking. Connecting the coils in series or subtractive series decreased the power output.

In [49], a tube-like harvester with magnetic springs was proposed. Two NdFeB N35 magnets were attached to both ends of the printed hollow tube with a coil (480 turns) wrapped around the middle of tube. A stack consisting of two magnets was able to move inside the tube. The available length for the stack's movement as well as the stack's mass were adjustable and therefore allowed to tune the resonance frequency. For the stack, the magnets were orientated with opposite poles facing each other which was shown to provide a higher magnetic flux. The authors investigated combinations of lengths (140 mm or 200 mm) and magnet stacks consisting of either three or six magnets under different conditions. The harvester was attached to the leg of a participant either vertically or transversely while exposed to different walking speeds. The maximum power output of 10.66 mW was under running conditions with the combination of 200 mm length and a six magnets-stack. The authors mentioned the higher mass of the stack shifted the resonance frequency more towards the human motion frequency.

In [50], a magneto-mechanical device was reported. The harvester was based on a straight tube design, composed of a printed cylindrical housing with magnetic springs located at the top and bottom end and a coil wrapped around the centre. The end-magnets were mounted on bolts to adjust their distance. The moving NdFeB disc-magnet was fixed with four oblique springs (made of rubber) to the housing in order to vertically align it and prevent flipping. The oblique springs reduced the energy loss due to friction between magnet and walls as well as increased the non-linearity of the harvester. The housing had air vents for a better airflow and thus reduction in viscous damping. Three-dimensional printing material HIPS was applied via FFF. The maximum power was around 7 mW at 15 Hz with a 1.14 kΩ load and 0.75 g.

In [24], a vibration harvester was reported. It consisted of a group of magnets (adhesive force: 17 kg; weight: 18.8 g) on a bar, attached to a printed spring (diameter: 50 mm). These components were placed inside a cone-like structure with a coil (300 turns, 2.39 Ω) wrapped around its top. The spring caused the magnet to move up and down along the cone-structure's axis. The used material was ABS and the printer a X400 PRO 3D. Human motions with low frequency and high amplitude (8–16 Hz; 2.842–31.5827 m/s^2) were targeted—walking or running in particular. Therefore, the harvester was intended to be attached in the area under the knee. Different spirals were tested. Power outputs of 12–76 mW$_{rms}$ at resonance with 8–15 Hz and a 2 Ω load were achieved. A decreasing output voltage during the 30 min test-cycle and a shift of the resonance frequency towards lower values were shown by the results. The geometry of the spring or the ABS's mechanical properties were pointed out as possible areas to be investigated in this regard.

4.2. Cantilever

In [51], a spiral micro array coil, a multipole magnet thin-plate and a 3D printed cantilever were utilised to create a micro-electromechanical system (MEMS) harvester. The magnet (8.9 mm × 8.9 mm, thickness 0.5 mm) had 16 poles with a chess pattern for an alternating magnetisation-direction. The coil array had 24 coils connected in series with a total resistance of 5.38 Ω and 144 turns. Both the multipole magnet and the coil array were fabricated by the authors. In one experiment, they 3D printed a 65 mm long cantilever made of PLA via FFF. The multipole magnet was attached to the cantilever while the coil remained fixed. An amount of 3.34 µW was reported as power output and 5.22 µW cm^{-3} as power density at 38 Hz and 11.6 g.

In [52], a coreless, non-linear, resonant vibration harvester was presented. The authors intended to improve the efficiency of coreless harvesters. They optimised the distribution of the coil's windings, resulting in fewer turns required and thus a higher power output. Figure 5 shows the harvester's magnets and coils. The two coils were racetrack-shaped with two times three magnets arranged between them. The magnets were split into four fixed (outer ones) and two movable ones (inner magnets). The latter were mounted on a cantilever, being deflected under vibrations and thus resulting in a varying magnetic flux. Due to optimisation, the coil's design had 746 turns (optimised) compared to 2000 turns (non optimised, uniformly distributed). An increase of 300% regarding the power output was achieved. The maximum power output was 6.46 mW at 35.5 Hz. A printing process or material were not mentioned. Based on the provided pictures, FFF was assumed as the process.

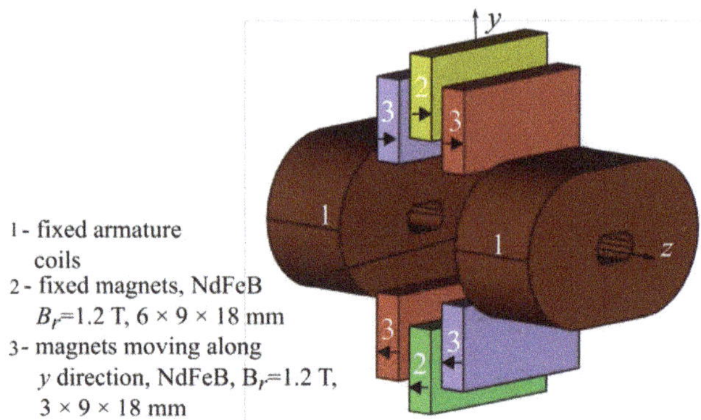

1 - fixed armature
coils
2 - fixed magnets, NdFeB
B_r=1.2 T, 6 × 9 × 18 mm
3 - magnets moving along
y direction, NdFeB, B_r=1.2 T,
3 × 9 × 18 mm

Figure 5. Magnets and coils from proposed harvester [52].

4.3. Pendulum

In [53], a non-linear pendulum-harvester was proposed. A cylindrical geometry with 12 symmetrically mounted coils (six on each side made of 28 AWG) and a pendulum with a permanent NdFeB N37 magnet in the middle was presented. Figure 6 shows the schematic of the pendulum. The authors connected the coils in series in order to achieve a high voltage-response of the harvester. Additional repulsive magnets were installed in some coils. Three-dimensional printing with ABS material was utilised to avoid magnetic coupling between the housing-geometries and magnets. As mentioned in the paper, a power output around 10 mW at 15 Hz was achieved. The authors pointed out that additional tests have to be carried out with higher values for the load resistor or excitation-frequency, but that was hindered by design constraints.

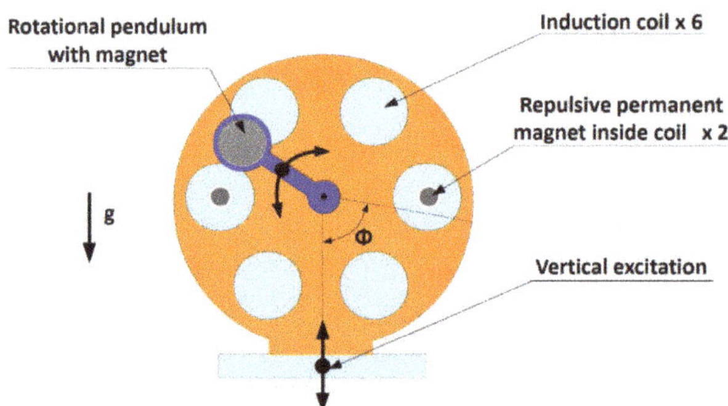

Figure 6. Pendulum-harvester schematic [53].

In [54], an inkjet printed pendulum-based harvester in a wrist watch housing was reported. The harvester was composed of the pendulum with eight integrated neodymium magnets, a disk with mounted SMD coils and a housing. The pendulum was attached to the disk and housing with a bearing. A Projet 3510 printer was utilised with Visijet M3 Crystal and S300 as model material and support material, respectively. The device provided a maximum peak-to-peak voltage of 1.8 V.

4.4. Others

In [55], a 3 degree of freedom (3-DOF) serial manipulator-type harvester was presented. The device was composed of a coil holder and three serially connected leaf hinge joints with a proof mass attached to the free end. The proof mass had three pairs of magnets integrated and the coil holder contained a pair of concentric coils (1400 turns each). The harvester was printed with a Dimension SST1200ES 3D printer and ABSplus. In order to harvest multiple frequencies the device had three resonance frequencies at 23.4, 29.2, and 34.8 Hz. Based on the measured peak-to-peak voltages across the optimal load of 1.01 kΩ the peak powers were 1.28, 0.89 and 1.32 mW at an acceleration of 1.5 m/s^2. The normalized power density was reported as 2.85 kg s m^{-3}.

In [41], a non linear in-plane vibration harvester with a leaf isosceles trapezoidal flexural (LITF) pivot topology was presented. The out-of-plane motion was restricted by the LITF topology. FFF was applied to print the harvester's topology. The used printer was a Stratasys Mojo with ABS. NdFeB N35 magnets were glued to the movable tip and on the opposite of the frame. Between this arrangement, a coil (7060 turns, air core) was fixed to the stationary part. Mechanical energy was converted to the electrical domain once the structure was exposed to vibrations. Maximum power output was achieved at the resonance frequency. A power output of 2.9 mW (down sweep) was shown at 1 g. The authors emphasised the superiority of 3D printing for the monolithic fabrication of LITF pivot topologies. A review on 3D printed vibration harvesters as well as a review on (non 3D printed) in-plane harvesters were presented too.

In [56], the vibrations of a flying moth were harvested. The moth showed an available amplitude of 1 mm at 25 Hz measured on its back surface. A EM multi-phase AC generator was applied to harness the available energy. On the back of the moth, an FFF-printed ABS pedestal-structure was attached. Three NdFeB magnets (each 9.5 mm $\times$ 1.8 mm, 0.3 mm thick) and a core (0.2 mm thick) were held on the tip of the structure. Two magnet-core-assemblies faced each other with an air gap between. The PCB-windings utilised Kapton as substrate and were located in the air gap. Each of the three coils had two windings in a figure eight design on the double sided substrate. The generator had a total mass of 1.28 g. Tests were carried out with the generator attached on a shaker. A 1.7 mW power output into the load was reported with a shaker-amplitude of $\pm$0.37 mm at 25.8 Hz and 0.9 mW with $\pm$0.23 mm at 25.8 Hz.

In [57], a non-resonant vibration harvester was presented. It consisted of a printed housing with a spherical cavity containing a NdFeB magnet ball. The parameters (e.g., coil turns, cavity diameter) were varied for testing. Depending on the variant one or two copper coils were wound around the device. The goal was to harness human motion in different conditions (walking/running). The housing was printed via binder jetting with powder/resin on a Spectrum Z510 3D printer. A better power output was shown with two coils placed with an offset to the cavity's equator. A power output of 1.44 mW was achieved by a variant placed in the pocket (offset wound coil with 500 turns). The highest shown power density was 0.5 mW/cm^3.

In [58], a resonance-based vibration harvester was printed via FFF and ABS material. The first generator was based on assembled aluminium parts fixed with screws. Screws were not suitable in a vibrating environment the authors pointed out. Therefore, FFF (Dimension BST) was utilised to print ABS-parts for the generator. Epoxy glue was applied to connect the printed parts. A lever with a magnetic circuit at its tip induced a voltage in a fixed coil due to Faraday's law. The lever's stiffness was tuned via permanent magnets repelling a magnet on the lever. Printing offers parts with low weight and durability according to the authors. The generator operated at 17 Hz with an amplitude of 0.1–1 g. The maximum power output was 26 mW with 9 V (DC). The harvester's size was 80 $\times$ 60 $\times$ 60 mm^3. It is not clear if the lever was printed or made of non-printed material. The latter would meet the criteria for exclusion of this paper.

Table 2 summarises the non-hybrid harvesters for later discussion.

Table 2. 3D printed electromagnetic non-hybrid vibration harvesters in the literature.

Ref.	Year	Structure	Power in mW	Energy Source	Special Feature
[51]	2021	Cantilever	3.34×10^{-3}	Low frequency vibrations	MEMS harvester
[43]	2020	Straight tube	1.4	Low frequency vibrations	Self-powered accelerometer
[53]	2020	Pendulum	10	Mechanical vibrations (shaker)	Oscillations and rotations possible
[44]	2020	Cubic tube	2.37×10^{-2}	Mechanical vibrations	Total volume less than 1 cm^3
[52]	2020	Cantilever	6.46	Shaker	Optimised coil windings
[45]	2019	Straight tube [(1)]	6×10^{-2}	Environment with low frequency + high amplitude	Ferrofluid as lubricant
[46]	2019	Cycloid tube	8.8	Human motion wrist or foot	Optimised shape of harvester
[47]	2019	Straight tube	6.1	Shaker	2-DOF harvester
[55]	2019	Manipulator type	$1.28_{(23.4\,\text{Hz})}$ [(p)] $0.89_{(29.2\,\text{Hz})}$ [(p)] $1.32_{(34.8\,\text{Hz})}$ [(p)]	Shaker	3-DOF harvester with leaf hinge joints
[54]	2019	Pendulum	-	Wrist movement	Inkjet printed harvester
[48]	2018	Straight tube	63.9	Human hand shaking	Coils all over surface
[49]	2017	Straight tube	10.66	Human leg motion	Magnetic spring
[50]	2017	Straight tube	7 [(p)]	Shaker	Oblique springs for alignment
[41]	2016	LITF pivot	2.9 [(ds)(p)]	Mechanical vibrations	Topology restricts out of plane movement
[24]	2016	Straight tube	76	Human motion (walking)	Printed spring
[56]	2009	3D spring	1.7	Vibrations from flying moth	Harvesting energy from moth
[57]	2008	Spherical	1.44	Random human motion	Multi-direction harvesting
[58]	2008	Moving arm	26	Ambient mechanical vibration	Repelling magnets as oscillator

[(1)] final prototype not 3D printed [(ds)] down sweeps values [(p)] peak power.

5. 3D Printed Hybrid Vibration Harvester

Hybrid harvester combine multiple energy sources and/or harvesting principles, for instance solar and kinetic sources or piezo and electromagnetic transducers. Figure 1 shows possible combinations of energy sources.

5.1. Tube-Shaped

In [59], a 3D printed tube contained a magnet attached to a spring, an electromagnetic generator (EMG), a contact separation mode triboelectric nanogenerator (CTENG) and a sliding mode triboelectric nanogenerator (STENG). An aluminium (Al) strip was wrapped around the magnet for the STENG. Two coils were mounted around and the STENG's electrodes inside the tube. The CTENG was located in the bottom. As explained by the authors, once the magnet started oscillating a voltage was induced in the coil by Faraday's law, a triboelectric charge transfer was induced in the STENG and the CTENG was compressed in the magnets lowest position. The maximum peak power outputs were 717 mW at 600 Ω for the EMG, 18.9 mW at 2 MΩ for the CTENG and 1.7 mW at 6 MΩ for the STENG. The device's regulated DC power output was 34.11 mW.

In [60], a non resonant impact based harvester with a similar structure as in [59] was presented. A magnet with two springs on its top and bottom was inside a printed rectangular tube with a coil mounted in the centre as EMG. Two CTENGs were integrated with their positive and negative triboelectric material on the open ends of the springs as well as the tube's bottom and ceiling, respectively. The magnet started oscillating once the device was under excitation. According to the paper a soft magnetic film was utilised as flux-concentrator for the EMG on the coil's outside. This increased the induced voltage in the coil by 1.39 times. Furthermore a dry lubricant was applied to the coils frame for easier magnet-movement. The harvester's power output was 144.1 mW with a load of 1.5 kΩ at 6 Hz and 1 g.

In [61], a springless hybrid, non resonant harvester was proposed. An EMG and four triboelectric nanogenerators (TENG) (two contact mode and two sliding mode) were combined. The EMG utilised a dual-Halbach array (NdFeB N52) for an increased magnetic flux density and a coil (460 turns). The TENG-materials were PTFE and Al. The harvester's structure was a bobbin with the coil inside a hollow, 3D printed frame for the Halbach-arrays with a distance of 1.2 mm between coil and magnets. On the outside of the Halbach-frame Al-electrodes were mounted. Around the bobbin and Halbach-frame another frame was located with PTFE-film and cooper electrode attached. Bobbin and outer frame were attached, the Halbach-frame could freely slide inside. Excitation of the device caused the Halbach-frame to slide back and forth along the axis. Thus, electromagnetic induction in the coil as well as electrostatic induction in the sliding mode TENG occurred. The contact mode TENGs were active when the Halbach-frame slid back after touching the ends of the outer frame. The transformer was attached to the TENGs and a multiplier circuit to the EMG in order to achieve a combined, regulated output for the device. Since no spring was applied in the harvester, it showed different power outputs when applied horizontally or vertically due to gravity. The horizontal power output on a Shaker at 6 Hz and 1 g was 5.41 mW with a 1.1 KΩ load. Vertically, 3.28 mW was achieved. The device was also tested under human motions such as handshaking, walking or slow running and delivered up to 2.9 mW. The authors mentioned wearable and portable devices as applications for the harvester.

In [62], a 3D printed wrist-wearable hybrid harvester for human frequencies ≤5 Hz was reported. A hollow curved tube, printed with ABS contained an EMG and TENG. The former consisted of a 10 mm NdFeB N52 magnet ball and two serially connected coils wrapped around the tube. The latter utilised the ABS-tube and the ball as triboelectric material with a PTFE-film between, inside the tube. Al-electrodes were wrapped outside around the tube. The PTFE-ABS composite was applied because the voltage generated just with ABS was not high enough and more suitable materials such as PTFE were not available for 3D printing. Only ABS resulted in a peak-to-peak voltage around 7.5 V. Applying a nano-structured PTFE-film achieved around 20 V. The power density was reported with 5.14 mW cm^{-3} with 49.2 Ω load for the EMG and 0.22 µW cm^{-3} with 13.9 MΩ for the TENG between 3 and 4 Hz. The harvested energy from 5 s running was enough to power an electric wrist watch for 410 s.

In [63], an improved version of [62] was reported. A broader range of human motion patterns could be exploited to harvest energy. The hybrid device combined electromagnetic and triboelectric generators to harvest human wrist motion such as shaking, twisting or waving with ≤5 Hz. The harvester's structure was a 3D printed circular hollow tube made of PLA. The EMG composed of four serially connected coils (300 turns each) evenly distributed and wound around the tube as well as a 10 mm magnet ball inside. A FeSiCr/PDMS-film was wrapped around the coils as flux concentrator to enhance the EMG. Simulations showed that the concentrator increased the flux density from 0.577 T to 1.168 T. The TENG composed of a PTFE-film inside the PLA-tube, four Al-electrodes wrapped around the tube and the magnet ball. The TENG's capabilities were also enhanced; for example, the ball was treated with grit abrasive paper to achieve a micro-structured surface. At 4–5 Hz excitation frequency and a 100 Ω load the EMG had an average power output of 3.46 mW without and 4.98 mW with the flux concentrator. The TENG achieved 0.0058 µW at 12 MΩ on average without and 93 nW with the PTFE-film.

5.2. Spherical

In [64], a multi-direction, low-frequency hybrid device to harvest the motion of waves was presented. The showed harvester with a cylindrical shape was on based on an EMG and a TENG. Four S-shaped TENGs in the bottom supported a printed curved surface above. A magnetic ball was enclosed inside the housing while being able to freely move around on the curved surface. Above the ball's area a coil (1500 turns) was located. Inside the curved surface a free-standing mode TENG was embedded which utilised the ball as positive tribolayer. The authors exploited this to measure the direction and amplitude based on the balls movement. Furthermore, the moving ball caused deformation of the supporting S-shaped contact-separation TENGs (CTENG) as well as a varying magnetic flux through the coil. The maximum power output at 1 Hz were 3.65 mW for the S-shaped TENGs and 22.4 mW for the EMG. The authors also demonstrated a self-powered seawater splitting system and a self-powered electrochemical cathodic protection system with the harvester.

In [65], a spherical hybrid harvester was presented as self-powered six axis inertial sensor for hand motion recognition as well as monitoring human activity. The authors combined electromagnetic, piezoelectric and triboelectric transducers in the device along with multiple magnetic balls inside the hollow 3D printed sphere. All generators were implemented in the shell of the sphere with wound wire or films. The device was characterised by vibrating it periodically along its x-axis with an amplitude of 26 cm. The EMG's two coils with 91 buckyballs had a power output of 13.8 nW and 22.4 nW with loads of 25 Ω and 20 Ω at a frequency around 2.5 Hz. The two piezoelectric films had a power output of 0.12 µW and 0.19 µW with loads of 10 MΩ at circa 2.5 Hz. The triboelectric generator's films showed 0.22 µW and 0.72 µW at 12 MΩ and 15 MΩ at approximately 3 Hz. The overall power output was not mentioned and is, therefore, not shown in Table 3.

5.3. Pendulum

In [66], a non-resonant hybrid EMG-TENG harvester for ultralow frequencies was proposed. The harvester had a printed cylindrical shell (white resin) with a vertical pendulum inside. The pendulum was similar to a metronome but flexible into all directions. The pendulum's shaft consisted of a spring with a printed, hollow rod around to increase its stiffness. The shaft was fixed to the shell at the bottom and on the top end a magnet ($\varnothing$ 22 mm $\times$ 20 mm) was attached. Above the magnet a coil was attached onto the shell's ceiling and four TENGs were attached to the inner walls of the cylindrical shell. The TENGs were designed with a double helix structure made of copper-foil and FEP-film. When exposed to external excitation the pendulum with the magnet started swinging around, inducing a voltage into the coil due to Faraday's law of induction. When the magnet collided with the TENG's the layers were pressed together, achieving temporary contact, thus resulting in triboelectrification. At 2.2 Hz, the peak power output for the EMG was 523 mw at 280 Ω and for the TENG 470 µW at 0.5 MΩ. A wireless temperature sensor was driven by the harvester and a LTC3106 as power management.

In [67], a piezo-electromagnetic hybrid-harvester with frequency up-conversion was reported. Figure 7 (left) illustrates the harvester's structure. A rotating mass block with integrated magnets was attached via a bearing to a shaft with four cantilevers. Piezo-elements were glued on the cantilevers surface. On the cantilevers' free end, NdFeB magnets were mounted, facing coils (200 turns each) attached to the fixed base. The mass block and the base were 3D printed with PLA. Under excitation the mass block started swinging. When a magnet on a cantilever's free end was passed by a magnet in the mass block, the cantilever was excited in its natural frequency of 42 Hz. Due to this structure frequency up-conversion was provided. The magnets mounted on the cantilevers caused a varying magnetic flux through the coils, resulting in an induced voltage due to Faraday's law. The reported peak power outputs were 1.28 mW for the piezo, 30 µW for the EMG and 1.31 mW when both were coupled. Figure 7 right shows the experimental setup with the printed harvester attached to a shaker.

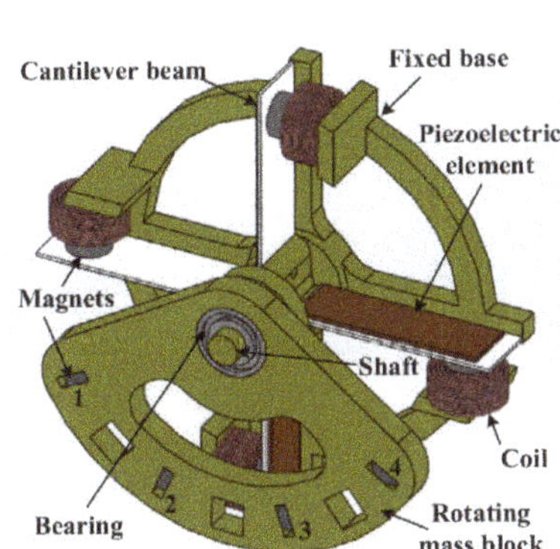

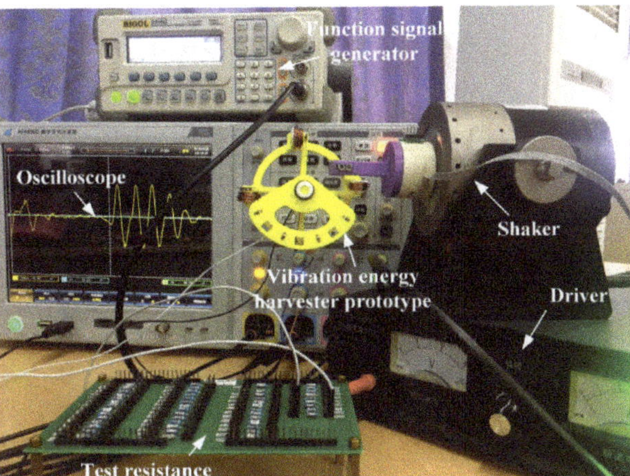

Figure 7. (**left**) Structure of the hybrid piezo-electromagnetic multi-directionalharvester [67], (**right**) experimental setup with printed harvester [67].

5.4. Others

In [68], a hybrid electromagnetic (EMG) and triboelectric nanogenerator (TENG) with three axis motion sensing capabilities was proposed. The vibration harvester's geometry consisted of a central cylinder with four arms symmetrically distributed around made of printed PLA. Inside the centre and each arm, slidable magnets were integrated. At the outer end of the arms, fixed magnets were attached. Coils were located on the top and bottom (1200 turns each) of the centre as well as wound around each arm (500 turns each). Due to repulsion from the centre-magnet the movable magnets were pushed towards the outer end. The TENG, made of micro-structured PTFE film, Al film and copper electrodes, was implemented in the centre's bottom. Due to the moving magnets, a voltage was induced in the coils (EMG) and a charge transfer occurred due to the sliding frictional effect (TENG). The peak power output were 18 mW with 193 Ω for the EMG and 3.25 µW with 10.5 MΩ load for the TENG. The reported resonance-frequency was 7 Hz. The Combination of EMG and TENG increased the voltage of a charged capacitor by around 50% compared to the EMG alone. A version printed with PLA with a higher power output (27 mW EMG and 56 µW TENG) was shown by the same authors in [69].

Table 3. 3D printed electromagnetic hybrid vibration harvesters in the literature.

Ref.	Year	Structure	Combination	Power in mW	Energy Source	Special Feature
[66]	2021	Pendulum	EMG	523 (p)	Blue energy	Multi-direction pendulum
[69]	2020	Cross shape	TENG	0.47 (p)	In-plane motions and vibrations	Linear/rotational motion sensor
			EMG	27 (p)		
			TENG	5.6×10^{-2} (p)		
[64]	2020	Cylinder/	EMG	22.4 (p)	Random low frequency	Vibration amplitude sensor
		spherical	TENG	3.65 (p)	vibrations	
[59]	2020	Straight tube	EMG, TENG	34.11	Human motion, ocean waves, automotive vibration	Universal self-chargeable power module

Table 3. Cont.

Ref.	Year	Structure	Combination	Power in mW	Energy Source	Special Feature
[60]	2020	Straight tube	EMG, TENG	144.1	Human motion	Universal power source
[67]	2020	Pendulum	Piezo	1.28 [(p)]	Human motion,	Frequency up-conversion
			EMG	3×10^{-2} [(p)]	low frequency vibrations	
[68]	2019	Cross shape	EMG	18 [(p)]	In-plane motions and vibrations	Linear/rotational motion sensor
			TENG	3×10^{-3} [(p)]		
[65]	2019	Spherical	EMG, Piezo, TENG	-	Human motion	Six axis inertial sensor
[61]	2018	Straight tube	EMG, TENG	5.41	Human induced vibration	Springless harvester
[62]	2018	Curved tube	EMG, TENG	-	Human motion	Wrist wearable harvester
[63]	2018	Circular tube	EMG	4.98	Human wrist motion	Flux concentrator
			TENG	9.3×10^{-5}		

Note: If a dedicated power output was given for EMG, TENG or piezo they were separated by a line break. [(p)] peak power.

6. Considerable Fields of Investigation for Future Vibration Harvesters

In this section, a few compression harvesters as well as some other interesting approaches are listed. For instance, multi-direction-harvesting, functional materials or micro-organism-harvesting. These might provide inspiration and impulses to adapt functionalities in order to create new approaches or improvements for vibration harvesting.

6.1. Compression

In [70], a spherical multi-direction harvester with a moulded silicone-shell was presented. The devices core was printed with PLA with guiding tubes for two integrated magnet balls. Coils were mounted around the tubes. Figure 8 shows the harvester's internal structure. The authors applied a compressible silicone-shell. Under compression, the shell actuated olive oil, which then forced the magnet balls to move, thus passing a coil and inducing a voltage. Once the deformation of the shell was gone, the fluid flowed back into its original location and pushed the balls into the opposite direction through the coil. This combination of shell and fluid allowed the multi-directional actuation. Olive oil was applied due to its high viscosity to move the balls. A less viscous fluid such as water would flow around the balls without moving them. The harvester was tested under frequencies of 4–15 Hz and achieved between 17 and 44 mV.

In [71], a harvester was presented, which translated compression from human walking (linear movement) into a rotary motion. Once compressed, the harvester kept rotating inertially. The housing of the device was 3D printed. The device applied a twisted steel rod. Due to compression, the rod actuated a ratchet-pawl-combination and the rotor started spinning. A spring pulled the rotor back up to the starting position. While rotating the eight NdFeB N52-magnets on the rotor and four fixed coils (400 turns each) converted mechanical energy to the electrical domain according to Faraday's law. A power output of 11 mW at 0.84 Hz and 85 Ω load was achieved. The harvester was also integrated into a shoe to harvest human motion. Under running conditions (9 km/h), a power output of 85 mW was observed.

In [72], a rectangular piece of magnetostrictive material (2826 MB) was embedded inside an FFF printed bone, made of PLA. Around the bone a collector coil (4000 turns, 0.385 mm wire diameter) was attached along with a bias NdFeB magnet ($150 \times 20 \times 10 \, \text{mm}^3$). The external magnet was necessary to provide a working magnetic field. Applied axial stress applied to the bone caused a compression of the magnetostrictive material and therefore a change of its permeability due to the Villari effect. Thus, the magnetic flux density changed and induced a voltage in the coil according to Faraday's law. Various human walking-conditions were simulated for testing. The maximum harvested power was 0.1 mW during simulated quick running.

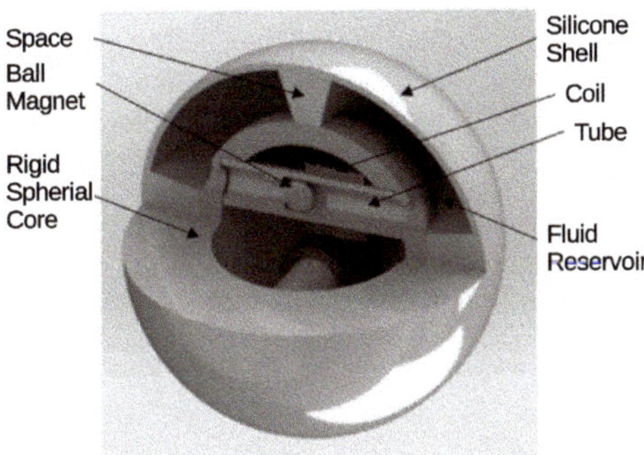

Figure 8. Image showing the harvester's internal structure [70].

6.2. Others

In [73], a rotational harvester with a 3D printed magnet was proposed for energy harvesting from power transmission lines. The NdFeB-magnet was made of compound material Neofer 25/60p. It was printed via FFF on a low-cost 3D printer. The magnet was mounted on a shaft with two bearings, which were fixed in a frame. A coil (750 turns) was wound around the frame. Two devices were built and tested in a two axis Helmholtz coil platform. A DC bias magnetic field was required for centring the magnet in its equilibrium position. An AC magnetic field then induced a rotary movement of the magnet. One device's magnet has been topologically optimised in order to achieve a more homogeneous radial magnetic field. The optimised magnet was recessed on two sides of its cylindrical shape. As a consequence, the distortion power factor of the optimised harvester increased by 55%. However, the power output decreased by 25% due to the lower magnet volume. At resonance, a power output of 93 mW with a 10 Ω load was achieved. The applied DC magnetic field had an impact on the device's resonance frequency. Applications mentioned were wireless sensor networks and IoT.

An interesting approach was presented in [74], where micro-organisms (phytoplankton and zooplankton) were utilised to power 3D printed actuation mechanisms. The movement of the micro-organisms was controlled by geometric forms as well as external stimuli. Phototaxis as one kind of stimulus caused micro-organisms to move towards (positive) or away from (negative) a light source. A linear movement was been achieved with a printed float with a fin beneath and matured Artemia sallina. The fin separated a channel with blue LEDs on both ends. If the LED was turned on, the negative phototaxis organisms escaped from it. They collided with the fin and pushed it to the opposite direction. With 50 Artemia, an average speed of 0.21 mm/s and a driving force of 0.537 mN were observed. This equalled 0.11 mW per organism. For practical application, a higher number of Artemia would be necessary. A rotary movement of 0.4 rpm was achieved with a printed ratchet and 300 Artemia in their larva-stage (positive phototaxis). The ratchet was printed on KEYENCE AGILISTA-3100 (inkjet) and AR-M2 (UV cureable resin) as material. A mask was applied to cause positive phototaxis movement around the ratchet. A ratchet made by photolithography was also tested with another organism (Volvox). Two ratchet-designs were tested and obtained—0.86 rpm and 2.01 rpm. A conversion from the kinetic into the electric domain has not been implemented.

7. Discussion

Table 2 summarises the reviewed non-hybrid harvesters. The first printed vibration harvesters (based on the criteria of this review) were presented in 2008. Figure 9 shows the distribution regarding the shape. Tube-like harvesters (straight or circular) are the biggest category while other categories are rather evenly distributed. The power outputs of the harvesters vary from 3 µW to 76 mW. The power density would provide a good comparison, taking the size into account. However, the harvester's dimensions were not provided in most papers. Therefore, size or power density could not be investigated properly and were not listed in the tables. Figure 10 shows the output power grouped into <1 mW, from 1–10 mW and >10 mW with average and peak power separated. It can be seen that the majority of the non-hybrid harvesters fall into the range of 1–10 mW. The harvested energy sources are mostly human motions or mechanical vibrations. Regarding the special feature-column, there are various interesting topics, from miniaturisation to coil optimisation or expanding the harvester's degree of freedoms.

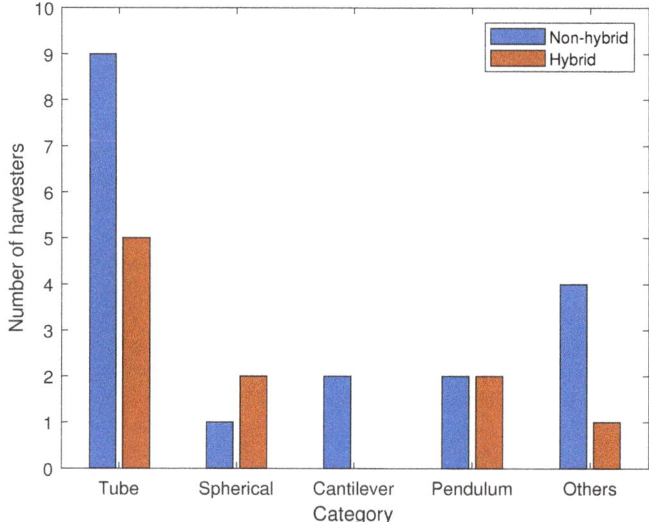

Figure 9. Bar graph with 3D printed EM hybrid and non-hybrid vibration harvesters grouped into categories.

Hybrid harvesters are presented in Table 3. A frequently applied shape is tube-like, as shown in Figure 9. EMG were mostly combined with TENG. Piezo-transducer were only utilised two times. Regarding the average power output, which was not given in some papers, 5 mW to 144 mW were found. The peak power outputs vary from around 1.3 mW to over 520 mW. The common targeted energy sources are human motions as well as low frequency vibrations such as blue energy. The special features are the exploitation of the harvester as sensor, utilisation of flux concentrators or multi-direction-harvesting.

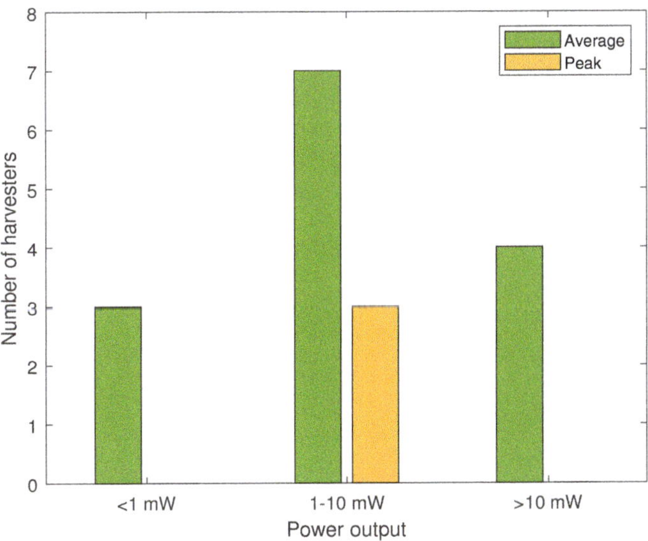

Figure 10. Bar graph with power outputs of 3D printed EM non-hybrid vibration harvesters grouped into three power classes.

7.1. Utilisation of 3D Printing

Tables 4 and 5 show the reviewed harvesters, the applied printing technology, the utilised material, the printed parts as well as the utilisation of 3D printing categorised by [41]. Binder jetting and FFF were applied first in 2008, while FFF was the most common process in the following years. Harvesters printed via inkjet came up quite recently. Inkjet offers better printing resolutions than FFF and therefore potential for miniaturised harvesters as seen in [44]. In some cases, the technology was not mentioned, but for ABS/PLA, it was most likely that FFF was utilised. As for materials ABS and PLA, these were printed most commonly for hybrid and non-hybrid, followed by resin-materials. It can be seen that none of the harvesters utilised functional materials, indicating a field worth investigating. Based on [41] utilising 3D printing for vibration harvesters can be categorised into fabricating support structures or suspension mechanisms. In [73], (rotational harvester) magnetic material was 3D printed for an optimised magnet topology in order to achieve a more homogeneous radial magnetic field. Other functional materials such as carbon fibre filaments or conductive materials could also offer benefits for vibration harvesters, their longevity or transducer mechanisms.

Table 4. Comparison of applied 3D printing techniques for electromagnetic vibration harvesters.

Ref.	Year	Structure	Process	Material	Printed Parts	Utilisation 3D Printing
[51]	2021	Cantilever	FFF	PLA	Cantilever	Suspension
[43]	2020	Straight tube	FFF	PLA	Tube	Support
[53]	2020	Pendulum	FFF (assumed)	ABS	Housing, pendulum	Support
[44]	2020	Cubic tube	Inkjet	Visijet M3 Crystal, S300	Spring, frame	Suspension
[52]	2020	Cantilever	FFF (assumed)	-	Cantilever, coil holder	Suspension
[45]	2019	Straight tube [1]	-	-	Tube	Support
[46]	2019	Cycloid tube	-	PLA	Tube	Support
[47]	2019	Straight tube	-	UV curable resin	Tube	Support
[55]	2019	Manipulator type	FFF	ABSplus	Main body, proof masses	Suspension
[54]	2019	Pendulum	Inkjet	Visijet M3 Crystal, S300	Housing, pendulum	Support
[48]	2018	Straight tube	-	Resin	Tube	Support
[49]	2017	Straight tube	-	-	Tube	Support
[50]	2017	Straight tube	FFF	HIPS	Housing	Support
[41]	2016	LITF pivot	FFF	ABS	Main body	Suspension
[24]	2016	Straight tube	FFF	ABS	Spiral/spring, housing	Suspension
[56]	2009	3D spring	FFF	ABS	3D spring	Suspension
[57]	2008	Spherical	Binder jet	Powder/resin	Housing	Support
[58]	2008	Moving arm	FFF	ABS	Immovable parts	Support

[1] final prototype not 3D printed.

Table 5. Comparison of applied 3D printing techniques for hybrid electromagnetic vibration harvesters.

Ref.	Year	Structure	Process	Material	Printed Parts	Utilisation 3D Printing
[66]	2021	Pendulum	-	Resin	Shell, magnet support, adjusting stud, rod	Support
[69]	2020	Cross shape	-	PLA	Body, covers	Support
[64]	2020	Cylinder	-	-	Disc, circular surface	Support
[59]	2020	Tube	-	PLA	Main hollow body	Support
[60]	2020	Tube	FFF (assumed)	ABS	Frame/Tube	Support
[67]	2020	Pendulum	-	PLA	Mass block, base	Support
[68]	2019	Cross shape	-	PLA	Body	Support
[65]	2019	Spherical	-	-	Shell	Support
[61]	2018	Tube	-	-	Housing	Support
[62]	2018	Curved tube	-	ABS	Tube	Support
[63]	2018	Circular tube	-	PLA	Tube	Support

7.2. Advantages of 3D Printing

Three-dimensional printing offers many advantages such as a faster, easier and cheaper process compared to traditional fabrication [46,54,62]. It is possible to create monolithically fabricated 3D structures in a single process [41] and, in general, complex structures are more feasible and cheaper to fabricate [62]. Furthermore, unintentional electromagnetic coupling can be avoided with 3D printing materials [53]. Miniaturisation-opportunities have been shown for inkjet [44]. Taking a look outside the box, in [75] a triboelectric generator was realised with polyamide and the rubber-like material TangoBlack. Based on the work covered in this paper, the utilisation of multi-material-printing with such materials or others has not been found. Therefore, multi-material-printing might be a field worth investigating for novel vibration generator designs.

7.3. Disadvantages of 3D Printing

In [45], small ridges from printing resulted in an increase in the friction coefficient. Based on the printing technique, material, layer resolution, orientation in the build camber as well as the printed geometry itself, the characteristic transitions from layer to layer could be an issue. Possible solutions are presented in the following section. Regarding hybrid harvesters/TENGs, certain materials (such as PTFE) offering better performance are not available yet [62]. In [24], a performance-decrease in the harvester's printed spring during testing was observed. Therefore, the long-term behaviour of printed materials in regard to their application should be considered. Simulations regarding stress and fatigue for 3D printed harvester parts should be investigated as well. Regarding inkjet, the high absorption of electromagnetic waves for photoactive compounds mentioned in [33] should be examined.

7.4. Enhancing Performance

In order to increase the harvester's power output in [63] a FeSiCr/PDMS-film was wrapped around the coils as flux concentrator. Simulations showed that the concentrator increased the flux density from 0.577 T to 1.168 T. In [45], ridges caused by the layer-stacking when printing were a drawback of 3D printing. In [47], intermittent sliding contact might have caused differences between experiment and simulation. Three-dimensional printing was not mentioned as reason, but reducing the surface's friction might improve the harvesters performance. In [60], a dry lubricant was applied to counter this issue. In [49], graphite powder was used as lubricant to reduce the friction and in [43] guide rails were utilised. In [50], a reduction in viscous damping was achieved with holes in the housing to improve the airflow. Furthermore, oblique springs were applied to align the magnet [50].

7.5. Applications

There are various applications mentioned; specific applications as well as potential fields of application. The most common fields are WSNs [44,52,58,66] as well as wearables/portable devices [24,46,54,62,63]. Health/fitness monitoring [62,63] are a subcategory of wearables. For WSNs, animal monitoring [54], forest monitoring [54] or environmental monitoring [64] are named. Specific applications in the field of portable devices are a self-chargeable power module [59] and a universal power source [60]. Furthermore, a self-powered seawater splitting system [64] and a self-powered electrochemical cathodic protection system [64] are mentioned. A very promising development is the exploitation of the harvester directly as sensor [64,65,68,69].

8. Conclusions

This review summarised the 3D printed EM vibration harvesters over the past 13 years. An introduction with the different energy harvesting categories and their combination was presented. The applied methodology for the review was explained. A total of 18 different 3D printed non-hybrid and 11 different hybrid harvesters were reviewed, compared and discussed. Maximum average power outputs of 76 mW (non-hybrid) and 144 mW (hybrid) were found. The tube-shape was the most applied geometry. A dedicated analysis on the utilisation of 3D printing for vibration harvesters was conducted. It has been shown that 3D printing is suitable to manufacture vibration harvesters. FFF was the most applied printing technology for non-hybrid and hybrid vibration harvesters. Research gaps were derived with multi-material-printing, functional materials or multi-directional harvesting as well as movement–conversion approaches. Furthermore, micro-organism-harvesting could be a future field of research. Applications for printed harvesters were gathered and listed. Finally, the remaining challenges and opportunities are summarised; challenges are the following:

- Finding miniaturisation strategies for harvesters utilising FFF.
- Identifying more strategies for FFF to overcome the drawback of the characteristic surface-ridges.
- Investigating the long-term behaviour of 3D printed suspensions and potential differences between the printing techniques in this regard.
- Examining the effect of photoactive compounds on electromagnetic waves for inkjet.

  Arising opportunities are the following:

- The monolithic printing of harvesters.
- No dependency on additional tools or moulds, thus quick realisation of variants.
- Fabricating complex and filigree structures.
- Faster and easier production than with traditional techniques and, therefore, mass customisation capabilities. This allows for tailoring the harvester specifically towards the application's surrounding conditions.
- Printing multiple materials in one process as well as the availability of functional materials and materials without electromagnetic coupling.
- Miniaturisation with inkjet (volumes less than 1 cm^3).

Author Contributions: P.G. is the lead author of this review. He conceptualised, investigated and wrote the original version of this paper. T.M.W., L.S., N.H. and U.B.H. contributed by reviewing the paper and T.M.W. by project administration. All authors have read and agreed to the published version of the manuscript.

Funding: The article processing charge was funded by the Baden-Württemberg Ministry of Science, Research and Culture and the Offenburg University of Applied Sciences in the funding programme Open Access Publishing.

Institutional Review Board Statement: Not applicable.

Informed Consent Statement: Not applicable.

Data Availability Statement: The raw data of graphs presented in this review are available on request from the corresponding author.

Conflicts of Interest: The authors declare no conflict of interest.

Abbreviations

The following abbreviations are used in this manuscript:

ABS	Acrylonitrile Butadiene Styrene
AM	Additive Manufacturing
AWG	American Wire Gauge
BJT	Binder Jetting
CAD	Computer-Aided Design
CTENG	Contact-separation Mode Triboelectric Nanogenerator
DC	Direct Current
DOF	Degree Of Freedom
EH	Energy Harvesting
EM	Electromagnetic
EMG	Electromagnetic Generator
FDM	Fused Deposition Modeling
FEP	Fluorinated Ethylene Propylene

FFF	Fused Filament Fabrication
HIPS	High Impact Polystyrene
LITF	Leaf Isosceles Trapezoidal Flexural
MEMS	Microelectromechanical System
NFC	Near Field Communication
PDMS	Polydimethylsiloxane
PLA	Polylactic Acid
PTFE	Polytetrafluoroethylene
RF	Radio Frequency
STENG	Sliding Mode Triboelectric Nanogenerator
TEG	Thermoelectric Generator
TENG	Triboelectric Nanogenerator
WSN	Wireless Sensor Node

References

1. Ku, M.L.; Li, W.; Chen, Y.; Ray Liu, K.J. Advances in Energy Harvesting Communications: Past, Present, and Future Challenges. *IEEE Commun. Surv. Tutor.* **2016**, *18*, 1384–1412. [CrossRef]
2. Elvin, N.; Erturk, A. *Advances in Energy Harvesting Methods*; Springer: New York, NY, USA, 2013. [CrossRef]
3. Tang, L.; Yang, Y.; Soh, C.K. Toward Broadband Vibration-based Energy Harvesting. *J. Intell. Mater. Syst. Struct.* **2010**, *21*, 1867–1897. [CrossRef]
4. Chien, L.J.; Drieberg, M.; Sebastian, P.; Hiung, L.H. A simple solar energy harvester for wireless sensor networks. In Proceedings of the 2016 6th International Conference on Intelligent and Advanced Systems (ICIAS), Kuala Lumpur, Malaysia, 15–17 August 2016; pp. 1–6. [CrossRef]
5. Senivasan, S.; Drieberg, M.; Singh, B.S.M.; Sebastian, P.; Hiung, L.H. An MPPT micro solar energy harvester for wireless sensor networks. In Proceedings of the 2017 IEEE 13th International Colloquium on Signal Processing Its Applications (CSPA), Penang, Malaysia, 10–12 March 2017. [CrossRef]
6. Kokert, J.; Beckedahl, T.; Reindl, L.M. Medlay: A Reconfigurable Micro-Power Management to Investigate Self-Powered Systems. *Sensors* **2018**, *18*, 259. [CrossRef] [PubMed]
7. Pubill, D.; Serra, J.; Verikoukis, C. Harvesting artificial light indoors to power perpetually a Wireless Sensor Network node. In Proceedings of the 2018 IEEE 23rd International Workshop on Computer Aided Modeling and Design of Communication Links and Networks (CAMAD), Barcelona, Spain, 17–19 September 2018; pp. 1–6.
8. Ma, X.; Bader, S.; Oelmann, B. Characterization of Indoor Light Conditions by Light Source Classification. *IEEE Sens. J.* **2017**, *17*, 3884–3891. [CrossRef]
9. Verma, G.; Sharma, V. A Novel Thermoelectric energy harvester for Wireless Sensor Network Application. *IEEE Trans. Ind. Electron.* **2019**, *66*, 3530–3538. [CrossRef]
10. Moser, A.; Erd, M.; Kostic, M.; Cobry, K.; Kroener, M.; Woias, P. Thermoelectric Energy Harvesting from Transient Ambient Temperature Gradients. *J. Electron. Mater.* **2012**, *41*, 1653–1661. [CrossRef]
11. Yedavalli, P.S.; Riihonen, T.; Wang, X.; Rabaey, J.M. Far-Field RF Wireless Power Transfer with Blind Adaptive Beamforming for Internet of Things Devices. *IEEE Access* **2017**, *5*, 1743–1752. [CrossRef]
12. Boada, M.; Lazaro, A.; Villarino, R.; Gil, E.; Girbau, D. Near-Field Soil Moisture Sensor with Energy Harvesting Capability. In Proceedings of the 2018 48th European Microwave Conference (EuMC), Madrid, Spain, 23–27 September 2018; pp. 235–238. [CrossRef]
13. Liu, H.; Hou, C.; Lin, J.; Li, Y.; Shi, Q.; Chen, T.; Sun, L.; Lee, C. A non-resonant rotational electromagnetic energy harvester for low-frequency and irregular human motion. *Appl. Phys. Lett.* **2018**, *113*, 203901. [CrossRef]
14. Tan, Y.; Dong, Y.; Wang, X. Review of MEMS Electromagnetic Vibration Energy Harvester. *J. Microelectromechanical Syst.* **2017**, *26*, 1–16. [CrossRef]
15. Wang, L.; Todaria, P.; Pandey, A.; O'Connor, J.; Chernow, B.; Zuo, L. An Electromagnetic Speed Bump Energy Harvester and Its Interactions With Vehicles. *IEEE/ASME Trans. Mechatronics* **2016**, *21*, 1985–1994. [CrossRef]
16. Pourshaban, E.; Karkhanis, M.U.; Deshpande, A.; Banerjee, A.; Ghosh, C.; Kim, H.; Mastrangelo, C.H. Flexible Electrostatic Energy Harvester Driven by Cyclic Eye Tear Wetting and Dewetting. In Proceedings of the 2021 IEEE International Conference on Flexible and Printable Sensors and Systems (FLEPS), virtually, 20–23 June 2021; pp. 1–4. [CrossRef]
17. Qian, Y.; Yu, J.; Zhang, F.; Kang, Y.; Su, C.; Pang, H. Facile synthesis of sub-10 nm ZnS/ZnO nanoflakes for high-performance flexible triboelectric nanogenerators. *Nano Energy* **2021**, *88*, 106256. [CrossRef]
18. Qian, Y.; Lyu, Z.; Kim, D.H.; Kang, D.J. Enhancing the output power density of polydimethylsiloxane-based flexible triboelectric nanogenerators with ultrathin nickel telluride nanobelts as a co-triboelectric layer. *Nano Energy* **2021**, *90*, 106536. [CrossRef]
19. Zhang, Y.; Wang, T.; Luo, A.; Hu, Y.; Li, X.; Wang, F. Micro electrostatic energy harvester with both broad bandwidth and high normalized power density. *Appl. Energy* **2018**, *212*, 362–371. [CrossRef]
20. Aljadiri, R.T.; Taha, L.Y.; Ivey, P. Electrostatic Energy Harvesting Systems: A Better Understanding of Their Sustainability. *J. Clean Energy Technol.* **2017**, *5*, 409–416. [CrossRef]

21. Yuan, X.; Gao, X.; Yang, J.; Shen, X.; Li, Z.; You, S.; Wang, Z.; Dong, S. The large piezoelectricity and high power density of a 3D-printed multilayer copolymer in a rugby ball-structured mechanical energy harvester. *Energy Environ. Sci.* **2020**, *13*, 152–161. [CrossRef]

22. Thakare, N.S.; Thakare, S.S.; Shahakar, R.S. Recent Advancement and Comparative Performance Analysis of Energy Harvesting Technique. In Proceedings of the 2018 Second International Conference on Intelligent Computing and Control Systems (ICICCS), Madurai, India, 14–15 June 2018; pp. 1631–1634. [CrossRef]

23. Zhao, J.; You, Z. A shoe-embedded piezoelectric energy harvester for wearable sensors. *Sensors* **2014**, *14*, 12497–12510. [CrossRef] [PubMed]

24. Garcia-Moreno, P.; Perez, M.E.; Estevez, F.J.; Gloesekoetter, P. Study of Wearable and 3D-Printable Vibration-Based Energy Harvesters. In Proceedings of the 2016 15th International Conference on Ubiquitous Computing and Communications and 2016 International Symposium on Cyberspace and Security (IUCC-CSS), Granada, Spain, 14–16 December 2016; pp. 101–108. [CrossRef]

25. Herawan, S.G.; Syahputra, S.A.; Tokit, E.M.; Sa'at, F.A.Z.M.; Rosli, M.A.M. Effect of number of permanent magnetic poles on 3D printed coreless generator rotor. *IOP Conf. Ser. Mater. Sci. Eng.* **2021**, *1082*, 012009. [CrossRef]

26. Herawan, S.G.; Syahputra, S.A.; Tokit, E.M.; Sa'at, F.A.Z.M.; Rosli, M.A.M. Energy harvesting applications using 3D-printed coreless generator. *IOP Conf. Ser. Mater. Sci. Eng.* **2021**, *1082*, 012004. [CrossRef]

27. Soemphol, C.; Angkawisittpan, N. 3D-printed materials based low-speed permanent magnet generator for energy harvesting applications. *Mater. Today Proc.* **2020**, *22*, 180–184. [CrossRef]

28. Adamski, K.T.; Adamski, J.W.; Urbaniak, L.; Dziuban, J.A.; Walczak, R.D. 3D Printed Miniature Water Turbine with Integrated Discrete Electronic Elements for Energy Harvesting and Water Flow Measurement. *J. Phys. Conf. Ser.* **2018**, *1052*, 012086. [CrossRef]

29. Lee, J.; Jeon, G.; Kim, S. Magnetically axial-coupled propeller-based portable electromangetic energy-harvesting device using air and water stream. In Proceedings of the 2018 IEEE International Magnetics Conference (INTERMAG), Singapore, 23–27 April 2018; p. 1. [CrossRef]

30. Han, N.; Zhao, D.; Schluter, J.U.; Goh, E.S.; Zhao, H.; Jin, X. Performance evaluation of 3D printed miniature electromagnetic energy harvesters driven by air flow. *Appl. Energy* **2016**, *178*, 672–680. [CrossRef]

31. Maamer, B.; Boughamoura, A.; Fath El-Bab, A.M.; Francis, L.A.; Tounsi, F. A review on design improvements and techniques for mechanical energy harvesting using piezoelectric and electromagnetic schemes. *Energy Convers. Manag.* **2019**, *199*, 111973. [CrossRef]

32. Gibson, I.; Rosen, D.; Stucker, B.; Khorasani, M. *Additive Manufacturing Technologies*, 3rd ed.; Springer: Cham, Swizterland, 2021. [CrossRef]

33. Calignano, F.; Manfredi, D.; Ambrosio, E.P.; Biamino, S.; Lombardi, M.; Atzeni, E.; Salmi, A.; Minetola, P.; Iuliano, L.; Fino, P. Overview on Additive Manufacturing Technologies. *Proc. IEEE* **2017**, *105*, 593–612. [CrossRef]

34. Chen, T.C.T.; Lin, Y.C. A three-dimensional-printing-based agile and ubiquitous additive manufacturing system. *Robot. Comput.-Integr. Manuf.* **2019**, *55*, 88–95. [CrossRef]

35. Darwish, L.R.; El-Wakad, M.T.; Farag, M.M. Towards sustainable industry 4.0: A green real-time IIoT multitask scheduling architecture for distributed 3D printing services. *J. Manuf. Syst.* **2021**, *61*, 196–209. [CrossRef]

36. Junk, S.; Gawron, P.; Schröder, W. Development of an Additively Manufactured Adaptive Wing Using Digital Materials. In *Sustainable Design and Manufacturing*; Ball, P., Huaccho Huatuco, L., Howlett, R.J., Setchi, R., Eds.; Springer: Singapore, 2019; pp. 49–59.

37. Wendt, T.; Hangst, N.; Gawron, P.; Junk, S. 3D-Druck von leitfähigen Materialien bei gedruckter Sensorik in intelligenten und multifunktional aufgebauten Mensch-Roboter-Kollaborations-Greifsystemen: 3D-Printing of Conductive Materials by Printed Sensors in Intelligent and Multifunctional Human-Robot-Collaboration-Grippers. In *Sensors and Measuring Systems*; VDE: Frankfurt am Main, 2018; pp. 135–138.

38. Wendt, T.; Gawron, P.; Hangst, N. Conductive materials and 3D printing—An overview. In *LOPEC Conference Proceedings 2019*; Messe München GmbH: München, 2019.

39. Flowers, P.F.; Reyes, C.; Ye, S.; Kim, M.J.; Wiley, B.J. 3D printing electronic components and circuits with conductive thermoplastic filament. *Addit. Manuf.* **2017**, *18*, 156–163. [CrossRef]

41. Constantinou, P.; Roy, S. A 3D printed electromagnetic nonlinear vibration energy harvester. *Smart Mater. Struct.* **2016**, *25*, 095053, doi:10.1088/0964-1726/25/9/095053.

41. Constantinou, P.; Roy, S. A 3D printed electromagnetic nonlinear vibration energy harvester. *Smart Mater. Struct.* **2016**, *25*, 095053. [CrossRef]

42. Pautasso, M. Ten simple rules for writing a literature review. *PLoS Comput. Biol.* **2013**, *9*. [CrossRef]

43. Beato-López, J.J.; Royo-Silvestre, I.; Algueta-Miguel, J.M.; Gómez-Polo, C. A Combination of a Vibrational Electromagnetic Energy Harvester and a Giant Magnetoimpedance (GMI) Sensor. *Sensors* **2020**, *20*, 1873. [CrossRef] [PubMed]

44. Kawa, B.; Śliwa, K.; Lee, V.C.; Shi, Q.; Walczak, R. Inkjet 3D Printed MEMS Vibrational Electromagnetic Energy Harvester. *Energies* **2020**, *13*, 2800. [CrossRef]

45. POROBIC, I.; GONTEAN, A. Electromagnetic energy harvester. In Proceedings of the 2019 IEEE 25th International Symposium for Design and Technology in Electronic Packaging (SIITME), Cluj-Napoca, Romania, 23–26 October 2019; pp. 151–154. [CrossRef]

46. Maharjan, P.; Bhatta, T.; Salauddin Rasel, M.; Salauddin, M.; Toyabur Rahman, M.; Park, J.Y. High-performance cycloid inspired wearable electromagnetic energy harvester for scavenging human motion energy. *Appl. Energy* **2019**, *256*, 113987. [CrossRef]

47. Fan, K.; Liang, G.; Zhang, Y.; Tan, Q. Hybridizing linear and nonlinear couplings for constructing two–degree–of–freedom electromagnetic energy harvesters. *Int. J. Energy Res.* **2019**, *5*, 041306. [CrossRef]

48. Zhao, X.; Cai, J.; Guo, Y.; Li, C.; Wang, J.; Zheng, H. Modeling and experimental investigation of an AA-sized electromagnetic generator for harvesting energy from human motion: ACCEPTED MANUSCRIPT. *Smart Mater. Struct.* **2018**, *27*, 085008. [CrossRef]

49. Wang, W.; Cao, J.; Zhang, N.; Lin, J.; Liao, W.H. Magnetic-spring based energy harvesting from human motions: Design, modeling and experiments. *Energy Convers. Manag.* **2017**, *132*, 189–197. [CrossRef]

50. Nammari, A.; Caskey, L.; Negrete, J.; Bardaweel, H. Design and investigation of an enhanced magneto-mechanical nonlinear energy harvester. In *Active and Passive Smart Structures and Integrated Systems 2017*; Park, G., Ed.; SPIE: Bellingham, WA, USA, 2017; p. 101642K. [CrossRef]

51. Han, D.; Shinshi, T.; Kine, M. Energy Scavenging From Low Frequency Vibrations Through a Multi-Pole Thin Magnet and a High-Aspect-Ratio Array Coil. *Int. J. Precis. Eng. Manuf.-Green Technol.* **2021**, *8*, 139–150. [CrossRef]

52. Kulik, M.; Jagieła, M.; Łukaniszyn, M. Surrogacy-Based Maximization of Output Power of a Low-Voltage Vibration Energy Harvesting Device. *Appl. Sci.* **2020**, *10*, 2484. [CrossRef]

53. Ambrożkiewicz, B.; Litak, G.; Wolszczak, P. Modelling of Electromagnetic Energy Harvester with Rotational Pendulum Using Mechanical Vibrations to Scavenge Electrical Energy. *Appl. Sci.* **2020**, *10*, 671. [CrossRef]

54. Adamski, K.; Walczak, R. Pendulum base 3D printed electromagnetic energy harvester. *J. Phys. Conf. Ser.* **2019**, *1407*, 012114. [CrossRef]

55. Kim, H.S.; Ryu, W.; Park, S.b.; Choi, Y.J. 3-Degree-of-freedom electromagnetic vibration energy harvester with serially connected leaf hinge joints. *J. Intell. Mater. Syst. Struct.* **2019**, *30*, 308–322. [CrossRef]

56. Chan, S.C.; Yaul, F.M.; Dominguez-Garcia, A.; O'Sullivan, F.; Otten, D.M.; Lang, J.H. Harvesting energy from moth vibrations during flight. In Proceedings of the PowerMEMS 2009, Washington, DC, USA, 1–4 December 2009; pp. 57–60.

57. Bowers, B.J.; Arnold, D. Spherical Magnetic Generators for Bio-Motional Energy Harvesting. In Proceedings of the PowerMEMS 2008, Sendai, Japan, 9–12 November 2008; pp. 281–284.

58. Hadas, Z.; Zouhar, J.; Singule, V.; Ondrusek, C. Design of Energy Harvesting Generator Base on Rapid Prototyping Parts. In Proceedings of the 2008 13th International Power Electronics and Motion Control Conference, Poznań, Poland, 1–3 September 2008; pp. 1665–1669.

59. Maharjan, P.; Bhatta, T.; Cho, H.; Hui, X.; Park, C.; Yoon, S.; Salauddin, M.; Rahman, M.T.; Rana, S.S.; Park, J.Y. A Fully Functional Universal Self–Chargeable Power Module for Portable/Wearable Electronics and Self–Powered IoT Applications. *Adv. Energy Mater.* **2020**, *10*, 2002782. [CrossRef]

60. Rahman, M.T.; Rana, S.S.; Salauddin, M.; Maharjan, P.; Bhatta, T.; Park, J.Y. Biomechanical Energy–Driven Hybridized Generator as a Universal Portable Power Source for Smart/Wearable Electronics. *Adv. Energy Mater.* **2020**, *10*, 1903663. [CrossRef]

61. Salauddin, M.; Toyabur, R.M.; Maharjan, P.; Rasel, M.S.; Kim, J.W.; Cho, H.; Park, J.Y. Miniaturized springless hybrid nanogenerator for powering portable and wearable electronic devices from human-body-induced vibration. *Nano Energy* **2018**, *51*, 61–72. [CrossRef]

62. Maharjan, P.; Cho, H.; Rasel, M.S.; Salauddin, M.; Park, J.Y. A fully enclosed, 3D printed, hybridized nanogenerator with flexible flux concentrator for harvesting diverse human biomechanical energy. *Nano Energy* **2018**, *53*, 213–224. [CrossRef]

63. Maharjan, P.; Toyabur, R.M.; Park, J.Y. A human locomotion inspired hybrid nanogenerator for wrist-wearable electronic device and sensor applications. *Nano Energy* **2018**, *46*, 383–395. [CrossRef]

64. Yang, H.; Deng, M.; Zeng, Q.; Zhang, X.; Hu, J.; Tang, Q.; Yang, H.; Hu, C.; Xi, Y.; Wang, Z.L. Polydirectional Microvibration Energy Collection for Self-Powered Multifunctional Systems Based on Hybridized Nanogenerators. *ACS Nano* **2020**, *14*, 3328–3336. [CrossRef]

65. Koh, K.H.; Shi, Q.; Cao, S.; Ma, D.; Tan, H.Y.; Guo, Z.; Lee, C. A self-powered 3D activity inertial sensor using hybrid sensing mechanisms. *Nano Energy* **2019**, *56*, 651–661. [CrossRef]

66. Xie, W.; Gao, L.; Wu, L.; Chen, X.; Wang, F.; Tong, D.; Zhang, J.; Lan, J.; He, X.; Mu, X.; et al. A Nonresonant Hybridized Electromagnetic-Triboelectric Nanogenerator for Irregular and Ultralow Frequency Blue Energy Harvesting. *Research* **2021**, *2021*, 5963293. [CrossRef] [PubMed]

67. Shi, G.; Chen, J.; Peng, Y.; Shi, M.; Xia, H.; Wang, X.; Ye, Y.; Xia, Y. A Piezo-Electromagnetic Coupling Multi-Directional Vibration Energy Harvester Based on Frequency Up-Conversion Technique. *Micromachines* **2020**, *11*, 80. [CrossRef] [PubMed]

68. Bhatta, T.; Maharjan, P.; Park, J.Y. All-Direction In-Plane Magnetic Repulsion-Based Self-Powered Arbitrary Motion Sensor and Hybrid Nanogenerator. In Proceedings of the 2019 19th International Conference on Micro and Nanotechnology for Power Generation and Energy Conversion Applications (PowerMEMS), Kraków, Poland, 2–6 December 2019; pp. 1–4. [CrossRef]

69. Bhatta, T.; Maharjan, P.; Salauddin, M.; Rahman, M.T.; Rana, S.S.; Park, J.Y. A Battery-Less Arbitrary Motion Sensing System Using Magnetic Repulsion–Based Self–Powered Motion Sensors and Hybrid Nanogenerator. *Adv. Funct. Mater.* **2020**, *30*, 2003276. [CrossRef]

70. Hall, R.G.; Rashidi, R. Multi-Directional Universal Energy Harvesting Ball. *Micromachines* **2021**, *12*, 457. [CrossRef] [PubMed]

71. Luo, A.; Zhang, Y.; Xu, W.; Lu, Y.; Wang, F. Electromagnetic Energy Harvester with Inertial Rotary Structure for Human Motion Application at Ultra-Low Frequency. In Proceedings of the 2020 IEEE 33rd International Conference on Micro Electro Mechanical Systems (MEMS), Vancouver, BC, Canada, 18–22 January 2020; pp. 536–539. [CrossRef]
72. Tan, Y.; Zhang, Y.; Ren, L. Energy Harvesting From an Artificial Bone. *IEEE Access* **2019**, *7*, 120065–120075. [CrossRef]
73. Wang, Z.; Huber, C.; Hu, J.; He, J.; Suess, D.; Wang, S.X. An electrodynamic energy harvester with a 3D printed magnet and optimized topology. *Appl. Phys. Lett.* **2019**, *114*, 013902. [CrossRef]
74. Hatsuzawa, T.; Yanagida, Y.; Nisisako, T. Microorganisms driven micro actuation mechanisms for the kinetic energy harvesting. In Proceedings of the 2017 19th International Conference on Solid-State Sensors, Actuators and Microsystems (TRANSDUCERS), Kaohsiung, Taiwan, 18–22 June 2017; pp. 2067–2070. [CrossRef]
75. Haque, R.I.; Farine, P.A.; Briand, D. 3D Printed Materials Based Triboelectric Device for Energy Harvesting and Sensing. *Proceedings* **2017**, *1*, 580. [CrossRef]

 **MDPI**

Review

Triboelectric Nanogenerators for Energy Harvesting in Ocean: A Review on Application and Hybridization

Ali Matin Nazar [1], King-James Idala Egbe [1], Azam Abdollahi [2] and Mohammad Amin Hariri-Ardebili [3,4,*]

[1] Institute of Port, Coastal and Offshore Engineering, Ocean College, Zhejiang University, Zhoushan 316021, China; ali.matinnazar@zju.edu.cn (A.M.N.); ekjames@zju.edu.cn (K.-J.I.E.)

[2] Department of Civil Engineering, University of Sistan and Baluchestan, Zahedan 45845, Iran; azam.abdollahi@pgs.usb.ac.ir

[3] Department of Civil Environmental and Architectural Engineering, University of Colorado, Boulder, CO 80309, USA

[4] College of Computer, Mathematical and Natural Sciences, University of Maryland, College Park, MD 20742, USA

[*] Correspondence: mohammad.haririardebili@colorado.edu; Tel.: +1-303-990-2451

Abstract: With recent advancements in technology, energy storage for gadgets and sensors has become a challenging task. Among several alternatives, the triboelectric nanogenerators (TENG) have been recognized as one of the most reliable methods to cure conventional battery innovation's inadequacies. A TENG transfers mechanical energy from the surrounding environment into power. Natural energy resources can empower TENGs to create a clean and conveyed energy network, which can finally facilitate the development of different remote gadgets. In this review paper, TENGs targeting various environmental energy resources are systematically summarized. First, a brief introduction is given to the ocean waves' principles, as well as the conventional energy harvesting devices. Next, different TENG systems are discussed in details. Furthermore, hybridization of TENGs with other energy innovations such as solar cells, electromagnetic generators, piezoelectric nanogenerators and magnetic intensity are investigated as an efficient technique to improve their performance. Advantages and disadvantages of different TENG structures are explored. A high level overview is provided on the connection of TENGs with structural health monitoring, artificial intelligence and the path forward.

Keywords: triboelectric nanogenerators; ocean wave; energy harvesting; artificial intelligence; structural health monitoring

Citation: Matin Nazar, A.; Idala Egbe, K.-J.; Abdollahi, A.; Hariri-Ardebili, M.A. Triboelectric Nanogenerators for Energy Harvesting in Ocean: A Review on Application and Hybridization. *Energies* **2021**, *14*, 5600. https://doi.org/10.3390/en14185600

Academic Editor: Dibin Zhu

Received: 19 July 2021
Accepted: 1 September 2021
Published: 7 September 2021

Publisher's Note: MDPI stays neutral with regard to jurisdictional claims in published maps and institutional affiliations.

1. Introduction

Ocean energy is one of the most powerful energy resources in the environment. Waves and tides have a considerable amount of mechanical energy that can be harvested and used toward technological development [1–8]. However, harvesting the ocean's energy is a challenging task because the difference of ions inside the water may damage electronic devices [9]. During the last decade, triboelectric nanogenerators (TENG) have played a remarkable role in ocean energy storage development with numerous benefits [7,8,10]. For example, wind farms have usually been built according to electromagnetism systems and a turbine structure produces environmental noise which is classified as an ecological problem [11–19]. There are several limitations associated with wind farms: They need to be operated under high wind speed, their equipment is very large and the installation price is typically high [20–22]. On the other hand, the TENG can resolve some of the mentioned issues by performing well with lightweight equipment and a low vibration system [23], which allows its application in various situations [24–30]. TENGs have multiple benefits in the context of hybrid energy collection, for instance, the electromagnetism compound with TENG to produce energy based on the ocean waves [31,32]. The principles of TENG were inspired from triboelectrification [33–36]. Triboelectrification is the process by which two originally uncharged bodies become charged when brought into contact and then separated.

When two dielectric materials are physically in contact and producing a triboelectric charge, a potential difference is generated by dividing two surfaces within the mechanical motion [37–39], making electrons flow in the external circuit and therefore, sets with the electrostatic operation. TENGs can be classified into four modes: Vertical contact separation, lateral sliding, single electrode and freestanding triboelectric layer [31,32,40–49]. Unlike traditional energy harvester devices, which are large [23], costly [50] and complicated to fabricate [51–57], TENGs are low damage and lightweight and they frequently have been applied to create ocean energy power [58,59].

The TENG's principles in ocean energy harvesting can be approximately separated into two stages: (1) Direct contact between the tribo-surface and water and an encapsulated design relying on solid–solid contact [60–77] and (2) direct contact between the tribo-surface and water is like raindrop energy harvesting. The method founded on solid–solid contact is influenced by the surface roughness and the contact/triboelectric field [78,79]. Investigating TENG-based ocean energy harvesting includes various perspectives such as structural design, operation optimization and meteorological regulation [80–97]. Furthermore, the large-scale combination of a TENG network relies on its design for improved flexibility and independence, forming self-powered wireless sensor networks [98–101]. The potential ocean energy utilization includes, but is not restricted to, long-term environmental monitoring, navigation at the ocean, decomposition of the hydrogen fuel water and purification of the polluted water [102]. The TENG networks floating on the ocean shore can be created to combine with another energy harvester [19,103,104].

This review article aims to provide an overview and summarizes the main research trends of TENG in the ocean. Section 2 states an overview of the structure and fundamentals of the ocean and sea waves including some discussion on the advantages and disadvantages of the current energy harvesters. Furthermore, a summary is provided on the physics of TENGs and the parameters that affect its trend. Section 3 introduces the hybridization of TENGs with other types of energy harvester including solar cells and electromagnetic and magnetic intensity. The applications of TENGs and the future challenges are discussed in Section 4 and the paper is concluded with a short summary in Section 5.

2. Fundamentals of Ocean Wave Energy and Different TENG Structures

Devices designed to absorb energy from sea waves are typically installed near shores. This choice of location reduces the risk of damage in stormy conditions [105]. However, due to their proximity to the shore, the amount of generated energy is not significant [106]. The majority of the devices in shallow water are attached to the seabed as a support base. However, their performance is improved if they are installed in deep water. On the other hand, there is an argument about the meaning of "deep" water. Some researchers consider tens of meters to be the definition [107], others say the depth is greater than 40 m [106] and another group take a profundity surpassing one-third of the wavelength [108]. However, all of them are agreed that the energy-absorbing devices from the ocean waves show a supreme performance if they are placed in deep water. The reason is pretty simple: Deep water generates more massive waves and subsequently more energy can be extracted. Nevertheless, it should not be forgotten that access to devices in deep waters is more complicated and the maintenance cost is also high. In addition, since they are located far from the shore and also they are exposed to large waves, it is necessary to consider a special design for them (probably with higher safety factor) [109] which increases the overall cost of the project.

2.1. Global Ocean Wave

A wave farm is a set of wave energy-extracting devices working together in a given location in harmony to produce more energy [110,111]. As a result of their numbers, the regional climate condition and the strategies needed to control them, these devices interact to generate more energy and reduce costs [112–117]. Many companies have invested in this technology and have set up a wave farm in different dimensions over the years [118–122]. Wave characteristics such as wavelengths and velocity at the water surface are fundamental features because they analyze how the waves break on the shores. Several countries including the United States, the United kingdom, Portugal and Australia operate wave farms and kinds of power take-off systems, including elastomeric hose pump, hydraulic ram, pimp-to-shore, hydroelectric turbine, air turbine and linear electrical generator.

2.2. Characteristics of Ocean Wave

The wave energy flux formula is formulated in deep waters where the depth of the water is more than half the wavelength. This equation represents the wave power in terms of the wave energy period, T and the significant wave height, H_{m0}, [123]:

$$P = \frac{\rho g^2}{64\pi} H_{m0}^2 T \tag{1}$$

where P is the wave energy flux per unit of wave-crest length. ρ and g are water density and gravitational acceleration, respectively.

On the other hand, the total potential of an ocean wave can be presented as [124]:

$$E = \frac{1}{2}\rho g A^2 \tag{2}$$

where A is the wave amplitude.

The average energy flux or wave power, P_w, is calculated by multiplying the energy term, E, by wave propagation speed, $V_g = \frac{L}{2T}$. In this formula, T is the wave period and L is the wavelength [125].

$$P_w = \frac{1}{2}\rho g A^2 \frac{L}{2T} \tag{3}$$

The dispersion relationship which describes the connection between the wave period and the wavelength, $L = \frac{g T^2}{2\pi}$, is combined with Equation (3) and results in [125]:

$$P_w = \frac{\rho g^2 T A^2}{8\pi} \quad or \quad P_w = \frac{\rho g^2 T H^2}{32\pi} \tag{4}$$

where H is the wave height.

Wave energy generation is an emerging commercial technology, in comparison with the other renewable energy sources. Since the 1890s, many countries located next to the oceans have been trying to harness this high-potential energy source [126]. In general, the energy harvesting devices from ocean waves can be classified according to their location (including shoreline, nearshore and offshore) and the power take-off system (including hydraulic ram, elastomeric hose pump, pump-to-shore, hydroelectric turbine, air turbine and linear electrical generator) [127]. Similar to other renewable energy sources, the energy from the ocean waves is a significant and endless energy source. While it has been used in many countries like China, the United States, Scotland and Australia, there are several disadvantages associated with this type of energy. Table 1 summarizes the pros and cons of this energy source.

Table 1. Comparing various features related to the wave energy harvesting from the ocean.

Advantage	Disadvantage
Renewable: Unlike fossil fuels, which we see running out every day, the energy from waves is renewable and vast [128]	**Not applicable everywhere:** Just like most natural resources, it is location-specific. Thus, this type of energy is only useful for countries with access to the ocean and sea [128].
Environment Friendly: These days, energy production is a significant problem in the world. On the other hand, pollution from energy production is another concern that human beings pay special attention to because pollution from the production of fossil fuels causes global warming. In contrast, the energy from the waves is environmentally friendly [128].	**Danger to the marine ecosystem:** This type of energy poses a threat to aquatic habitats. Some of these devices fixed to the ocean floor, which can damage habitats and sometimes cause sea creatures to collide with turbine blades or even get electrocuted [129].
Easy access: The advantage of easy access arises for nations with borders along the coast with high wave intensity [128]	**Disruption of ship traffic:** Energy-efficient devices from the ocean waves located near shores and in the direction of the wave, a transit point for cargo ships and cruise ships. These devices make it a bit difficult to get around.
Technology growth in this area: : Many devices have been designed and implemented to extract energy in this [128].	**Poor performance in stormy weather:** One of the problems with wave energy extraction devices is poor performance in rough weather, which can even cause severe damage.
Predictable: One of the essential advantages of this type of energy is calculating the amount of energy production and its predictability.	**Noise pollution:** Another downside of these devices is the noise, which can significantly reduce real estate value for areas near the coast [128].
Less dependence on fossil fuels: It reduces dependence on fossil fuels. It can help reduce pollution around the globe by reducing the dependence on fossil-generated energy [128].	**Dependence on wind:** This wave category is driven by wind, so when the wind is not stable, it is not possible to extract significant energy.
Earth protection: Unlike fossil fuels that require deep drilling for extraction, there is no need to damage the earth to extract this type of energy [130]	

Figure 1a illustrates the absorption point device which works by floating on the water's surface. It is maintained by the cables connected to the seafloor. Generally, the perfect point absorber has the same features as a useful wave-maker. The system of absorption points is ocean-level floating structures whose structural physics is designed to have slightly horizontal dimensions relative to their vertical dimensions. The technology has been hailed by researchers working to obtain energy from ocean waves. When the vessel is excited by the waves (point absorption), it works so that the current moves relative to the fixed reference point. They use linear generators. At the top of this device, the degree of freedom glass considered to create the best necessary performance and also this device use the movement in the heave axis [131–134]. The wave energy converter(WEC) system has three main parts: Buoy, power take-off and heave plate. The buoy part consists of components placed in a series and a vertical direction, the most important hydraulic cylinder components and a spring. To prevent the device from drowning, the empty volume inside the device is filled with urethane foam. With this method, if water enters the system through the cracks in the device, it will not cause the system to drown. Another innovative feature of this device is that a urethane-shaped ring is attached to the device's outer wall, increasing buoyancy. The power take-off device uses a spring to maintain linear stress. The WEC's hydraulic system also uses four low-pressure valves to control the movement between the inlet and outlet of the end of the cylinder, which is installed at the outlet and before the flow limiter. The primary purpose of this placement is to control the pressure as a function of the current. The heave plate comprises a steel rod fastened to a steel plate. Cast press weights slide over the steel bar, allowing straightforward modification of the common heave plate mass. A parabolic bowl with a center hole slides over the center shaft and clamps onto the barbell weights. The reason for the heave plate is to supply a counter constrain to the buoy.

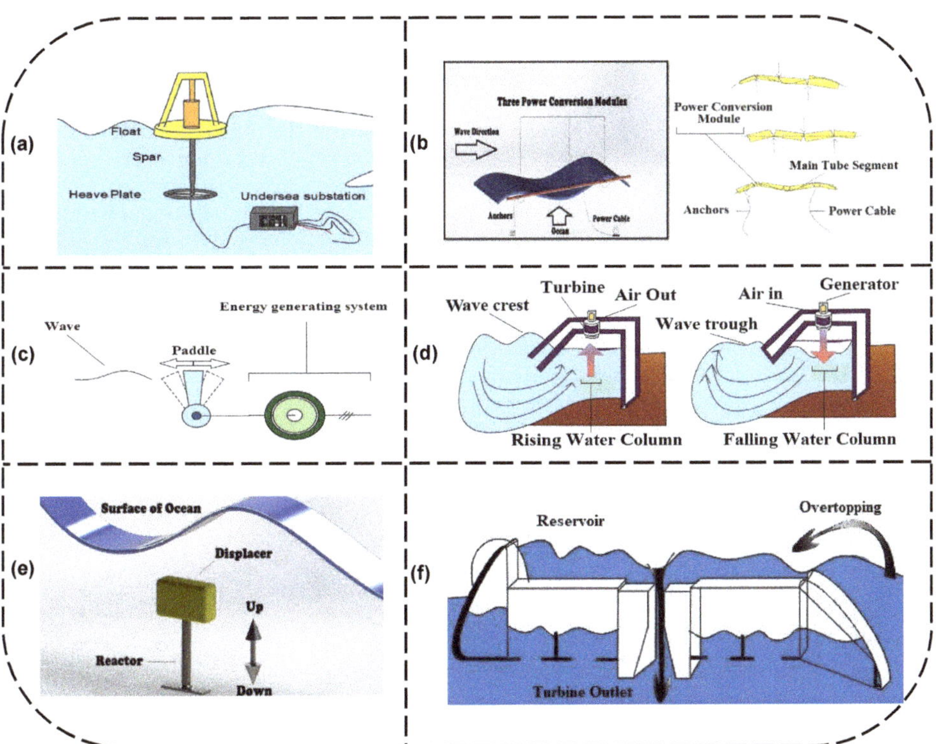

Figure 1. Summary of wave energy converters: (**a**) Point absorber buoy, (**b**) Surface attenuator, (**c**) Oscillating wave surge converter, (**d**) Oscillating water column, (**e**) Submerged pressure differential, (**f**) Overtopping device.

Figure 1b Shows the surface attenuators that are used to convert the energy from ocean waves. They typically move parallel to the ocean waves' direction or parallel to the movement of the ocean waves. In this system, two long pieces are usually connected. The attenuators rely on the flexibility of the joints to produce the power. Examples include surge, sway and heave. One of the most popular and used types of attenuators is WEC Pelamis [134]. Figure 1c demonstrates the oscillating wave surge converter (OWSCs) whose energy is produced using the movement of ocean waves. The structure of this device is made of a central arm or flap, such that, when ocean waves are created, these waves stimulate the designed arm. The flaps are located in a device connected to a generator and its operation is such that it acts as an arm for a large lever and rotates the generator, which generates electricity or, in other words, converts wave energy into electricity. It is possible. These devices are designed to be used frequently in shallow waters [134–136].

Figure 1d shows the air chamber located at the top of the water surface in the device configuration. In this system when a wave travels across the device, the water level rises and falls with standard performance. The system is designed so that the top of the compartment is a turbine as well as a duct. The primary function of the channel is to enter and exit the air. When the water level in the chamber drops, the pressure in the chamber also decreases, which results in a vacuum. In this case, the air comes from outside the device compartment to the inside. This process occurs back and forth. When the water level rises, excess air escapes from the designed turbine. This water movement puts pressure on the turbine blades, causing them to move and generate electricity [137–140].

Figure 1e presents an overtopping device, also called a stabilizer, that uses the ocean waves to generate energy. The device's function is to simulate the same wave motion seen on the shores and, through that simulation, the electricity generated. The Wave Dragon Project is an example of a conceptual design of an overflow hydraulic power plant. The mechanism of this device's operation is such that the two sloping obstacles are concentrated towards the center of the device. This is to optimize energy harvesting. As a result, spilled water moves the turbine to the center, which is a low-pressure turbine. After that, the spilled water is temporarily stored in a reservoir and returned to the ocean. The system acts as a floating marine power plant not connected to the shore. These types of devices are mostly located near the shores. In order to be most efficient, it adjusts its surface height to the height of the waves that occur in the ocean [141,142].

Figure 1f illustrates submerged pressure differential. This system produces energy based on the differential pressure of immersion in the ocean's depths. It is made up of flexible but amplified membranes to extract energy from ocean waves. The device generally uses the pressure difference under the wave at different locations to create a pressure difference in a liquid system that rises from closed power. This pressure difference directs the turbine, which results in the production of electrical energy. There are two different systems, one located near the coast and on the seabed that relies on pressure fluctuations. The other model is similar to point absorption but immersed in water, which floats back and forth in a wave motion, moving a linear generator to convert energy [143–151].

In summary, Table 2 compares and contrasts the advantages and disadvantages of the six above-discussed modern technologies for energy harvesting from the ocean waves.

Table 2. Comparison of modern technologies used for energy harvesting from ocean waves.

Technology	Advantage	Disadvantage
Point absorber buoy	Simple structure	Habitat destruction
Oscillating wave surge converter	Generates electricity in both forward and reverse directions	Habitat destruction
Submerged pressure differential	Can be used on the seabed and as a float	Danger of electricity for aquatic animals
Surface attenuator	High efficiency	Best performance when it is aligned in the wave direction
Overtopping device	Independence of the beach	Danger of sea animals being trapped in a water storage tank
Oscillating water column	Innovative structure	Danger of turbine blades for aquatic animals

2.3. Physics of Triboelectric Nanogenerators

According to Figure 2, several models have been proposed for the TENGs. The first category is the formal physical model, which is implemented according to the classical electromagnetic theory [152–162]. It should be noted that the 3D mathematical models and the distance-dependent electric field model are created according to the quasi-electrostatic model [141,154–156]. In the second category, an equivalent electrical circuit model is notable. The circuit includes the CA model as well as the Norton equivalent circuit model [153–156].

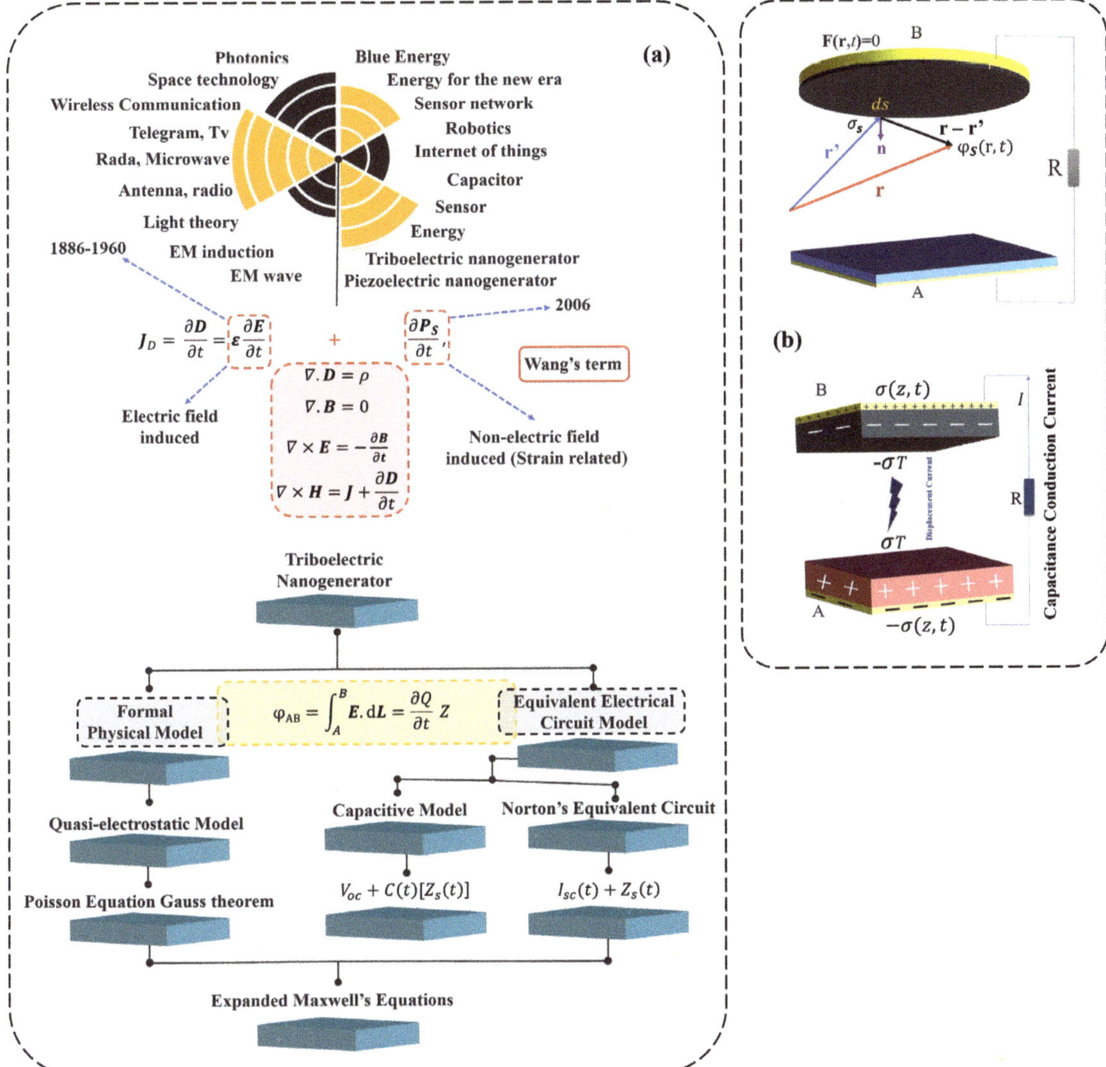

Figure 2. Summary of underpinning theories related to the physics of the TENGs. (a) Comparison of displacement current between Maxwell's equation and Wang's expanded equation and relationships between the formal physical model and the equivalent electrical circuit model (TENG models). (b) Schematic simulation of nano generator and illustration of a vertical contact-separation mode of TENG.

Figure 2 summarizes a transport equation describing the formal physical model and an equivalent electrical circuit model. It should be noted that both models are linked to each other. As it turns out, ϕ_{AB} is a potential drop for the TENG system which appears in the left side of the equation. Moreover, $V = \frac{\partial Q}{\partial t} \times Z$ presents the voltage across the external load (appearing on the right side). According to Kirchhoff's voltage law, the potential difference between two TENG electrodes is equal to the load resistance voltage. The final product is the transportation equation. The physics of TENGs is determination by the variation of potential, ϕ, electric field, E, polarization of the dielectric material, P and the Maxwell's displacement current, I_D. The circuit models determine the outputs from the external circuit, e.g., variation of voltage, V, current, I, power P and extracted electrical energy, E [142].

According to Figure 2, Maxwell's equations, known as Wang's term, are added by the term P_S [157]. It should be noted that Wang's term is not the result of moderate polarization due to the P electric field. Wang's term derives from the existence of electrostatic surface charges:

$$D = \varepsilon_0 E + P + P_s \tag{5}$$

The corresponding displacement current density, J_D, is given by

$$J_D = \frac{\partial D}{\partial t} = \varepsilon_0 \frac{\partial E}{\partial t} + \frac{\partial P}{\partial t} + \frac{\partial P_s}{\partial t} = \varepsilon \frac{\partial E}{\partial t} + \frac{\partial P_s}{\partial t} \tag{6}$$

where ε_0 and ε are permittivity of free space (vacuum) and permittivity of the material (or medium), respectively. These two terms are connected as $\varepsilon \equiv \varepsilon_0(1 + \gamma_e)$, where γ_e presents the electric susceptibility of the medium. Knowing that $P = (\varepsilon - \varepsilon_0)E$, the volume charge density (Equation (7)) and the density of current density (Equation (8)) are defined by

$$\rho' = \rho - \nabla \cdot P_s \tag{7}$$

$$J' = J + \frac{\partial P_s}{\partial t} \tag{8}$$

Satisfying the charge conversion and continuation equation [157]:

$$\nabla \cdot J' + \frac{\partial \rho'}{\partial t} = 0 \tag{9}$$

As a result, Maxwell's equations are rewritten as [157]:

$$\begin{aligned}
\nabla \cdot D' - \rho' &= 0 \\
\nabla \cdot B &= 0 \\
\nabla \times E + \frac{\partial B}{\partial t} &= 0 \\
\nabla \times H - J' - \frac{\partial D'}{\partial t} &= 0
\end{aligned} \tag{10}$$

It is noteworthy that the self-consistent equations mentioned above describe the relationships between electromagnetic fields and charges as well as the current distribution in TENGs [157], where:

$(\varepsilon \partial E/\partial t)$: Well-known contribution to Maxwell's displacement current.

$(\varepsilon \partial P_s/\partial t)$: Displacement current due to the presence of surface charges

$$\phi_{AB} = \int_A^B E \cdot dL = \frac{\partial Q}{\partial t} Z \tag{11}$$

The equation mentioned in this section is very important. Because this equation is a link between the internal circuit and the external circuit. It is also noteworthy that, by calculating the surface integral J_D, the displacement current I_D is obtained [142,157,163].

$$I_D = \int J_D \cdot ds = \int \frac{\partial D}{\partial t} \cdot ds = \frac{\partial}{\partial t} \int (\nabla \cdot D) dr = \frac{\partial}{\partial t} \int \rho \, dr = \frac{\partial Q}{\partial t} \tag{12}$$

The following results are based on the fact that Q is a free charge of the electrode [142]:

- The displacement current is the internal driving force in TENGs. While conducting current, the received current is on the load.
- Ideally, the conduction current is equal to the displacement current.
- The conduction current and the displacement current form a complete loop in the TENG electrodes (where they are connected). Moreover, using formal physical and equivalent electrical circuit models, the TENG outputs are fully predictable [157].

In general, there are four main modes for TENG, each of them with unique features and benefits, see Figure 3, i.e., freestanding triboelectric layer mode, single electrode mode, lateral sliding mode and contact separation mode. The general basis of the TENG function, in all modes, is the transfer of electrostatic charges to the electrodes. All TENG modes, except single electrode mode, use two electrodes. When a displacement is applied to one of the TENG layers, the state is out of electrostatic mode and a potential difference occurs. The current from the external charge is driven by such a potential difference to balance the electrostatic state. It should be mentioned that moving in the contrary direction of the TENG layer will induce an inverse potential difference between the electrodes. Hence, by having a reciprocating motion, an AC output can be received from TENG [164–166].

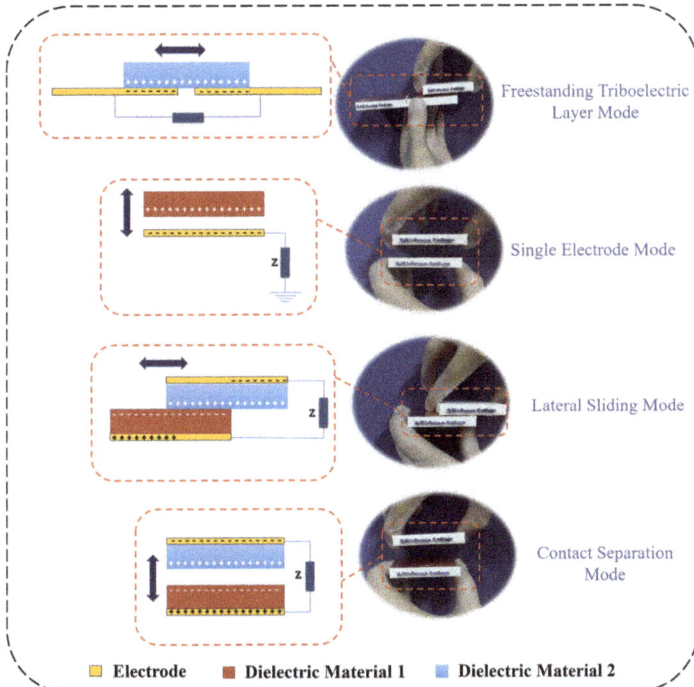

Figure 3. Four principal modes of TENGs: Contact separation, lateral sliding, single electrode and freestanding triboelectric layer (Note: Photo by authors).

The contact separation mode consists of two electrodes, which are located behind the TENG-layers. In this case, a potential difference occurs when the contact and separation processes takes place. The output voltage can be measured using a voltmeter by connecting it to one electrode and the other end to the other electrode. Then, periodically contact and disconnect operation and the output voltage can observed [167–169].

The lateral sliding mode consists of two electrodes, which, like the contact separation mode and freestanding triboelectric layer mode, are located behind the TENG layers. The TENG-layers' relative slip creates the lateral sliding mode, the state removed from the electrostatic state and a potential difference. In this case, using a simple voltmeter, the resulting voltage can be measured. It is enough to connect one end of the voltmeter to one electrode and the other end of the voltmeter to another electrode. With the reciprocating slip of the TENG layer, the voltage created by the potential difference can be seen [169]. One of the most significant drawbacks of the lateral sliding mode and contact separation mode is that both electrodes must have an output wire, limiting their applications. A solution to solve this problem is to provide a single electrode mode. In this case, only one electrode is used. When the TENG layer comes in contact with the electrode, the condition goes out of the electrostatic state and the potential difference causes an electric current [168–173]. In this case, known as freestanding triboelectric layer mode, the TENG layer is self-moving without connecting to the electrode and the two electrodes are spaced apart. When the TENG layer slips from the first electrode to the second electrode, it causes a potential difference. In this case, to display the output voltage, it is enough to use a simple voltmeter [174]. Connect one end of the voltmeter to one electrode and the other end of the voltmeter to the other electrode. Then, by moving the TENG layer back and forth from one electrode to another, the voltage can be display by the voltmeter [175].

Figure 4a demonstrates the trends of publications about the energy harvesting system base on the TENG. Once the TENG technology was discovered around 2012, many researchers turned to work in this field and the number of published articles has increased significantly. This can be considered a turning point in the field of mechanical energy harvesting [176]. Figure 4b shows various applications of TENGs. After a decade of working on nanogenerators, researchers have demonstrated many applications and potentials in this field. TENGs are not only useful in the production of hydropower and wind energy; they can also be used in medical, civil engineering (such as structural health monitoring (SHM) systems [177] and self-power sensors as an energy source for some structures like a bridge) and other fields to protect the environment and reduce fossil fuel production [176] (Note: Various images are adopted from [176,177]).

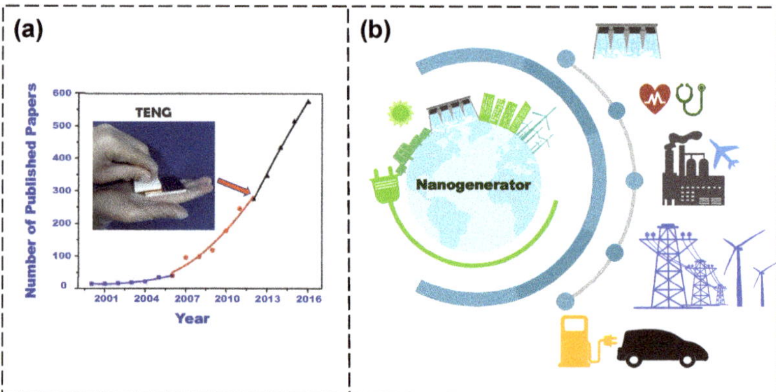

Figure 4. Trends of TENG-related publications and the applications: (**a**) Trends of publications on the energy harvesting concept and TENG's impact [176]. (**b**) Various applications of TENG.(Note: Various images are adopted from [176,177] and photo by authors)

2.4. Spherical Triboelectric Nanogenerator Networks

Figure 5a illustrates blue energy, including solar panels, wind turbines and TENGs. This design provides a concept to combine all three sources of energy as networks. However, the main problem is transporting the voltage and power to a port to be used in the city. On the other hand, this structure must be fixed to islands or underwater mountains to prevent its movement and protect marine life [104]. Figure 5b illustrates the split ball–shell structured TENG principle with silicone rubber balls and outer shells. By rotating, silicone rubber balls from one layer to another can transfer electrons and generate a high and efficient voltage from ocean waves. This design is a network that can be located on the surface of the water [178]. Figure 5e also supports this concept with different materials and different connections [19].

Figure 5c shows spherical TENG networks with different operations. In this design, when the ocean waves come from any direction, the pendulum can rotate freely and transfer electrons between layers to harvest energy [179–181]. Figure 5d illustrates another concept of spherical TENG networks with a different structure. The principle of this design is based on the movement of polyacrylate balls between several layers of spherical TENG [182]. The main advantage of the all-spherical design is that it can work in any ocean wave direction.

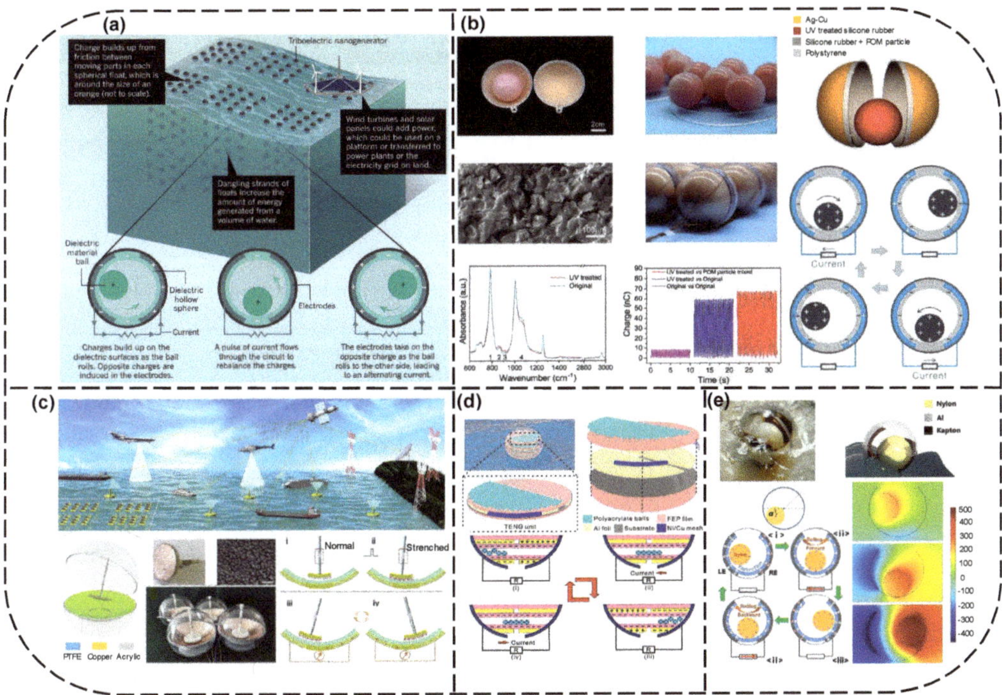

Figure 5. Spherical TENG Networks: (**a**) Harvest renewable energy from the ocean using floating nets of nanogenerators [104]. (**b**) Design and working principle of the split ball–shell structured TENG with silicone rubber balls and outer shells [178]. (**c**) Principle of the ES-TENG networks for collecting energy from ocean waves [179]. (**d**) Theory of the spherical TENG with dense point contacts [182]. (**e**) Freestanding-triboelectric-layer nanogenerator's design and working principles (RF-TENG) with a rolling nylon ball enclosed [19]. Note: Various images are adopted from [19,104,178,179,182].

2.5. Spring-Assisted Triboelectric Nanogenerator

TENG without help from other components such as spring and magnet are more productive for harvesting energy. Nevertheless, ocean waves have a lower frequency, which may cause some difficulties in generating a very restricted electrical energy. Some gadget TENGs use springs to collect unstable dynamic energy that perform poorly concerning high-frequency oscillations and enable energy transformation efficiency. Figure 6a presents a conceptual framework of tandem TENG with a cascade impact structure (CIT-TENG) for generating energy from different layers of TENG (TENG-1, TENG-2, TENG-3 and TENG-4) with various frequencies for each TENG layer. The tribo-layer used in this design is PTEF and the electrode is aluminum. In this design, having a small vibration, all TENG layers move with different frequencies and generate energy based on the contact mode of TENGs [183].

Figure 6b presents a TENG built on a suspended 3D spiral structure that is assisted by mass and spring. In this method, the frequency can be increased by changing the mass and spring values to develop output voltage from ocean waves. Contact separation mode is also used in this structure. A triboelectric layer is Kapton and the electrode is aluminum [23]. Figure 6c illustrates the working principle of a spring-assisted TENG based on the vertical contact separation mode. In this method, the spring is used to increase the high-frequency oscillations and generate more dynamic energy. In this structure, the triboelectric layer is PTFE and the electrode is copper [184–186].

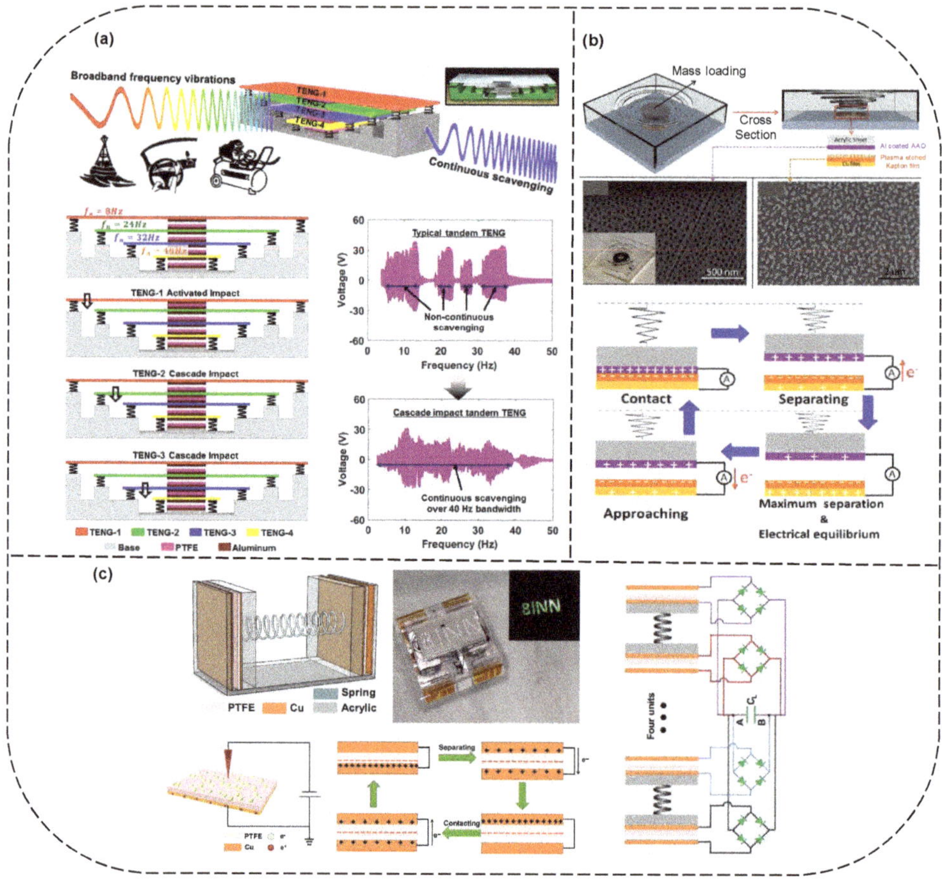

Figure 6. Spring-assisted TENGs: (**a**) Conceptual and cross-section schematic of the CIT-TENG design with output voltage [183]. (**b**) Principle of TENG built on a suspended 3D spiral structure [23]. (**c**) Principal and fabrication of spring-assisted TENG device with process for generating negative charges on the surface of PTFE [184]. Note: Various images are adopted from [23,183,184].

2.6. Liquid–Solid Interfacing Triboelectric Nanogenerator

TENGs based on liquid–solid contact have numerous benefits such as increasing output voltage from ocean wave energy, improving contact area between the FEP film and water and having a simple structure. Figure 7a shows the conceptual framework of a liquid–solid electrification-enabled generator (LSEG). In this design, the structure will be located in front of the wave and when the wave is contacted, water will come up and water will cross from one layer to another to transfer the electron [27,187–191]. Figure 7b presents the structure of a novel wave sensor based on a liquid–solid interfacing triboelectric nanogenerator (WS-TENG), which is useful for structural health monitoring of marine equipment. The triboelectric layer used in this design is PTEF and the electrode is copper. The principle of this method is based on the water which can come up and the water will cross from one layer to another for electron exchange [192]. As shown in Figure 7c, this structure can harvest energy by the network of the liquid–solid-contact buoy TENGs from ocean waves. The innovation of this design includes two external and internal layers of TENG, which can generate energy from the shaking and rotating movement of waves [26]. Figure 7d illustrates the mechanism of droplet-based TENGs for wave energy harvesting (DB-TENG) for marine vehicles. The principle of this design is also based on the water. The triboelectric layer used in this structure is FEP and the electrode is copper [193,194].

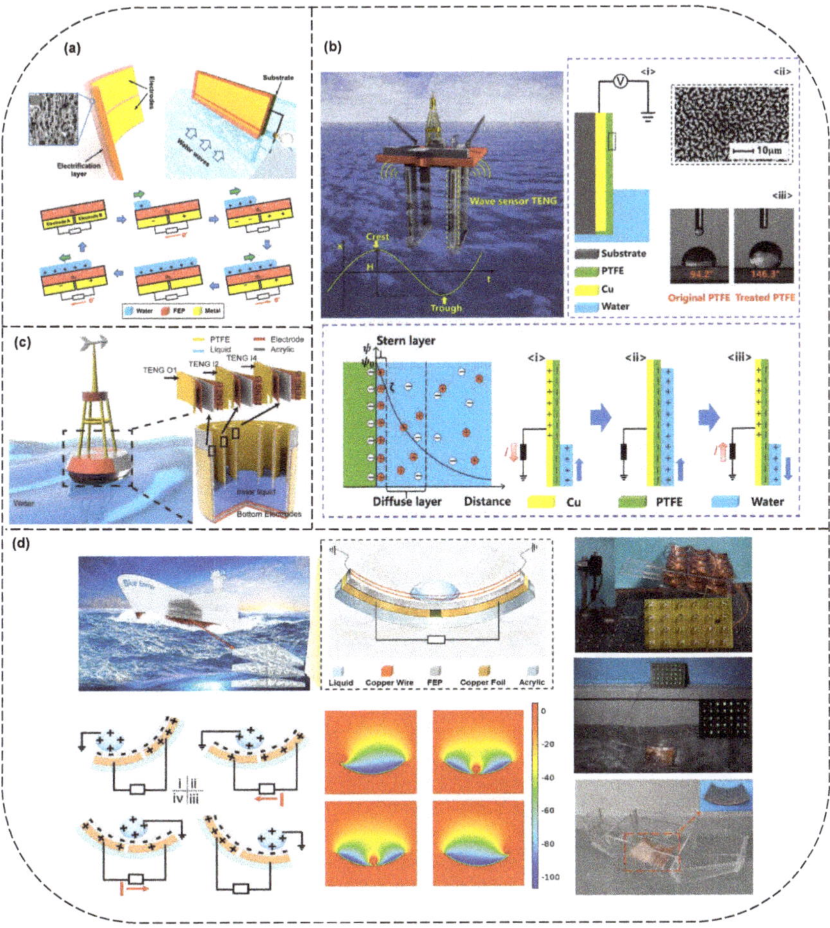

Figure 7. Liquid–solid interfacing TENGs: (**a**) Structural design of liquid–solid electrification enabled generator (LSEG) [27]. (**b**) Principal of a novel wave sensor based on liquid–solid interfacing triboelectric nanogenerator (WS-TENG) useful for monitoring waves around marine equipment [192]. (**c**) Structural design of the blue energy harvested by the network of the liquid–solid-contact buoy TENGs [26]. (**d**) Working mechanism of droplet-based TENG for wave energy harvesting (DB-TENG) [193]. Note: Various images are adopted from [26,27,192,193].

3. Hybridization of TENG with Other Energy Harvester Systems

So far, the concept of TENG, its applications, mechanism and different modes have been discussed. However, it is always challenging to improve the efficiency of TENGs by combining them with other energy-generating systems. The resulting hybrid system will benefit from advantages of both systems. In this section, the TENG is hybridized with solar cells, electromagnetic systems and some others.

3.1. Hybridized Piezoelectric and TENG

Figure 8a describes a high output piezo/triboelectric hybrid generator. This hybrid generator merges triboelectric output voltage and a high piezoelectric output current, which generates a peak voltage of ~370 V, a current density of ~12 μA cm^{-2} and the average power density ~4.44 mW cm^{-2}. The amount of power strongly turned on 600 LED bulbs by applying a mechanical force of 0.2 N and it is able to charge a 10 μF capacitor to 10 V during 25 s [195].

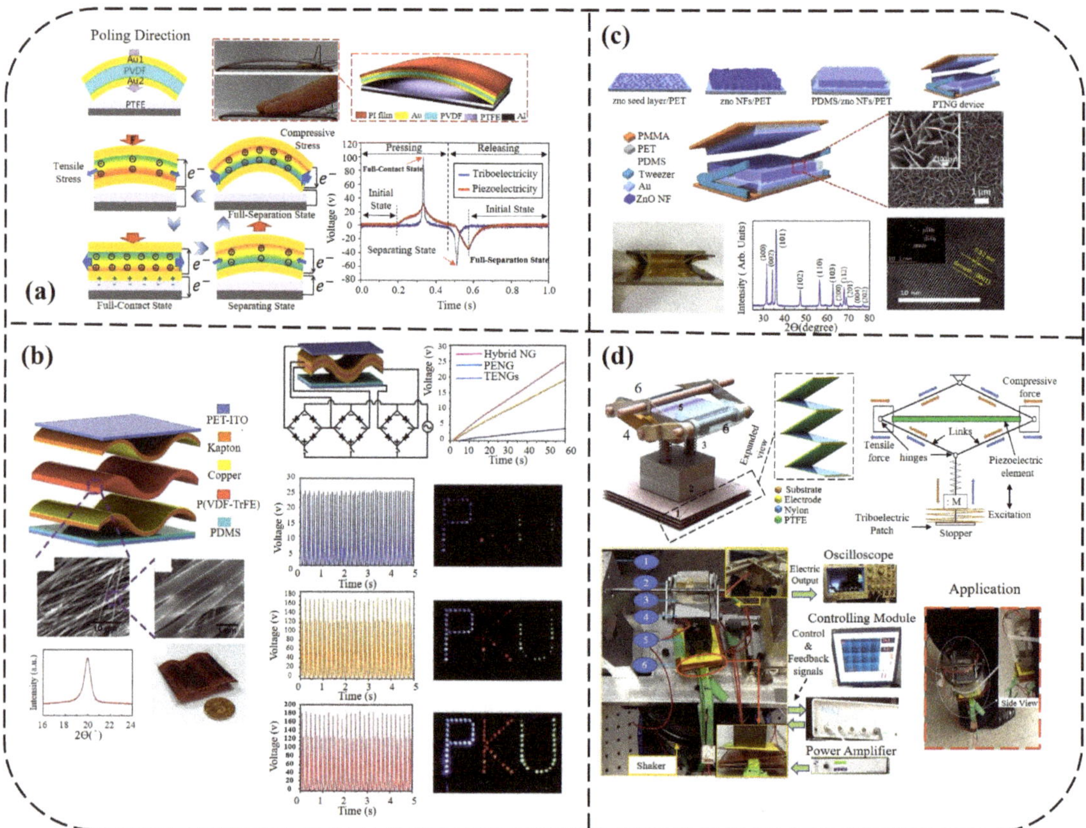

Figure 8. Different structures and their application of a piezoelectric and TENG: (**a**) Schematic of a working mechanism of the hybrid generator in a press and release cycle [195]. (**b**) Structure design of the wave-shaped hybrid piezoelectric and TENG based on P(VDF-TrFE) nanofibers [196]. (**c**) Schematic illustration and characterizations of the PTNG devices based on ZnO nanoflakes/polydimethylsiloxane composite films [197]. (**d**) Structural configuration of the generator for low-frequency and broad-bandwidth energy harvesting [198]. Note: Various images are adopted from [195–198].

In Figure 8b, a wave-shaped hybrid piezoelectric and TENG is illustrated founded on P(VDF-TrFE) nanofibers. In this technique, a piezoelectric P(VDF-TrFE) nanofiber is sandwiched between two wave-shaped Kapton films to form a three-layer pattern. This structure produces piezoelectric triboelectric outputs concurrently in one press and release cycle. Within well-organized experimental validation and situational analysis, the three-layer fabrication can produce noticeable output performance for both parts. Meanwhile, triggered with 4 Hz external force, the piezoelectric part produces a peak output and current of 96 V and 3.8 μA, which is $\sim$2 times higher than its first output. While the execution of triboelectric parts increases 8 V and 16 V with the support of piezoelectric potential [196].

Figure 8c presents ultrahigh output piezoelectric and triboelectric association nanogenerators based on ZnO nanoflakes/polydimethylsiloxane composite films. There is a hybrid nanogenerator (NG) utilizing both piezoelectric and triboelectric effects influenced from ZnO nanoflakes (NFs)/polydimethylsiloxane (PDMS) composite films by a facile, cost-effective fabrication design. This hybrid NG showed high piezoelectric output current owing to the improved surface piezoelectricity of the ZnO NFs, including high triboelectric output voltage owing to the notable triboelectrification of Au-PDMS contact, generating a voltage of $\sim$470 V, a current density of $\sim$60 μA. cm^{-2} and an average power density of $\sim$28.2 mW. cm^{-2}. The hybrid NGs with a space of 3 $\times$ 3 cm^2 immediately turned on 180 green light-emitting diodes within cyclic control compression [197].

In Figure 8d, a hybrid piezoelectric-triboelectric generator is depicted for low-frequency and broad-bandwidth energy harvesting. The generator is created using a piezoelectric energy harvester (PEH) patch, a TENG patch, a spring-mass method and an amplitude limiter. The spring-mass method takes energy from applied excitation and utilizes forces to the piezoelectric component and the triboelectric films. The novel amplitude limiter is deliberately organized inside the system, obtaining the effect of favorable frequency up-conversion increasing the voltage responses considerably. Moreover, the limiter creates hardening nonlinearity and dynamic bifurcation triggering super harmonic resonance due to resonation concerning the generator at a cycle of approximately 3 Hz. The implemented PEH adopts the extreme compressive performance mode and employs a truss mechanism to amplify the impact forces effectively. The open-circuit voltages following excitation of 1.0 g at resonance are 58.4 V and 60 V from PEH and TENG, respectively. The hybridized generator produces the highest power of 19.6 mW from two sources by matched impedances. The working bandwidths of the PEH and the TENG are increased to 5.39 Hz and 7.25 Hz, respectively. Meanwhile, the applications to charge capacitors, the high saturation voltage and the approximately short charging time validate the effectiveness of the power management technique. Besides, the generator is useful to efficiently scavenge energy from human body actions and charge a capacitor of 4.7 μF to 7.6 V in approximately 50 s. It shows a high potential of functional usages in wearable mechanisms [198].

3.2. Hybridized Solar Cells and TENG

Figure 9a describes integrating a TENG with a silicon solar cell by an electrode for harvesting energy from sunlight and raindrops. A heterojunction silicon (Si) solar cell is combined with a TENG using a mutual electrode of a poly (3.4-ethylene dioxythiophene):poly (styrene sulfonate) (PEDOT:PSS) film. The solar cell, printed PEDOT:PSS is applied to decrease light reflection, which begins to improve short-circuit current density. A single-electrode-setup water-drop TENG on the solar cell is made with merging imprinted polydimethylsiloxane (PDMS) as a triboelectric material merged with a PEDOT:PSS layer as an electrode. The expanding contact section among the imprinted PDMS and water drops considerably increases the voltage output of the TENG with a peak short-circuit current of $\sim$33.0 nA and a peak open-circuit voltage of $\sim$2.14 V, sequentially [67].

Figure 9b illustrates biomimetic anti-reflective TENGs for concurrent harvesting of solar and raindrop energies. A moth's eye mimicking TENG (MM-TENG) can perform the role of equivalent energy harvester to a standard solar cell due to its higher specular transmittance (maximum 91% for visible light). At the first time, strongly examine the visible effect of the MM-TENG on a solar cell by considering solar-weighted transmittance (SWT). The 0.01% developed SWT in the MM-TENG enhances the fill part and power transformation efficiency of the solar cell by 0.5% and 0.17%, respectively, compared with a traditional protective glass plate that constantly utilized a solar panel. Moreover, in addition to such prominent high transmittance, the self-cleaning property of the MM-TENG enables the long-term production of the solar panel. Besides, this design summarizes a unique electric circuit for effective control in a hybrid energy harvester through intermittently transferring the preserved electrical energy output of the MM-TENG [121].

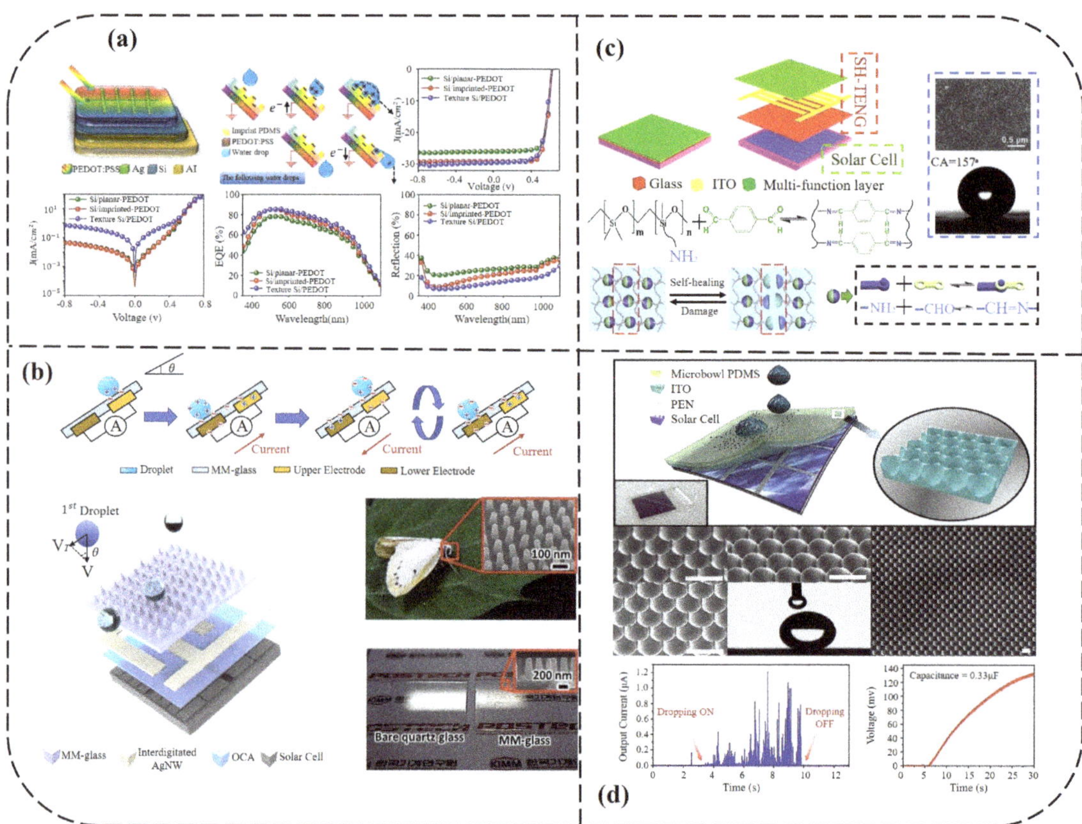

Figure 9. Various structures of integrating TENG with solar cells: (**a**) Schematic of a working mechanism of integrating a silicon solar cell with a TENG via a mutual electrode for harvesting energy from sunlight and raindrops [67]. (**b**) Structure design of the biomimetic anti-reflective TENG for concurrent harvesting of solar and raindrop energies [121]. (**c**) Schematic diagram of the hybrid energy system with the SH-TENG and solar cell [199]. (**d**) Structural configuration of the self-cleaning hybrid energy harvester to generate power from raindrop and sunlight [105]. Note: Various images are adopted from [67,105,121,199].

Figure 9c shows a hybrid energy system based on solar cells and a self-healing/self-cleaning TENG. This self-healable and transparent TENG presents an outstanding output execution on a rainy day with a peak short-circuit current of 0.8 μA and a peak open-circuit voltage of 6 V, separately. At the same time, it can work very strongly beside the solar cell on sunny days. Moreover, the self-healing TENG with elastic composition can function as a protection film for the solar cell, where the of the solar cell's risk of breakinfg can be significantly overcome [199].

Figure 9d shows a self-cleaning hybrid energy harvester to generate power from raindrops and sunlight. The transparent TENG made with superhydrophobic PDMS and ITO/PEN substrate. The amount of voltage current was 7 V and 128 nA, separately.

The highest output power is 0.27 µW. Additionally, output components of water TENG are studied with several solutions such as natural rainwater and flowing water to investigate the sensible potential of water TENG in harvesting actual raindrop energy [105].

3.3. Hybridized TENG with Electromagnetic

Figure 10a demonstrates a self-powered and self-functional tracking system-based triboelectric-electromagnetic hybridized blue energy harvesting module. A sophisticated designed rotating gyro structure, including a triboelectric-electromagnetic working principle, strongly established a battery-less tracking method. The unique gyro rolling mode solution of the TENG features its sensitivity, multiple directions and robustness. Eventually, the triboelectric-electromagnetic hybridized module was completely confirmed autonomously within the Huanghai Sea project [200].

Figure 10b shows a hybridized triboelectric-electromagnetic water wave energy harvester (WWEH) based on a magnetic sphere. A freely rolling magnetic sphere senses the water movement to drive the friction object sliding on a solid surface for TENG back and forth. Simultaneously, two coils convert the motion of the magnetic sphere into electricity based on the electromagnetic induction effect. Harvesting the blue energy from any place, the electrodes of the TENG are determined as the Tai Chi form, the effectiveness of which is analyzed and described. According to experimental results, the two friction films and two coils are given in parallel and series connection. A paper-based super-capacitor of ~1 mF is fabricated to save the generated energy. The WWEH is located on a buoy to examine in Lake Lanier. The super-capacitor can be charged to 1.84 V and the electric energy storage approximately 1.64 mJ during 162 s [201].

Figure 10c presents the complementary electromagnetic-triboelectric active sensor (ETAS) for detecting multiple mechanical triggerings. This work displays a combined ETAS for simultaneous detection of various mechanical triggering signals. The high-grade combination of a contact-separation mode TENG and an electromagnetic generator (EMG) recognizes the complements of their incomparable benefits. The logical consideration of EMG and TENG analysis are presented to explain the relation of output and the external mechanical signals. Corresponding to the experimental outcomes, the output voltage of the TENG part recognizes the magnitude of the external triggering force with a sensibility of around 2.01 VN^{-1}, while the output current of the EMG part is more suitable to showing that the triggering speed and the sensation is approximately 4.3 mA s/m. Furthermore, both the TENG and EMG parts display high stability after 20,000 cycles of force loading–unloading [202].

Figure 10d demonstrates an ultra-low-friction triboelectric-electromagnetic hybrid nanogenerator for generating energy as a self-powered sensor. A freestanding mode TENG and a rotating EMG are combined to recognize unique merits. The principal is based on contact between soft and flexible triboelectric elements in the TENG outputs with low friction force. The impressions of the model and the dimensions of the dielectric material on the performance of the TENG are regularly considered in hypotheses concerning experiments. According to the results, voltage output is improved by the rotation velocity, which is very different from a standard rotary TENG and is due to the contact area's increase. The optimized TENG has a maximal load voltage of 65 V and maximal load power per unit mass of 438.9 mW/kg under a velocity rotation of 1000 rpm. The EMG has a maximal load voltage of 7 V and a maximal load power density of 181 mW/kg. This illustrates that the hybrid NG can power various sensors such as humidity and temperature by converting wind energy within electric energy during wind speeds of 5.7 m/s. In contrast, it can be utilized as a self-powered wind velocity sensor to recognize wind rates as low as 3.5 m/s [203].

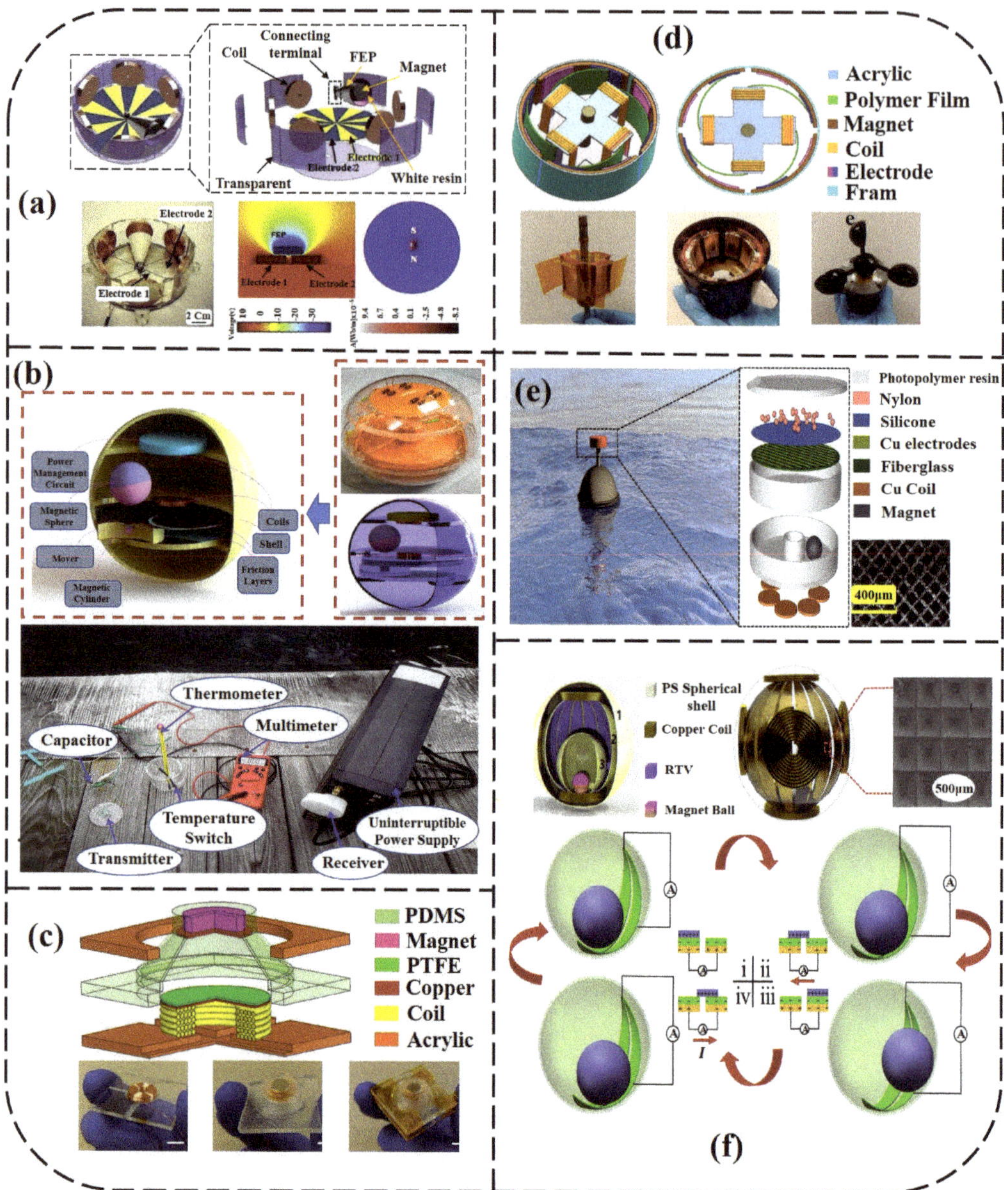

Figure 10. Various structures of hybridized electromagnetic-triboelectric nanogenerators: (**a**) Structure and working principle of a self-powered and self-functional tracking system based on a triboelectric-electromagnetic hybridized sort [200]. (**b**) Structure and illustration of the working principle of the hybridized triboelectric-electromagnetic water wave energy harvester (WWEH) [201]. (**c**) Structure design of the hybridized EMG-TENG active sensor [202]. (**d**) Working principle of an ultra-low-friction triboelectric-electromagnetic hybrid nanogenerator for rotation energy harvesting and self-powered wind speed sensor [203]. (**e**) Structural diagram of the pendulum hybrid generator used in a hydrophone-based system with the scanning electron microscopy (SEM) image of the silicone film [204]. (**f**) Schematic diagrams of the FSHG internal structure and SEM image of surface microstructure with the charge generation process in TENG modules [205]. Note: Various images are adopted from [200–205].

Figure 10e illustrates a pendulum hybrid generator for water wave energy harvesting and hydrophone-based wireless sensing. The aimed pendulum fabrication can harvest irregular water wave energy from random directions sensitively. Combining a freely rolling mode TENG and a magnetic sphere-based EMG produces complementary benefits and harvests wave energy in a wide frequency range. The hybrid generator is shown to drive 177 LEDs and power electronic equipment. At a wave driving rate of 1.4 Hz, the output power of the EMG and TENG is 6.7 mW and 8.01 μW, sequentially. A capacitor in this design can be charged to 26 V by the hybridized generator in 200 s by 1.8 Hz [204].

Figure 10f shows a 3D full-space triboelectric-electromagnetic hybrid nanogenerator toward high-efficient mechanical energy harvesting in a vibration method. The output voltage of the field test demonstrated the performance of TENG and EMGs, which can be influenced by the direction of external vibration and excitation frequency. The result of output voltage of TENG and EMGs develops as the excitation frequency rises. The outcome explains that the maximum output power of TENG is 18 mW at an external loading resistance of 200 MΩ and the maximum output power of EMG is 640 mW at an external loading resistance of 1000 Ω. The FSHG displays a fast-charging capacity for capacitor and the capability to power hundreds of LEDs. After saving energy within the capacitor, the DC signal can power a humidity/temperature sensor [205].

3.4. Hybridized TENG with Magnetic Intensity

Figure 11a demonstrates a self-powered multi-functional motion sensor enabled by a magnetic-regulated TENG. Typically, a self-powered multi-functional movement sensor (MFMS) is aimed in this design to detect the motion parameters such as area, velocity and acceleration of linear and rotary movements concurrently. The MFMS is made of a TENG module, a magnetic regulation module and an acrylic shell. The mode of TENG used in this structure is free standing with a polytetrafluorethylene (PTFE) plate and six copper electrodes. The working mechanism of the MFMS design is based on the sliding of the MD on the PTFE plate for electron exchange between layers. The precisely designed six copper electrodes can detect eight directions of action with the acceleration and determine the rotational velocity and direction.

Furthermore, the magnetic regulation module is utilized here by fixing a magnetic cylinder (MC) in the shell, right below the center of the PTFE plate. Because of the magnetic attraction utilized by the MC in this design, the MD will automatically return to the center to prepare for the next discovery round, making the proposed sensor much more suitable for practice [206].

Figure 11b illustrates wind energy and blue energy harvesting according to the magnetic-assisted noncontact TENG. An innovative approach to wind and blue energy harvesting based on a magnetic-assisted noncontact TENG has been described. With the compound of the magnetic responsive composite with the TENG design, the wind and water forces could be transformed toward the contact-separation movement between Al/Ni electrode and PDMS film. The influence of the related parameters (contact-separation frequency, wind speed and humidity) on the performances of the fabricated TENG has been regularly examined. The results explain the strong potential of magnetic-assisted noncontact TENG for wind and blue energy harvesting purposes [207].

Figure 11c shows a floating oscillator-embedded triboelectric generator (FO-TEG) for versatile mechanical energy harvesting. The FO-TEG is appropriate for impulse excitation and sinusoidal vibration, which are in the natural ecosystem. For the impulse excitation, the created current sustains and moderately decays by the residual oscillation of the floating oscillator. Resonance oscillation maximizes the output energy for sinusoidal vibration. The acting frequency range can be optimized via a high degree of freedom to satisfy different application requirements. Furthermore, the highest resistance to ambient humidity is empirically illustrated, which stems from the inherently packaged fabrication of FO-TEG. The prototype design presents a peak-to-peak open-circuit voltage of 157 V and an instantaneous short-circuit current of 4.6 μA, within sub-10 Hz of operating frequency [208].

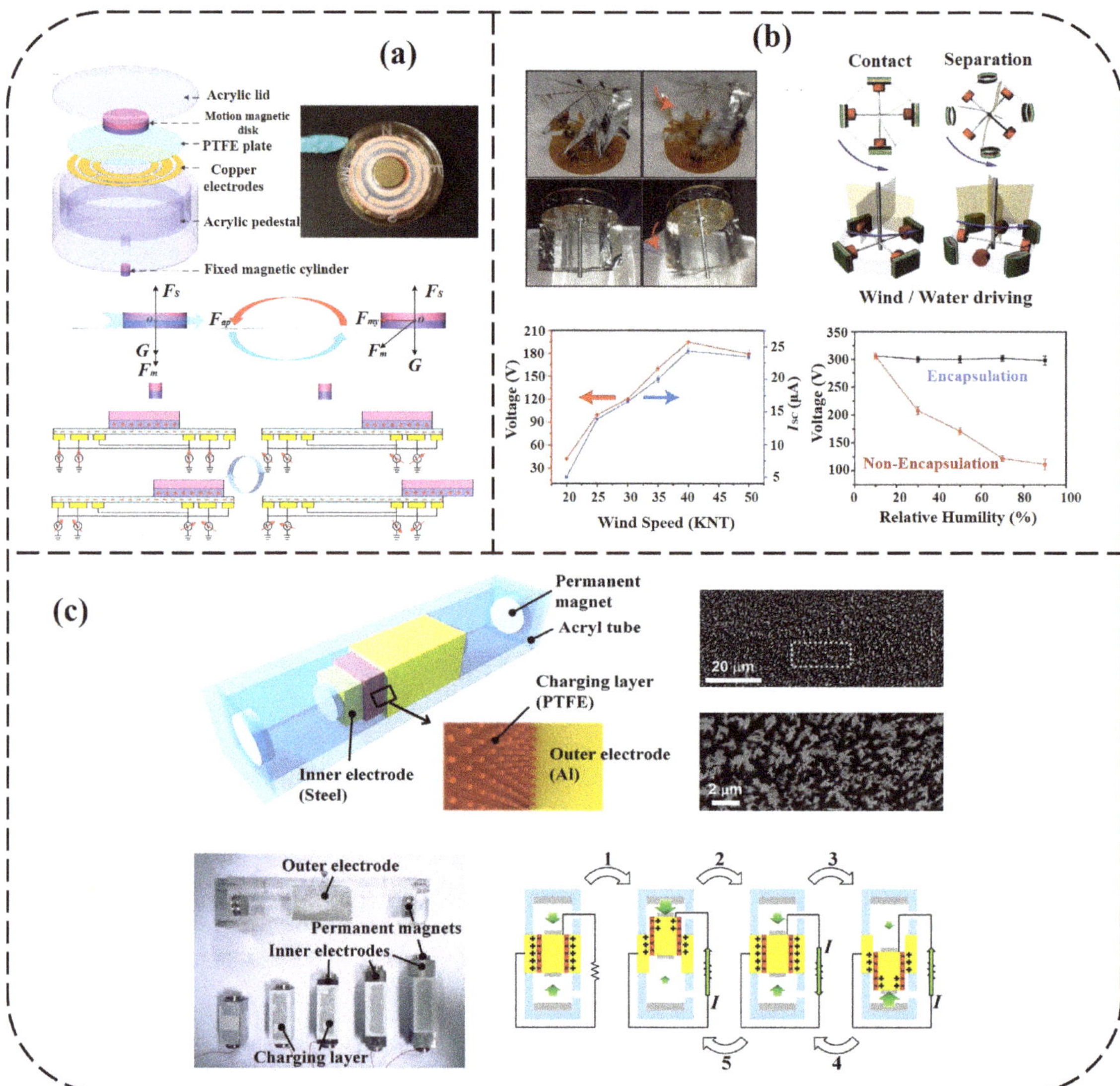

Figure 11. Various structures combining TENG with magnetic intensity: (**a**) Structure and illustration of the working principle of a self-powered multi-functional motion sensor (MFMS) [206]. (**b**) Demonstration and schematic illustration of magnetic-assisted noncontact TENG for harvesting energy from wind and water flow (four devices arranged with perpendicular angle) [207]. (**c**) Schematic illustration of the FO-TEG structure. The FO-TEG contains a tube part with a mobile oscillator part floating inside, suspended by magnetic repulsive forces. Vertical nanowire-like structures are formed on the sidewalls of the PTFE charging layer [208]. Note: Various images are adopted from [206–208].

Figure 12a demonstrates a magnetic switch structured TENG for continuous and regular wind energy harvesting. A magnetic switch structured triboelectric nanogenerator (MS-TENG) is developed, consisting of transmission gears, energy modulation modules and a creation unit. Meanwhile, wind falls intermittently on the wind scoop; the energy collected and released through the energy modulation do not depend on wind velocity. However, the magnetic force of the magnets allows the wind energy to be changed into continuous and normal electric energy. The experimental outcomes show that the MS-

TENG can succeed as a power supply, generating output characteristics of 410 V, 18 µA, 155 nC and peak power of 4.82 mW, satisfactory to power 500 LEDs in series [209].

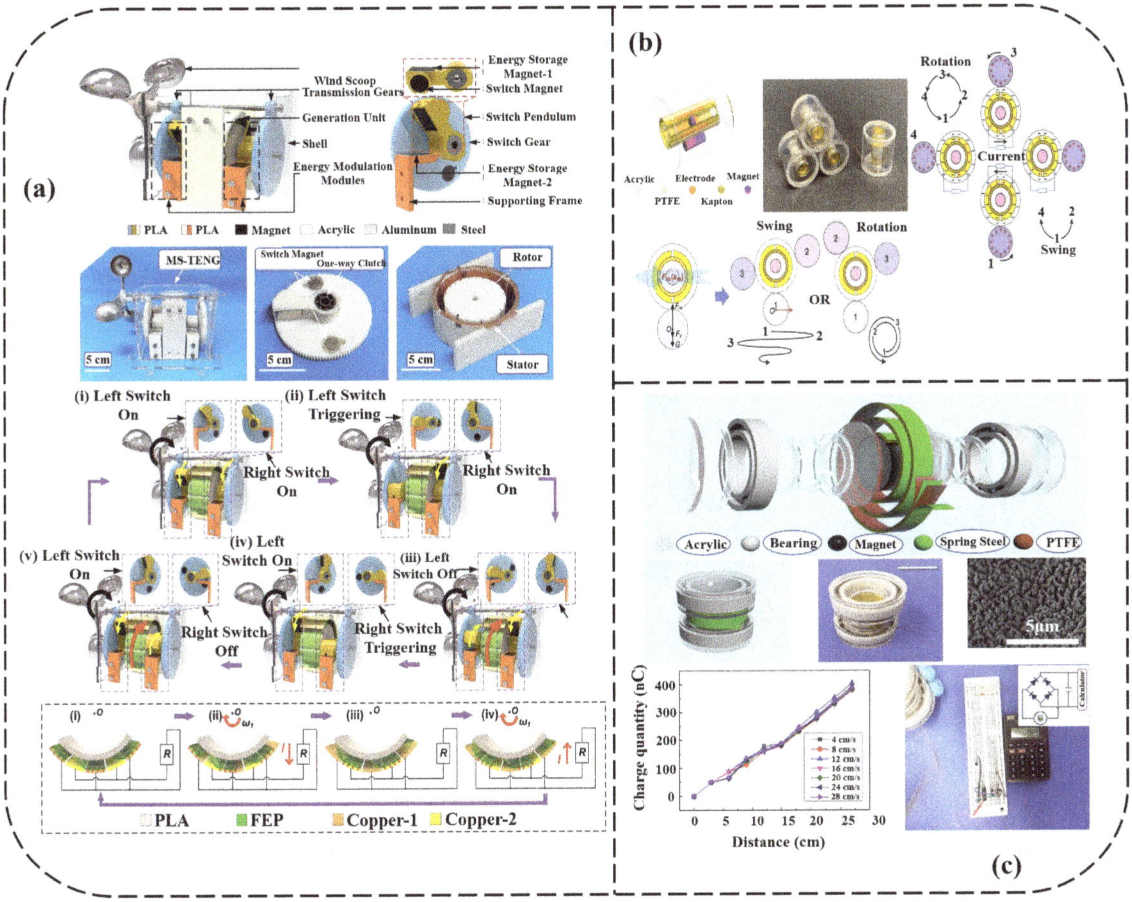

Figure 12. Various structures of hybridized TENG using magnetic intensity. (**a**) The magnetic switch structured triboelectric nanogenerator (MS-TENG) [209]. (**b**) Structural design and working principle of the multifunctional sensor based on translational-rotary triboelectric nanogenerator [210]. (**c**) Schematic illustrating the structure of assembled and digital photo of MR-TENG [211]. Note: Various images are adopted from [209–211].

Figure 12b explains a multi-functional sensor according to a translational rotary TENG. A cylindrical self-powered multi-functional sensor (MS) with a translational-rotary magnetic mechanism aims to recognize acceleration, force and rotational parameters. The MS can convert a translational movement into a swing motion or a multi-cycle rotational movement of a low damping magnetic cylinder around a friction layer, therefore driving the TENG to produce voltage output. To augment the TENG's output performance, an electrode material with a small work function, low resistance and suitable surface topography is the best choice. Based on the fabrication characteristic of the translational rotary magnetic mechanism, the MS can simply respond to a low striking and can measure the rotational parameters without the need for coaxial installation. According to the MS, some applications are installed [210].

Figure 12c shows a novel TENG relying on magnetically induced retractable spring steel tapes (MR-TENG) for efficient energy harvesting of large amplitude periodic motion.

An ingenious design utilizes a new material. The tape-like fundamental design ensures that the contact/separate direction of the friction layers is straight concerning the force direction, breaking the amplitude limitation of prior nanogenerators with vertical contact/separate movement. Conjunction of flexible spring steel tapes makes the structure of design enable portability, therefore widening its application. The outcomes present that the maximum short-circuit current, open-circuit voltage and instantaneous power are 21 µA, 342 V and 1.8 mW, sequentially [211].

3.5. Comparison of TENG Structures

Figure 13 illustrates the advantages and disadvantages of various design strategies for TENGs, including spherical TENG networks, spring-assisted TENGs, liquid–solid interfacing TENGs, hybridized piezoelectrics and TENGs, hybridized solar cells and TENGs, hybridized TENGs with electromagneticm and finally hybridized TENGs with magnetic intensity.

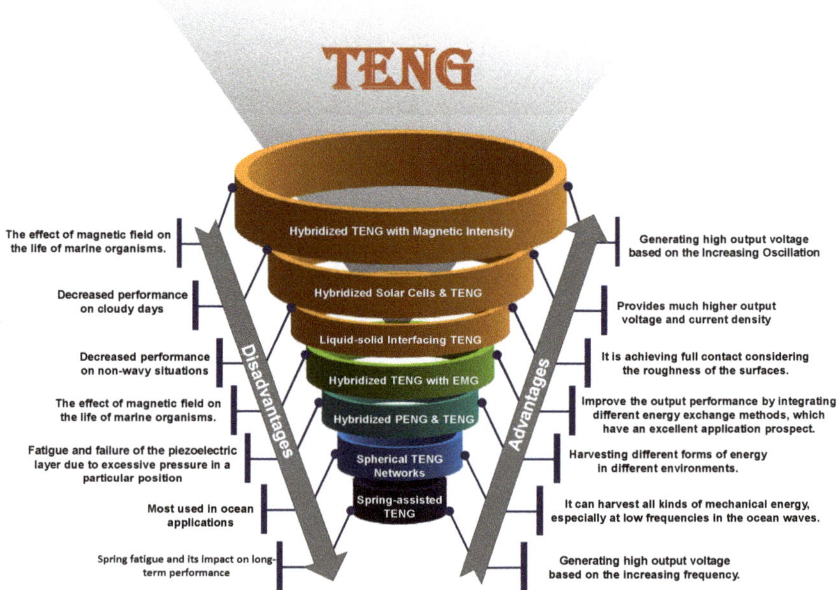

Figure 13. Advantages and disadvantages of different TENG structures.

4. Applications, Challenges and Future Trends of TENGs

The quest to find new (and also renewable) energy resources has always been a key point in modern history. Traditional sources like coal and oil are associated with a lot of pollution, as well as climate change. Therefore, the majority of recent efforts have been concentrated on clean energy resources. In the last decade, one of the most significant human discoveries in this field is the triboelectric nanogenerator. As discussed in Figure 4b, TENG has a broad domain of applications in human life, from medical sciences to engineering and technology.

Maybe two of the most interesting applications for TENG can be found in the civil engineering field. The first one is associated with the SHM of large bridges. For such infrastructure, not all parts are easily accessible by the technicians to install the sensors for damage detection purposes. These sensors require a continuous source of energy to be able to detect the vibrations, analyze them and transmit data to the center. Since the installation and maintenance of the batteries for these sensors are very difficult, an alternative can be to use a TENG as an energy source. On the other hand, TENGs can be used even in a larger scale to supply the required energy for smart cities.

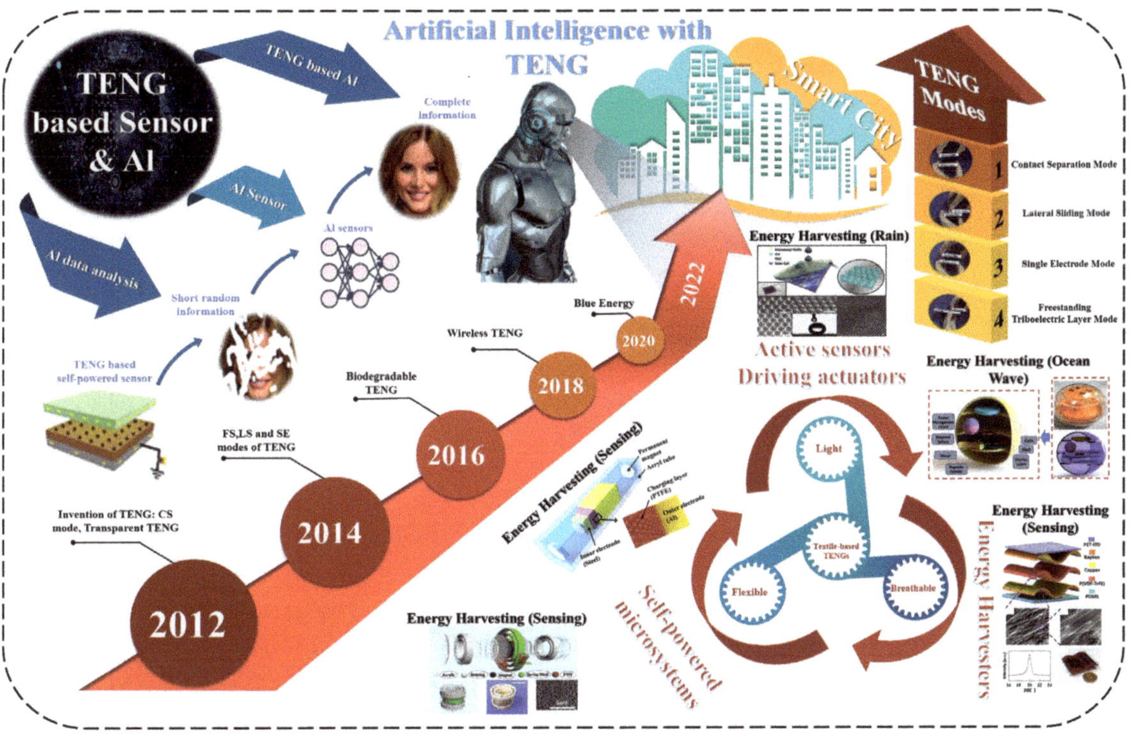

Figure 14. A summary of recent progress in development of TENGs from 2012 to 2022 including the future trends and their connection with AI.

Figure 14 shows the challenges and future trends related to TENGs. In January 2012, professor Zhong Lin Wang introduced the principle of another type of TENG, which can also harvest ambient mechanical energy by combining contact-electrification and the electrostatic phenomenon. The TENG concept has been expanded to biodegradable TENGs, wireless TENGs and blue energy and more recently in smart cities. TENGs also have a bright future in robotics and self-powered sensing technologies when they are combined with artificial intelligence (AI). TENGs have been proposed in ubiquitous computation, which presents the potential of applying them to increase an entire sensor network.

Table 3 presents the performance output comparison of various TENG devices to collect ocean wave energy such as spherical TENG networks, spring-assisted TENG, liquid–solid interfacing TENG, hybridized piezoelectrics and TENGs, hybridized solar cell sand TENGs, electromagnetic hybridized TENGs and hybridized TENGs with magnetic intensity. Most of the electrodes used in TENG are Al and Cu and also most of the triboelectric layers are PTFE.

Table 3. Summary of various hybridized TENG techniques to harvest ocean energy.

Structure	Year	Authors	Applied Tribo-Layer	Electrode Type	Max-Open-Circuit Voltage (V)	Max-Short-Circuit Current (μA)	Current Density	Surface Power Density	Power Density and Power	Load Resistance (Ω)
Spherical TENG Networks	2018	Xu et al. [178]	Polystyrene	Ag-Cu	1780	1.8	-	-	4.47 W/m^3	1 GΩ
	2021	Lin et al. [179]	PTFE	Cu	76	4.4	-	-	28 μW	100 MΩ
	2022	Yuan et al. [182]	FEP	Ni/Cu-Al	2103	15.5	-	-	20.57 W/m^3	300 MΩ
	2015	Wang [19]	Nylon 6/6 and Kapton	Al	903	1	-	-	10 mW	10 GΩ
Spring-assisted TENG	2019	Bhatia et al. [183]	PTFE	Al	14	0.75	-	-	-	40 GΩ
	2013	Hu et al. [23]	Kapton	Al/Cu	150	27	20 mA/m^2	2.76 W/m^2	-	6 MΩ
	2016	Jiang et al. [184]	PTFE	Cu	755.8	65	-	-	-	-
Liquid–solid interfacing TENG	2014	Zhu et al. [27]	FEP	Cu	3	-	-	-	0.12 μW	150 MΩ
	2019	Xu et al. [192]	PTFE	Cu	1.98	-	-	-	-	-
	2018	Li et al. [26]	PTFE/FEP/FET/PDMS	Cu/Al	600	9	-	-	-	-
	2021	Wei et al. [193]	FEP	Cu	237	746 nA	-	-	23.3 μW	500 MΩ
Hybridized PENG and TENG	2015	Jung et al. [195]	PTFE	Al/Au	370	12	-	4.4 mW/cm^2	-	-
	2018	He et al. [197]	PDMS/NFs	Au	470	60	-	28.2 mW/cm^2	-	-
	2017	Chen et al. [196]	Kapton/PDMS	Cu	361.27	10.85	-	-	441 μW	100 MΩ
	2018	Li et al. [198]	PTFE	Nylon	60	-	-	-	5.7 mW	1 MΩ
Hybridized TENG with solar cells	2018	Liu et al. [67]	PDMS/PEDOT:PSS	Al/Ag	2.14	33 nA	-	-	-	-
	2015	Jeon et al. [105]	PDMS	ITO/PEN	7	128 nA	-	27 μW	25 MΩ	
	2019	Yoo et al. [121]	MM-glass	AgNW	18.4	24.4	-	-	-	-
	2021	Yang et al. [199]	Silicone	ITO	6	0.8	-	-	-	-
Hybridized TENG with EMG	2020	Gao et al. [200]	FEP	PCB	23.76	-	-	-	-	-
	2019	Wu et al. [201]	PTFE	Cu	172.95	3.5	-	-	-	-
	2018	Wang et al. [203]	Polymer	Cu	65	-	-	-	438.9 mW/kg	-
	2020	Hao et al. [204]	Silicone	Cu/Nylon	72	1.7	-	-	8.01 μW	1 MΩ
	2018	Wang et al. [202]	PTFE/PDMS	Cu	13.1	0.25	-	-	-	-
	2020	He et al. [205]	Silicone	Cu	88	-	-	-	18 μW	200 MΩ
Hybridized TENG with magnetic intensity	2018	Wu et al. [206]	PTFE	Cu	~ 2	-	-	-	-	-
	2015	Seol et al. [208]	PTFE	Al	157	4.6	-	-	-	-
	2016	Huang et al. [207]	PDMS	Al/Ni	206	30	-	-	3 mW	10 MΩ
	2021	Liu et al. [209]	FEP	Cu	410	18	-	-	4.82 mW	50 MΩ
	2019	Wu et al. [210]	Kapton/PTFE	Ag/Au/Cu/ITO	1.75	70 nA	-	-	-	-
	2017	Liu et al. [211]	PTFE	Spring steel	342	21	-	-	1.8 mW	2 MΩ

5. Summary

This paper reviews and summarizes the concept of triboelectric nanogenerators (TENGs) for energy harvesting purposes from ocean waves. This review covers the TENGs from the fabrication of the design to the working principle, the recent improvements and various applications for ocean waves energy. The TENG technology that aims at numerous environmental energies is extensively utilized in the energy storage unit, which can be useful for self-power sensors. Furthermore, conjunction nanogenerators of a TENG and various types of power sources have been discussed for ocean as an innovative technology approach. Through this study, the emerging TENG technology has displayed a high potential in several fields from medical science to engineering.

It is expected that, with the current growth in TENG technology, as well as its hybridization with other energy sources, more efficient applications could be identified in the near future. This study provided a condensed (yet comprehensive) review of the current improvements in TENGs and the relevant technologies. First, we present the characteristics of ocean wave energy, followed by the underpinning principles of TENGs. Next, the function of hybridization of TENGs with another type of energy harvester and the various structural designs of TENGs as energy harvesters have been discussed. In addition, future improvements in TENG technology including a self-powered sort combined with sensors and actuators have been studied. Finally, the expansion of TENGs has been summarized and future challenges and opportunities have been briefly explained. Low motion frequency is the most important characteristic of ocean energy, so it can be considered an energy harvester application based on the TENG. Figure 15 shows a summary of the cloud words used in this review paper highlighting the most frequent words.

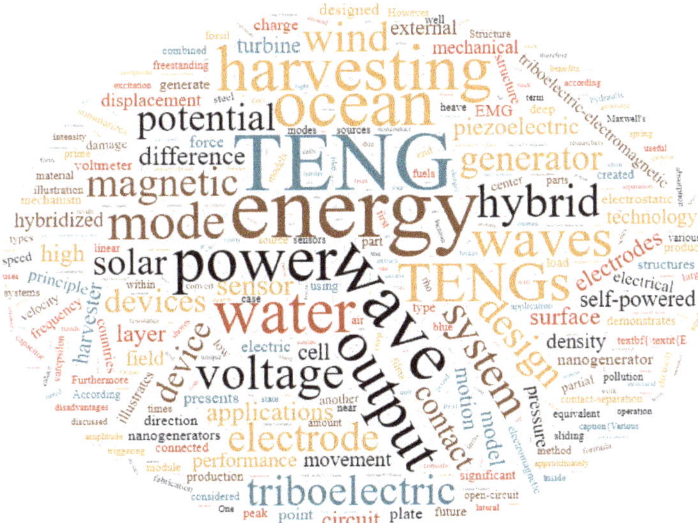

Figure 15. Word cloud summarizing the most frequent statements in this paper.

Author Contributions: Conceptualization, A.M.N.; investigation, A.M.N., K.-J.I.E., A.A., M.A.H.-A.; writing—original draft preparation, A.M.N.; writing—review and editing, A.M.N., K.-J.I.E., A.A., M.A.H.-A.; visualization, A.M.N., A.A., M.A.H.-A.; Supervision, M.A.H.-A.; project administration, M.A.H.-A. All authors have read and agreed to the published version of the manuscript.

Funding: This research received no external funding.

Conflicts of Interest: The authors declare no conflict of interest.

References

1. Atzori, L.; Iera, A.; Morabito, G. The internet of things: A survey. *Comput. Netw.* **2010**, *54*, 2787–2805. [CrossRef]
2. Santoro, G.; Vrontis, D.; Thrassou, A.; Dezi, L. The Internet of Things: Building a knowledge management system for open innovation and knowledge management capacity. *Technol. Forecast. Soc. Chang.* **2018**, *136*, 347–354. [CrossRef]
3. Wang, X.; Song, J.; Liu, J.; Wang, Z.L. Direct-current nanogenerator driven by ultrasonic waves. *Science* **2007**, *316*, 102–105. [CrossRef]
4. Wang, S.; Xie, Y.; Niu, S.; Lin, L.; Wang, Z.L. Freestanding triboelectric-layer-based nanogenerators for harvesting energy from a moving object or human motion in contact and non-contact modes. *Adv. Mater.* **2014**, *26*, 2818–2824. [CrossRef] [PubMed]
5. Wang, Z.L.; Jiang, T.; Xu, L. Toward the blue energy dream by triboelectric nanogenerator networks. *Nano Energy* **2017**, *39*, 9–23. [CrossRef]
6. Fan, F.R.; Tian, Z.Q.; Wang, Z.L. Flexible triboelectric generator. *Nano Energy* **2012**, *1*, 328–334. [CrossRef]
7. Wang, Z.L. Triboelectric nanogenerators as new energy technology for self-powered systems and as active mechanical and chemical sensors. *ACS Nano* **2013**, *7*, 9533–9557. [CrossRef]
8. Wang, Z.L.; Chen, J.; Lin, L. Progress in triboelectric nanogenerators as a new energy technology and self-powered sensors. *Energy Environ. Sci.* **2015**, *8*, 2250–2282. [CrossRef]
9. Qin, H.; Cheng, G.; Zi, Y.; Gu, G.; Zhang, B.; Shang, W.; Yang, F.; Yang, J.; Du, Z.; Wang, Z.L. High energy storage efficiency triboelectric nanogenerators with unidirectional switches and passive power management circuits. *Adv. Funct. Mater.* **2018**, *28*, 1805216. [CrossRef]
10. Kammen, D.M.; Sunter, D.A. City-integrated renewable energy for urban sustainability. *Science* **2016**, *352*, 922–928. [CrossRef]
11. Liu, J.; Cui, N.; Gu, L.; Chen, X.; Bai, S.; Zheng, Y.; Hu, C.; Qin, Y. A three-dimensional integrated nanogenerator for effectively harvesting sound energy from the environment. *Nanoscale* **2016**, *8*, 4938–4944. [CrossRef] [PubMed]
12. Javadi, M.; Heidari, A.; Darbari, S. Realization of enhanced sound-driven CNT-based triboelectric nanogenerator, utilizing sonic array configuration. *Curr. Appl. Phys.* **2018**, *18*, 361–368. [CrossRef]
13. Zhao, C.; Zhang, Q.; Zhang, W.; Du, X.; Zhang, Y.; Gong, S.; Ren, K.; Sun, Q.; Wang, Z.L. Hybrid piezo/triboelectric nanogenerator for highly efficient and stable rotation energy harvesting. *Nano Energy* **2019**, *57*, 440–449. [CrossRef]
14. Feng, A.S.; Narins, P.M. Ultrasonic communication in concave-eared torrent frogs (*Amolops tormotus*). *J. Comp. Physiol. A* **2008**, *194*, 159–167. [CrossRef]
15. Xi, Y.; Wang, J.; Zi, Y.; Li, X.; Han, C.; Cao, X.; Hu, C.; Wang, Z. High efficient harvesting of underwater ultrasonic wave energy by triboelectric nanogenerator. *Nano Energy* **2017**, *38*, 101–108. [CrossRef]
16. Painuly, J.P. Barriers to renewable energy penetration; a framework for analysis. *Renew. Energy* **2001**, *24*, 73–89. [CrossRef]
17. Reddy, S.; Painuly, J.P. Diffusion of renewable energy technologies—Barriers and stakeholders' perspectives. *Renew. Energy* **2004**, *29*, 1431–1447. [CrossRef]
18. Wang, Z.L. Catch wave power in floating nets. *Nat. News* **2017**, *542*, 159. [CrossRef] [PubMed]
19. Wang, Z.L. Triboelectric nanogenerators as new energy technology and self-powered sensors–Principles, problems and perspectives. *Faraday Discuss.* **2015**, *176*, 447–458. [CrossRef] [PubMed]
20. Gandomi, A.H.; Alavi, A.H.; Asghari, A.; Niroomand, H.; Matin Nazar, A. An innovative approach for modeling of hysteretic energy demand in steel moment resisting frames. *Neural Comput. Appl.* **2014**, *24*, 1285–1291. [CrossRef]
21. Jiao, P.; Egbe, K.J.I.; Xie, Y.; Matin Nazar, A.; Alavi, A.H. Piezoelectric sensing techniques in structural health monitoring: A state-of-the-art review. *Sensors* **2020**, *20*, 3730. [CrossRef] [PubMed]
22. Matin Nazar, A.; Jiao, P.; Zhang, Q.; Egbe, K.J.I.; Alavi, A.H. A New Structural Health Monitoring Approach Based on Smartphone Measurements of Magnetic Field Intensity. *IEEE Instrum. Meas. Mag.* **2021**, *24*, 49–58. [CrossRef]
23. Hu, Y.; Yang, J.; Jing, Q.; Niu, S.; Wu, W.; Wang, Z.L. Triboelectric nanogenerator built on suspended 3D spiral structure as vibration and positioning sensor and wave energy harvester. *ACS Nano* **2013**, *7*, 10424–10432. [CrossRef]
24. Lin, Z.H.; Cheng, G.; Lin, L.; Lee, S.; Wang, Z.L. Water–solid surface contact electrification and its use for harvesting liquid-wave energy. *Angew. Chem. Int. Ed.* **2013**, *52*, 12545–12549. [CrossRef] [PubMed]
25. Zhao, X.J.; Kuang, S.Y.; Wang, Z.L.; Zhu, G. Highly adaptive solid–liquid interfacing triboelectric nanogenerator for harvesting diverse water wave energy. *ACS Nano* **2018**, *12*, 4280–4285. [CrossRef]
26. Li, X.; Tao, J.; Wang, X.; Zhu, J.; Pan, C.; Wang, Z.L. Networks of high performance triboelectric nanogenerators based on liquid–solid interface contact electrification for harvesting low-frequency blue energy. *Adv. Energy Mater.* **2018**, *8*, 1800705. [CrossRef]
27. Zhu, G.; Su, Y.; Bai, P.; Chen, J.; Jing, Q.; Yang, W.; Wang, Z.L. Harvesting water wave energy by asymmetric screening of electrostatic charges on a nanostructured hydrophobic thin-film surface. *ACS Nano* **2014**, *8*, 6031–6037. [CrossRef]
28. Xu, L.; Pang, Y.; Zhang, C.; Jiang, T.; Chen, X.; Luo, J.; Tang, W.; Cao, X.; Wang, Z.L. Integrated triboelectric nanogenerator array based on air-driven membrane structures for water wave energy harvesting. *Nano Energy* **2017**, *31*, 351–358. [CrossRef]
29. Lee, K.; Lee, J.W.; Kim, K.; Yoo, D.; Kim, D.S.; Hwang, W.; Song, I.; Sim, J.Y. A spherical hybrid triboelectric nanogenerator for enhanced water wave energy harvesting. *Micromachines* **2018**, *9*, 598. [CrossRef]
30. An, J.; Wang, Z.M.; Jiang, T.; Liang, X.; Wang, Z.L. Whirling-folded triboelectric nanogenerator with high average power for water wave energy harvesting. *Adv. Funct. Mater.* **2019**, *29*, 1904867. [CrossRef]

31. Feng, Y.; Ling, L.; Nie, J.; Han, K.; Chen, X.; Bian, Z.; Li, H.; Wang, Z.L. Self-powered electrostatic filter with enhanced photocatalytic degradation of formaldehyde based on built-in triboelectric nanogenerators. *ACS Nano* **2017**, *11*, 12411–12418. [CrossRef]

32. Yang, Y.; Wang, Z.L. Hybrid energy cells for simultaneously harvesting multi-types of energies. *Nano Energy* **2015**, *14*, 245–256. [CrossRef]

33. Cha, S.N.; Seo, J.S.; Kim, S.M.; Kim, H.J.; Park, Y.J.; Kim, S.W.; Kim, J.M. Sound-driven piezoelectric nanowire-based nanogenerators. *Adv. Mater.* **2010**, *22*, 4726–4730. [CrossRef]

34. Huang, L.B.; Bai, G.; Wong, M.C.; Yang, Z.; Xu, W.; Hao, J. Triboelectric Nanogenerators: Magnetic-Assisted Noncontact Triboelectric Nanogenerator Converting Mechanical Energy into Electricity and Light Emissions (Adv. Mater. 14/2016). *Adv. Mater.* **2016**, *28*, 2843. [CrossRef]

35. Cheng, G.; Lin, Z.H.; Du, Z.l.; Wang, Z.L. Simultaneously harvesting electrostatic and mechanical energies from flowing water by a hybridized triboelectric nanogenerator. *ACS Nano* **2014**, *8*, 1932–1939. [CrossRef] [PubMed]

36. Cui, N.; Gu, L.; Liu, J.; Bai, S.; Qiu, J.; Fu, J.; Kou, X.; Liu, H.; Qin, Y.; Wang, Z.L. High performance sound driven triboelectric nanogenerator for harvesting noise energy. *Nano Energy* **2015**, *15*, 321–328. [CrossRef]

37. Miles, J.W. On the generation of surface waves by shear flows. *J. Fluid Mech.* **1957**, *3*, 185–204. [CrossRef]

38. Chen, J.; Huang, Y.; Zhang, N.; Zou, H.; Liu, R.; Tao, C.; Fan, X.; Wang, Z.L. Micro-cable structured textile for simultaneously harvesting solar and mechanical energy. *Nat. Energy* **2016**, *1*, 1–8. [CrossRef]

39. Ren, Z.; Ding, Y.; Nie, J.; Wang, F.; Xu, L.; Lin, S.; Chen, X.; Wang, Z.L. Environmental energy harvesting adapting to different weather conditions and self-powered vapor sensor based on humidity-responsive triboelectric nanogenerators. *ACS Appl. Mater. Interfaces* **2019**, *11*, 6143–6153. [CrossRef]

40. Chen, J.; Yang, J.; Li, Z.; Fan, X.; Zi, Y.; Jing, Q.; Guo, H.; Wen, Z.; Pradel, K.C.; Niu, S.; et al. Networks of triboelectric nanogenerators for harvesting water wave energy: A potential approach toward blue energy. *ACS Nano* **2015**, *9*, 3324–3331. [CrossRef] [PubMed]

41. Bai, P.; Zhu, G.; Liu, Y.; Chen, J.; Jing, Q.; Yang, W.; Ma, J.; Zhang, G.; Wang, Z.L. Cylindrical rotating triboelectric nanogenerator. *ACS Nano* **2013**, *7*, 6361–6366. [CrossRef] [PubMed]

42. Ren, X.; Fan, H.; Wang, C.; Ma, J.; Li, H.; Zhang, M.; Lei, S.; Wang, W. Wind energy harvester based on coaxial rotatory freestanding triboelectric nanogenerators for self-powered water splitting. *Nano Energy* **2018**, *50*, 562–570. [CrossRef]

43. Lin, Z.; Zhang, B.; Guo, H.; Wu, Z.; Zou, H.; Yang, J.; Wang, Z.L. Super-robust and frequency-multiplied triboelectric nanogenerator for efficient harvesting water and wind energy. *Nano Energy* **2019**, *64*, 103908. [CrossRef]

44. Ahmed, A.; Hassan, I.; Hedaya, M.; El-Yazid, T.A.; Zu, J.; Wang, Z.L. Farms of triboelectric nanogenerators for harvesting wind energy: A potential approach towards green energy. *Nano Energy* **2017**, *36*, 21–29. [CrossRef]

45. Liang, Q.; Yan, X.; Gu, Y.; Zhang, K.; Liang, M.; Lu, S.; Zheng, X.; Zhang, Y. Highly transparent triboelectric nanogenerator for harvesting water-related energy reinforced by antireflection coating. *Sci. Rep.* **2015**, *5*, 1–7. [CrossRef] [PubMed]

46. Liang, Q.; Yan, X.; Liao, X.; Zhang, Y. Integrated multi-unit transparent triboelectric nanogenerator harvesting rain power for driving electronics. *Nano Energy* **2016**, *25*, 18–25. [CrossRef]

47. Lai, Y.C.; Hsiao, Y.C.; Wu, H.M.; Wang, Z.L. Waterproof fabric-based multifunctional triboelectric nanogenerator for universally harvesting energy from raindrops, wind and human motions and as self-powered sensors. *Adv. Sci.* **2019**, *6*, 1801883. [CrossRef]

48. Xu, M.; Zhao, T.; Wang, C.; Zhang, S.L.; Li, Z.; Pan, X.; Wang, Z.L. High power density tower-like triboelectric nanogenerator for harvesting arbitrary directional water wave energy. *ACS Nano* **2019**, *13*, 1932–1939. [CrossRef]

49. Yang, J.; Chen, J.; Liu, Y.; Yang, W.; Su, Y.; Wang, Z.L. Triboelectrification-based organic film nanogenerator for acoustic energy harvesting and self-powered active acoustic sensing. *ACS Nano* **2014**, *8*, 2649–2657. [CrossRef]

50. Yang, Y.; Zhang, H.; Liu, R.; Wen, X.; Hou, T.C.; Wang, Z.L. Fully enclosed triboelectric nanogenerators for applications in water and harsh environments. *Adv. Energy Mater.* **2013**, *3*, 1563–1568. [CrossRef]

51. Zhang, D.; Shi, J.; Si, Y.; Li, T. Multi-grating triboelectric nanogenerator for harvesting low-frequency ocean wave energy. *Nano Energy* **2019**, *61*, 132–140. [CrossRef]

52. Zou, Y.; Tan, P.; Shi, B.; Ouyang, H.; Jiang, D.; Liu, Z.; Li, H.; Yu, M.; Wang, C.; Qu, X.; et al. A bionic stretchable nanogenerator for underwater sensing and energy harvesting. *Nat. Commun.* **2019**, *10*, 1–10. [CrossRef] [PubMed]

53. Cheng, P.; Guo, H.; Wen, Z.; Zhang, C.; Yin, X.; Li, X.; Liu, D.; Song, W.; Sun, X.; Wang, J.; et al. Largely enhanced triboelectric nanogenerator for efficient harvesting of water wave energy by soft contacted structure. *Nano Energy* **2019**, *57*, 432–439. [CrossRef]

54. Liu, W.; Xu, L.; Bu, T.; Yang, H.; Liu, G.; Li, W.; Pang, Y.; Hu, C.; Zhang, C.; Cheng, T. Torus structured triboelectric nanogenerator array for water wave energy harvesting. *Nano Energy* **2019**, *58*, 499–507. [CrossRef]

55. Mariello, M.; Guido, F.; Mastronardi, V.; Todaro, M.; Desmaële, D.; De Vittorio, M. Nanogenerators for harvesting mechanical energy conveyed by liquids. *Nano Energy* **2019**, *57*, 141–156. [CrossRef]

56. Cao, X.; Jie, Y.; Wang, N.; Wang, Z.L. Triboelectric nanogenerators driven self-powered electrochemical processes for energy and environmental science. *Adv. Energy Mater.* **2016**, *6*, 1600665. [CrossRef]

57. Khan, U.; Kim, S.W. Triboelectric nanogenerators for blue energy harvesting. *ACS Nano* **2016**, *10*, 6429–6432. [CrossRef]

58. Gu, L.; Cui, N.; Liu, J.; Zheng, Y.; Bai, S.; Qin, Y. Packaged triboelectric nanogenerator with high endurability for severe environments. *Nanoscale* **2015**, *7*, 18049–18053. [CrossRef]

116. Yang, Y.; Zhang, H.; Lee, S.; Kim, D.; Hwang, W.; Wang, Z.L. Hybrid energy cell for degradation of methyl orange by self-powered electrocatalytic oxidation. *Nano Lett.* **2013**, *13*, 803–808. [CrossRef] [PubMed]

117. Iglesias, G.; Carballo, R. Wave farm impact: The role of farm-to-coast distance. *Renew. Energy* **2014**, *69*, 375–385. [CrossRef]

118. Zhao, K.; Gu, G.; Zhang, Y.; Zhang, B.; Yang, F.; Zhao, L.; Zheng, M.; Cheng, G.; Du, Z. The self-powered CO2 gas sensor based on gas discharge induced by triboelectric nanogenerator. *Nano Energy* **2018**, *53*, 898–905. [CrossRef]

119. Liu, X.; Cheng, K.; Cui, P.; Qi, H.; Qin, H.; Gu, G.; Shang, W.; Wang, S.; Cheng, G.; Du, Z. Hybrid energy harvester with bi-functional nano-wrinkled anti-reflective PDMS film for enhancing energies conversion from sunlight and raindrops. *Nano Energy* **2019**, *66*, 104188. [CrossRef]

120. Xi, Y.; Guo, H.; Zi, Y.; Li, X.; Wang, J.; Deng, J.; Li, S.; Hu, C.; Cao, X.; Wang, Z.L. Multifunctional TENG for blue energy scavenging and self-powered wind-speed sensor. *Adv. Energy Mater.* **2017**, *7*, 1602397. [CrossRef]

121. Yoo, D.; Park, S.C.; Lee, S.; Sim, J.Y.; Song, I.; Choi, D.; Lim, H.; Kim, D.S. Biomimetic anti-reflective triboelectric nanogenerator for concurrent harvesting of solar and raindrop energies. *Nano Energy* **2019**, *57*, 424–431. [CrossRef]

122. Yang, X.; Xu, L.; Lin, P.; Zhong, W.; Bai, Y.; Luo, J.; Chen, J.; Wang, Z.L. Macroscopic self-assembly network of encapsulated high-performance triboelectric nanogenerators for water wave energy harvesting. *Nano Energy* **2019**, *60*, 404–412. [CrossRef]

123. Aydoğan, B.; Ayat, B.; Yüksel, Y. Black Sea wave energy atlas from 13 years hindcasted wave data. *Renew. Energy* **2013**, *57*, 436–447. [CrossRef]

124. Calisal, S. A note on the derivation of potential energy for two-dimensional water waves. *Ocean Eng.* **1983**, *10*, 133–138. [CrossRef]

125. Thorpe, T.W. *A Brief Review of Wave Energy*; Harwell Laboratory, Energy Technology Support Unit: London, UK, 1999.

126. Olsen, M.; Zhang, R.; Örtegren, J.; Andersson, H.; Yang, Y.; Olin, H. Frequency and voltage response of a wind-driven fluttering triboelectric nanogenerator. *Sci. Rep.* **2019**, *9*, 1–6. [CrossRef]

127. Suzuki, R.O.; Tanaka, D. Mathematical simulation of thermoelectric power generation with the multi-panels. *J. Power Sources* **2003**, *122*, 201–209. [CrossRef]

128. Prieto, L.F.; Rodríguez, G.R.; Rodríguez, J.S. Wave energy to power a desalination plant in the north of Gran Canaria Island: Wave resource, socioeconomic and environmental assessment. *J. Environ. Manag.* **2019**, *231*, 546–551. [CrossRef]

129. Leijon, M.; Bernhoff, H.; Agren, O.; Isberg, J.; Sundberg, J.; Berg, M.; Karlsson, K.E.; Wolfbrandt, A. Multiphysics simulation of wave energy to electric energy conversion by permanent magnet linear generator. *IEEE Trans. Energy Convers.* **2005**, *20*, 219–224. [CrossRef]

130. Pelc, R.; Fujita, R.M. Renewable energy from the ocean. *Mar. Policy* **2002**, *26*, 471–479. [CrossRef]

131. Chatzigiannakou, M.A.; Dolguntseva, I.; Leijon, M. Offshore deployments of wave energy converters by seabased industry AB. *J. Mar. Sci. Eng.* **2017**, *5*, 15. [CrossRef]

132. Nie, J.; Ren, Z.; Xu, L.; Lin, S.; Zhan, F.; Chen, X.; Wang, Z.L. Probing contact-electrification-induced electron and ion transfers at a liquid–solid interface. *Adv. Mater.* **2020**, *32*, 1905696. [CrossRef] [PubMed]

133. Moskvitch, K. News Briefing: In Num6ers-Sola Road. *Eng. Technol.* **2016**, *11*, 12–13. [CrossRef]

134. Zhao, X.; Chen, B.; Wei, G.; Wu, J.M.; Han, W.; Yang, Y. Polyimide/graphene nanocomposite foam-based wind-driven triboelectric nanogenerator for self-powered pressure sensor. *Adv. Mater. Technol.* **2019**, *4*, 1800723. [CrossRef]

135. McCormick, M.E.; Ertekin, R.C. Renewable sea power. *Mech. Eng.* **2009**, *131*, 36–39. [CrossRef]

136. Deane, J.; Dharmasena, R.; Gentile, G. Power computation for the triboelectric nanogenerator. *Nano Energy* **2018**, *54*, 39–49. [CrossRef]

137. Roscow, J.; Lewis, R.; Taylor, J.; Bowen, C. Corrigendum to "Modelling and fabrication of porous sandwich layer barium titanate with improved piezoelectric energy harvesting figures of merit" [Acta Mater. 128 (2017) 207-217]. *Acta Mater.* **2018**, *152*, 199. [CrossRef]

138. Yousry, Y.M.; Yao, K.; Chen, S.; Liew, W.H.; Ramakrishna, S. Mechanisms for enhancing polarization orientation and piezoelectric parameters of PVDF nanofibers. *Adv. Electron. Mater.* **2018**, *4*, 1700562. [CrossRef]

139. Harstad, S.; D'Souza, N.; Soin, N.; El-Gendy, A.A.; Gupta, S.; Pecharsky, V.K.; Shah, T.; Siores, E.; Hadimani, R.L. Enhancement of β-phase in PVDF films embedded with ferromagnetic Gd5Si4 nanoparticles for piezoelectric energy harvesting. *AIP Adv.* **2017**, *7*, 056411. [CrossRef]

140. Jbaily, A.; Yeung, R.W. Piezoelectric devices for ocean energy: A brief survey. *J. Ocean. Eng. Mar. Energy* **2015**, *1*, 101–118. [CrossRef]

141. Mo, J.; Zhang, C.; Lu, Y.; Liu, Y.; Zhang, N.; Wang, S.; Nie, S. Radial piston triboelectric nanogenerator-enhanced cellulose fiber air filter for self-powered particulate matter removal. *Nano Energy* **2020**, *78*, 105357. [CrossRef]

142. Gu, G.Q.; Han, C.B.; Lu, C.X.; He, C.; Jiang, T.; Gao, Z.L.; Li, C.J.; Wang, Z.L. Triboelectric nanogenerator enhanced nanofiber air filters for efficient particulate matter removal. *ACS Nano* **2017**, *11*, 6211–6217. [CrossRef]

143. Uchino, K. Piezoelectric actuators 2008: Key factors for commercialization. *Adv. Mater. Res.* **2008**, *55*, 1–9. [CrossRef]

144. Marino, A.; Genchi, G.G.; Mattoli, V.; Ciofani, G. Piezoelectric nanotransducers: The future of neural stimulation. *Nano Today* **2017**, *14*, 9–12. [CrossRef]

145. Gao, X.; Wu, J.; Yu, Y.; Chu, Z.; Shi, H.; Dong, S. Giant piezoelectric coefficients in relaxor piezoelectric ceramic PNN-PZT for vibration energy harvesting. *Adv. Funct. Mater.* **2018**, *28*, 1706895. [CrossRef]

146. Damjanovic, D. Ferroelectric, dielectric and piezoelectric properties of ferroelectric thin films and ceramics. *Rep. Prog. Phys.* **1998**, *61*, 1267. [CrossRef]

147. Messing, G.L.; Sabolsky, E.; Kwon, S.; Trolier-McKinstry, S. Templated grain growth of textured piezoelectric ceramics. *Key Eng. Mater.* **2001**, *206*, 1293–1296. [CrossRef]

148. Isarakorn, D.; Sambri, A.; Janphuang, P.; Briand, D.; Gariglio, S.; Triscone, J.M.; Guy, F.; Reiner, J.; Ahn, C.; De Rooij, N. Epitaxial piezoelectric MEMS on silicon. *J. Micromech. Microeng.* **2010**, *20*, 055008. [CrossRef]

149. Muralt, P.; Polcawich, R.G.; Trolier-McKinstry, S. Piezoelectric thin films for sensors, actuators and energy harvesting. *MRS Bull.* **2009**, *34*, 658–664. [CrossRef]

150. Roscow, J.; Taylor, J.; Bowen, C. Manufacture and characterization of porous ferroelectrics for piezoelectric energy harvesting applications. *Ferroelectrics* **2016**, *498*, 40–46. [CrossRef]

151. Martínez-Ayuso, G.; Friswell, M.I.; Adhikari, S.; Khodaparast, H.H.; Berger, H. Homogenization of porous piezoelectric materials. *Int. J. Solids Struct.* **2017**, *113*, 218–229. [CrossRef]

152. Wang, Z.L. Triboelectric nanogenerator (TENG)—Sparking an energy and sensor revolution. *Adv. Energy Mater.* **2020**, *10*, 2000137. [CrossRef]

153. Niu, S.; Wang, Z.L. Theoretical systems of triboelectric nanogenerators. *Nano Energy* **2015**, *14*, 161–192. [CrossRef]

154. Niu, S.; Zhou, Y.S.; Wang, S.; Liu, Y.; Lin, L.; Bando, Y.; Wang, Z.L. Simulation method for optimizing the performance of an integrated triboelectric nanogenerator energy harvesting system. *Nano Energy* **2014**, *8*, 150–156. [CrossRef]

155. Ma, P.; Zhu, H.; Lu, H.; Zeng, Y.; Zheng, N.; Wang, Z.L.; Cao, X. Design of biodegradable wheat-straw based triboelectric nanogenerator as self-powered sensor for wind detection. *Nano Energy* **2021**, *86*, 106032. [CrossRef]

156. Dai, K.; Wang, X.; Niu, S.; Yi, F.; Yin, Y.; Chen, L.; Zhang, Y.; You, Z. Simulation and structure optimization of triboelectric nanogenerators considering the effects of parasitic capacitance. *Nano Res.* **2017**, *10*, 157–171. [CrossRef]

157. Shao, J.; Jiang, T.; Wang, Z. Theoretical foundations of triboelectric nanogenerators (TENGs). *Sci. China Technol. Sci.* **2020**, *63*, 1087–1109. [CrossRef]

158. Shao, J.; Willatzen, M.; Jiang, T.; Tang, W.; Chen, X.; Wang, J.; Wang, Z.L. Quantifying the power output and structural figure-of-merits of triboelectric nanogenerators in a charging system starting from the Maxwell's displacement current. *Nano Energy* **2019**, *59*, 380–389. [CrossRef]

159. Shao, J.; Liu, D.; Willatzen, M.; Wang, Z.L. Three-dimensional modeling of alternating current triboelectric nanogenerator in the linear sliding mode. *Appl. Phys. Rev.* **2020**, *7*, 011405. [CrossRef]

160. Dharmasena, R.D.I.G.; Jayawardena, K.; Mills, C.; Deane, J.; Anguita, J.; Dorey, R.; Silva, S. Triboelectric nanogenerators: Providing a fundamental framework. *Energy Environ. Sci.* **2017**, *10*, 1801–1811. [CrossRef]

161. Dharmasena, R.; Jayawardena, K.; Mills, C.; Dorey, R.; Silva, S. A unified theoretical model for Triboelectric Nanogenerators. *Nano Energy* **2018**, *48*, 391–400. [CrossRef]

162. Shao, J.; Willatzen, M.; Wang, Z.L. Theoretical modeling of triboelectric nanogenerators (TENGs). *J. Appl. Phys.* **2020**, *128*, 111101. [CrossRef]

163. Shao, J.; Willatzen, M.; Shi, Y.; Wang, Z.L. 3D mathematical model of contact-separation and single-electrode mode triboelectric nanogenerators. *Nano Energy* **2019**, *60*, 630–640. [CrossRef]

164. Tang, W.; Jiang, T.; Fan, F.R.; Yu, A.F.; Zhang, C.; Cao, X.; Wang, Z.L. Liquid-metal electrode for high-performance triboelectric nanogenerator at an instantaneous energy conversion efficiency of 70.6%. *Adv. Funct. Mater.* **2015**, *25*, 3718–3725. [CrossRef]

165. Xie, Y.; Wang, S.; Niu, S.; Lin, L.; Jing, Q.; Yang, J.; Wu, Z.; Wang, Z.L. Grating-structured freestanding triboelectric-layer nanogenerator for harvesting mechanical energy at 85% total conversion efficiency. *Adv. Mater.* **2014**, *26*, 6599–6607. [CrossRef] [PubMed]

166. Wang, S.; Niu, S.; Yang, J.; Lin, L.; Wang, Z.L. Quantitative measurements of vibration amplitude using a contact-mode freestanding triboelectric nanogenerator. *ACS Nano* **2014**, *8*, 12004–12013. [CrossRef] [PubMed]

167. Fan, F.R.; Lin, L.; Zhu, G.; Wu, W.; Zhang, R.; Wang, Z.L. Transparent triboelectric nanogenerators and self-powered pressure sensors based on micropatterned plastic films. *Nano Lett.* **2012**, *12*, 3109–3114. [CrossRef] [PubMed]

168. Yang, Y.; Wang, S.; Zhang, Y.; Wang, Z.L. Pyroelectric nanogenerators for driving wireless sensors. *Nano Lett.* **2012**, *12*, 6408–6413. [CrossRef]

169. Zhu, G.; Wang, A.C.; Liu, Y.; Zhou, Y.; Wang, Z.L. Functional electrical stimulation by nanogenerator with 58 V output voltage. *Nano Lett.* **2012**, *12*, 3086–3090. [CrossRef]

170. Yang, Y.; Zhu, G.; Zhang, H.; Chen, J.; Zhong, X.; Lin, Z.H.; Su, Y.; Bai, P.; Wen, X.; Wang, Z.L. Triboelectric nanogenerator for harvesting wind energy and as self-powered wind vector sensor system. *ACS Nano* **2013**, *7*, 9461–9468. [CrossRef]

171. Zhu, G.; Bai, P.; Chen, J.; Wang, Z.L. Power-generating shoe insole based on triboelectric nanogenerators for self-powered consumer electronics. *Nano Energy* **2013**, *2*, 688–692. [CrossRef]

172. Xie, Y.; Wang, S.; Lin, L.; Jing, Q.; Lin, Z.H.; Niu, S.; Wu, Z.; Wang, Z.L. Rotary triboelectric nanogenerator based on a hybridized mechanism for harvesting wind energy. *ACS Nano* **2013**, *7*, 7119–7125. [CrossRef]

173. Zhou, Y.S.; Liu, Y.; Zhu, G.; Lin, Z.H.; Pan, C.; Jing, Q.; Wang, Z.L. In situ quantitative study of nanoscale triboelectrification and patterning. *Nano Lett.* **2013**, *13*, 2771–2776. [CrossRef]

174. Yang, W.; Chen, J.; Zhu, G.; Yang, J.; Bai, P.; Su, Y.; Jing, Q.; Cao, X.; Wang, Z.L. Harvesting energy from the natural vibration of human walking. *ACS Nano* **2013**, *7*, 11317–11324. [CrossRef]

175. Zhu, G.; Chen, J.; Zhang, T.; Jing, Q.; Wang, Z.L. Radial-arrayed rotary electrification for high performance triboelectric generator. *Nat. Commun.* **2014**, *5*, 1–9. [CrossRef]

176. Askari, H.; Khajepour, A.; Khamesee, M.B.; Saadatnia, Z.; Wang, Z.L. Piezoelectric and triboelectric nanogenerators: Trends and impacts. *Nano Today* **2018**, *22*, 10–13. [CrossRef]

177. Jung, Y.; Yu, J.; Hwang, H.J.; Bhatia, D.; Chung, K.B.; Choi, D. Wire-based triboelectric resonator for a self-powered crack monitoring system. *Nano Energy* **2020**, *71*, 104615. [CrossRef]

178. Xu, L.; Jiang, T.; Lin, P.; Shao, J.J.; He, C.; Zhong, W.; Chen, X.Y.; Wang, Z.L. Coupled triboelectric nanogenerator networks for efficient water wave energy harvesting. *ACS Nano* **2018**, *12*, 1849–1858. [CrossRef] [PubMed]

179. Lin, Z.; Zhang, B.; Xie, Y.; Wu, Z.; Yang, J.; Wang, Z.L. Elastic-Connection and Soft-Contact Triboelectric Nanogenerator with Superior Durability and Efficiency. *Adv. Funct. Mater.* **2021**, 2015237.

180. Liang, X.; Liu, Z.; Feng, Y.; Han, J.; Li, L.; An, J.; Chen, P.; Jiang, T.; Wang, Z.L. Spherical triboelectric nanogenerator based on spring-assisted swing structure for effective water wave energy harvesting. *Nano Energy* **2021**, *83*, 105836. [CrossRef]

181. Guan, D.; Cong, X.; Li, J.; Shen, H.; Zhang, C.; Gong, J. Quantitative Characterization of the Energy Harvesting Performance of Soft-contact Sphere Triboelectric Nanogenerator. *Nano Energy* **2021**, *87*, 106186. [CrossRef]

182. Yuan, Z.; Wang, C.; Xi, J.; Han, X.; Li, J.; Han, S.T.; Gao, W.; Pan, C. Spherical Triboelectric Nanogenerator with Dense Point Contacts for Harvesting Multidirectional Water Wave and Vibration Energy. *ACS Energy Lett.* **2021**, *6*, 2809–2816. [CrossRef]

183. Bhatia, D.; Hwang, H.J.; Huynh, N.D.; Lee, S.; Lee, C.; Nam, Y.; Kim, J.G.; Choi, D. Continuous scavenging of broadband vibrations via omnipotent tandem triboelectric nanogenerators with cascade impact structure. *Sci. Rep.* **2019**, *9*, 1–9.

184. Jiang, T.; Yao, Y.; Xu, L.; Zhang, L.; Xiao, T.; Wang, Z.L. Spring-assisted triboelectric nanogenerator for efficiently harvesting water wave energy. *Nano Energy* **2017**, *31*, 560–567. [CrossRef]

185. Xiao, T.X.; Liang, X.; Jiang, T.; Xu, L.; Shao, J.J.; Nie, J.H.; Bai, Y.; Zhong, W.; Wang, Z.L. Spherical triboelectric nanogenerators based on spring-assisted multilayered structure for efficient water wave energy harvesting. *Adv. Funct. Mater.* **2018**, *28*, 1802634. [CrossRef]

186. Wang, W.; Xu, J.; Zheng, H.; Chen, F.; Jenkins, K.; Wu, Y.; Wang, H.; Zhang, W.; Yang, R. A spring-assisted hybrid triboelectric–electromagnetic nanogenerator for harvesting low-frequency vibration energy and creating a self-powered security system. *Nanoscale* **2018**, *10*, 14747–14754. [CrossRef] [PubMed]

187. Wang, S.; Wang, Y.; Liu, D.; Zhang, Z.; Li, W.; Liu, C.; Du, T.; Xiao, X.; Song, L.; Pang, H.; et al. A robust and self-powered tilt sensor based on annular liquid–solid interfacing triboelectric nanogenerator for ship attitude sensing. *Sens. Actuators A Phys.* **2021**, *317*, 112459. [CrossRef]

188. Singh, H.H.; Khare, N. KPFM Study of Flexible Ferroelectric Polymer/Water Interface for Understanding the Working Principle of Liquid–Solid Triboelectric Nanogenerator. *Adv. Mater. Interfaces* **2021**, *8*, 2100032. [CrossRef]

189. Vo, C.P.; Shahriar, M.; Le, C.D.; Ahn, K.K. Mechanically active transducing element based on solid–liquid triboelectric nanogenerator for self-powered sensing. *Int. J. Precis. Eng. Manuf.-Green Technol.* **2019**, *6*, 741–749. [CrossRef]

190. Chung, J.; Cho, H.; Yong, H.; Heo, D.; Rim, Y.S.; Lee, S. Versatile surface for solid–solid/liquid–solid triboelectric nanogenerator based on fluorocarbon liquid infused surfaces. *Sci. Technol. Adv. Mater.* **2020**, *21*, 139–146. [CrossRef]

191. Vu, D.L.; Le, C.D.; Vo, C.P.; Ahn, K.K. Surface polarity tuning through epitaxial growth on polyvinylidene fluoride membranes for enhanced performance of liquid–solid triboelectric nanogenerator. *Compos. Part B Eng.* **2021**, *223*, 109135. [CrossRef]

192. Xu, M.; Wang, S.; Zhang, S.L.; Ding, W.; Kien, P.T.; Wang, C.; Li, Z.; Pan, X.; Wang, Z.L. A highly-sensitive wave sensor based on liquid–solid interfacing triboelectric nanogenerator for smart marine equipment. *Nano Energy* **2019**, *57*, 574–580. [CrossRef]

193. Wei, X.; Zhao, Z.; Zhang, C.; Yuan, W.; Wu, Z.; Wang, J.; Wang, Z.L. All-Weather Droplet-Based Triboelectric Nanogenerator for Wave Energy Harvesting. *ACS Nano* **2021**, online ahead of print. [CrossRef]

194. Nie, J.; Jiang, T.; Shao, J.; Ren, Z.; Bai, Y.; Iwamoto, M.; Chen, X.; Wang, Z.L. Motion behavior of water droplets driven by triboelectric nanogenerator. *Appl. Phys. Lett.* **2018**, *112*, 183701. [CrossRef]

195. Jung, W.S.; Kang, M.G.; Moon, H.G.; Baek, S.H.; Yoon, S.J.; Wang, Z.L.; Kim, S.W.; Kang, C.Y. High output piezo/triboelectric hybrid generator. *Sci. Rep.* **2015**, *5*, 1–6. [CrossRef]

196. Chen, X.; Han, M.; Chen, H.; Cheng, X.; Song, Y.; Su, Z.; Jiang, Y.; Zhang, H. A wave-shaped hybrid piezoelectric and triboelectric nanogenerator based on P (VDF-TrFE) nanofibers. *Nanoscale* **2017**, *9*, 1263–1270. [CrossRef]

197. He, W.; Qian, Y.; Lee, B.S.; Zhang, F.; Rasheed, A.; Jung, J.E.; Kang, D.J. Ultrahigh output piezoelectric and triboelectric hybrid nanogenerators based on ZnO nanoflakes/polydimethylsiloxane composite films. *ACS Appl. Mater. Interfaces* **2018**, *10*, 44415–44420. [CrossRef] [PubMed]

198. Li, Z.; Saadatnia, Z.; Yang, Z.; Naguib, H. A hybrid piezoelectric-triboelectric generator for low-frequency and broad-bandwidth energy harvesting. *Energy Convers. Manag.* **2018**, *174*, 188–197. [CrossRef]

199. Yang, D.; Ni, Y.; Su, H.; Shi, Y.; Liu, Q.; Chen, X.; He, D. Hybrid energy system based on solar cell and self-healing/self-cleaning triboelectric nanogenerator. *Nano Energy* **2021**, *79*, 105394. [CrossRef]

200. Gao, L.; Lu, S.; Xie, W.; Chen, X.; Wu, L.; Wang, T.; Wang, A.; Yue, C.; Tong, D.; Lei, W.; et al. A self-powered and self-functional tracking system based on triboelectric-electromagnetic hybridized blue energy harvesting module. *Nano Energy* **2020**, *72*, 104684. [CrossRef]

201. Wu, Z.; Guo, H.; Ding, W.; Wang, Y.C.; Zhang, L.; Wang, Z.L. A hybridized triboelectric–electromagnetic water wave energy harvester based on a magnetic sphere. *ACS Nano* **2019**, *13*, 2349–2356. [CrossRef]

202. Wang, P.; Liu, R.; Ding, W.; Zhang, P.; Pan, L.; Dai, G.; Zou, H.; Dong, K.; Xu, C.; Wang, Z.L. Complementary electromagnetic-triboelectric active sensor for detecting multiple mechanical triggering. *Adv. Funct. Mater.* **2018**, *28*, 1705808. [CrossRef]

203. Wang, P.; Pan, L.; Wang, J.; Xu, M.; Dai, G.; Zou, H.; Dong, K.; Wang, Z.L. An ultra-low-friction triboelectric–electromagnetic hybrid nanogenerator for rotation energy harvesting and self-powered wind speed sensor. *ACS Nano* **2018**, *12*, 9433–9440. [CrossRef]
204. Hao, C.; He, J.; Zhang, Z.; Yuan, Y.; Chou, X.; Xue, C. A pendulum hybrid generator for water wave energy harvesting and hydrophone-based wireless sensing. *AIP Adv.* **2020**, *10*, 125019. [CrossRef]
205. He, J.; Fan, X.; Mu, J.; Wang, C.; Qian, J.; Li, X.; Hou, X.; Geng, W.; Wang, X.; Chou, X. 3D full-space triboelectric-electromagnetic hybrid nanogenerator for high-efficient mechanical energy harvesting in vibration system. *Energy* **2020**, *194*, 116871. [CrossRef]
206. Wu, Z.; Ding, W.; Dai, Y.; Dong, K.; Wu, C.; Zhang, L.; Lin, Z.; Cheng, J.; Wang, Z.L. Self-powered multifunctional motion sensor enabled by magnetic-regulated triboelectric nanogenerator. *ACS Nano* **2018**, *12*, 5726–5733. [CrossRef] [PubMed]
207. Huang, L.b.; Xu, W.; Bai, G.; Wong, M.C.; Yang, Z.; Hao, J. Wind energy and blue energy harvesting based on magnetic-assisted noncontact triboelectric nanogenerator. *Nano Energy* **2016**, *30*, 36–42. [CrossRef]
208. Seol, M.L.; Han, J.W.; Jeon, S.B.; Meyyappan, M.; Choi, Y.K. Floating oscillator-embedded triboelectric generator for versatile mechanical energy harvesting. *Sci. Rep.* **2015**, *5*, 1–10. [CrossRef]
209. Liu, S.; Li, X.; Wang, Y.; Yang, Y.; Meng, L.; Cheng, T.; Wang, Z.L. Magnetic switch structured triboelectric nanogenerator for continuous and regular harvesting of wind energy. *Nano Energy* **2021**, *83*, 105851. [CrossRef]
210. Wu, Z.; Zhang, B.; Zou, H.; Lin, Z.; Liu, G.; Wang, Z.L. Multifunctional sensor based on translational-rotary triboelectric nanogenerator. *Adv. Energy Mater.* **2019**, *9*, 1901124. [CrossRef]
211. Liu, G.; Chen, J.; Guo, H.; Lai, M.; Pu, X.; Wang, X.; Hu, C. Triboelectric nanogenerator based on magnetically induced retractable spring steel tapes for efficient energy harvesting of large amplitude motion. *Nano Res.* **2018**, *11*, 633–641. [CrossRef]

Article

Vibration Converter with Passive Energy Management for Battery-Less Wireless Sensor Nodes in Predictive Maintenance

Sonia Bradai, Ghada Bouattour, Dhouha El Houssaini and Olfa Kanoun *

Measurement and Sensor Technology, Faculty of Electrical Engineering, Chemnitz University of Technology, 09111 Chemnitz, Germany; sonia.bradai@etit.tu-chemnitz.de (S.B.); ghada.bouattour@etit.tu-chemnitz.de (G.B.); dhouha.el houssaini@etit.tu-chemnitz.de (D.E.H.)
* Correspondence: olfa.kanoun@etit.tu-chemnitz.de

Abstract: Predictive maintenance is becoming increasingly important in industry and requires continuous monitoring to prevent failures and anticipate maintenance processes, resulting in reduced downtime. Vibration is often used for failure detection and equipment conditioning as it is well correlated to the machine's operation and its variation is an indicator of process changes. In this context, we propose a novel energy-autonomous wireless sensor system that is able to measure without the use of batteries and automatically deliver alerts once the machine has an anomaly by the variation in acceleration. For this, we designed a wideband electromagnetic energy harvester and realized passive energy management to supply a wireless sensor node, which does not need an external energy supply. The advantage of the solution is that the designed circuit is able to detect the failure without the use of additional sensors, but by the Analog Digital Converter (ADC) of the Wireless Sensor Nodes (WSN) themselves, which makes it more compact and have lower energy consumption. The electromagnetic converter can harvest the relevant energy levels from weak vibration, with an acceleration of 0.1 g for a frequency bandwidth of 7 Hz. Further, the energy-management circuit enabled fast recharging of the super capacitor on a maximum of 31 s. The designed energy-management circuit consists of a six-stage voltage multiplier circuit connected to a wide-band DC-DC converter, as well as an under-voltage lock-out (UVLO) circuit to connect to the storage device to the WSN. In the failure condition with a frequency of 13 Hz and an acceleration of 0.3 g, the super capacitor recharging time was estimated to be 24 s. The proposed solution was validated by implementing real failure detection scenarios with random acceleration levels and, alternatively, modus. The results show that the WSN can directly measure the harvester's response and decide about the occurrence of failure based on its characteristic threshold voltage without the use of an additional sensor.

Keywords: energy harvesting; autonomous wireless sensor; passive energy management; weak vibration; electromagnetic converter; wideband; planar spring; voltage multiplier; rectifier; predictive maintenance; failure detection; WSN

Citation: Bradai, S.; Bouattour, G.; El Houssaini, D.; Kanoun, O. Vibration Converter with Passive Energy Management for Battery-Less Wireless Sensor Nodes in Predictive Maintenance. *Energies* **2022**, *15*, 1982. https://doi.org/10.3390/en15061982

Academic Editor: Dibin Zhu

Received: 18 January 2022
Accepted: 3 March 2022
Published: 8 March 2022

Publisher's Note: MDPI stays neutral with regard to jurisdictional claims in published maps and institutional affiliations.

1. Introduction

Predictive maintenance has become one of the key factors developed in modern industries to detect failure before it leads to catastrophic damage, which helps to reduce maintenance costs [1]. It can be associated with motors, conveyor chain failure, and mechanical component deterioration. Predictive maintenance can be achieved classically based on the expected lifetime of devices, which does not ensure the recognition of sudden damage. With industrial evolution, reliable and continuous system monitoring, especially potential failure prediction, is still quite challenging. It can be ensured through the continuous measurement of several parameters, such as temperature, vibration, pressure, and speed. To this end, various sensors are required, which lead to a large number of cables for data and power transmission [2]. Further, using such types of sensors limits their implementation, in particular, in inaccessible and harsh environments. Therefore, wireless sensor nodes (WSNs) in machine condition monitoring (MCM) are highly required to avoid catastrophic damages as well as enable easy control for the operating status of equipment in industry. Hence, to develop a reliable and autonomous WSN, several challenges occur nowadays. One of the most important challenges is the availability of a continuous and robust power source for the WSN. Battery technology is still used, but is not reliable and requires human intervention. Therefore, ambient energy harvesting from the environment surrounding the WSN presents an interesting alternative. This includes solar, vibration, and thermoelectric sources. Nevertheless, the challenge is to ensure a continuous and relevant energy amount. Even though solar sources can provide a considerable energy amount, it becomes limited, especially in indoor applications [2,3]. For thermoelectric sources, several research works have shown that the conversion efficiency is very limited [4,5]. To this end, the vibration source is most promising in industrial applications due to its presence in machines and its relatively high energy density [6]. In this case, several principles can be used to harvest energy from vibration, including electrostatic [7,8], piezoelectric [9,10], electromagnetic [11,12], and magnetoelectric [13,14] principles. Electromagnetic converters are the most considered in industrial applications due to their robustness, relatively relevant energy output, and their ease of integration compared to the other principles [6]. The main challenge for electromagnetic converters is generating relevant energy outcomes from weak vibration sources with low acceleration limited to 0.1 g to 0.2 g and low frequencies up to 30 Hz. Until now, in the state of the art, the developed harvesters for such vibration sources have several limitations. In [15], an eccentric pendulum-based electromagnetic converter was developed to harvest energy from 0.1 g acceleration and a frequency of 2 Hz, which was successful, but the total volume of the harvester was 153.9 cm^3, which results in a bulky system. Further, in [16], a compact solution with a total volume of ~27.4 cm^3 was developed for a low frequency of 10 Hz, but was unable to generate energy under 1 g of applied acceleration. In [17], a nonlinear magnetic rolling pendulum-based electromagnetic principle was developed, which could work for low frequencies in the range lower than 30 Hz and an acceleration of 0.2 g; nevertheless, the proposed solution was difficult to be implemented easily in industrial applications. To this end, some researchers focused on rotational electromagnetic converters for low frequencies, as in [18], where, unfortunately, a high displacement of 20 mm was required to generate energy, and in [19], where the solution could work at low displacement, but presented a bulky solution with a total volume of 97 cm^3.

Further, most of the electromagnetic harvesters generate low AC voltage, where rectification is required, and sometimes the required energy for that is not provided by the harvester. Moreover, matching the internal resistance of the energy harvester and the energy-management circuit is required to achieve the maximum power output for the WSN. Several works investigated such challenges; nevertheless, most of them reached only up to 20–80% of efficiency [20] in the case of passive energy management solutions or up to 90% [20] using an active solution with a wide-band input power level, which is not recommended for an autonomous WSN due to their large size and complexity.

In this context, few solutions have realized an autonomous solution-based energy harvesting and wireless sensor node for IIoT devices in industry and predictive maintenance. In [20–22], a sensor network powered through an electromagnetic harvester was developed for the diagnosis of ball bearings; nevertheless, the results showed that the system required 100 mW to measure and transmit the information, which was quite high. Further, in [22], a wireless sensor node was developed for structural health monitoring and had the advantage of low power consumption. Nevertheless, the solution had a limitation in terms of the photovoltaic source, which led to a long charging time of 30 min. In other research work [23], 5 min was required for the wireless sensor node to transmit the information, and the solution was based on the use of a thermoelectric harvester. In this case, the placement of the harvester is crucial and specific to ensure the required energy, which limits its use.

To conclude this research, there is always a challenge in ensuring a system with low power consumption and with a continuous energy source, which is the aim of this paper.

This paper proposes a passive solution for predictive maintenance through an autonomous WSN powered by a vibration energy harvester with its passive energy management solution. The proposed solution consists of continuous measurement of data to predict failures and to avoid catastrophic damage to the equipment through the converter response. In particular, it enables the continuous measurement of acceleration and alerting through the WSN in case of high acceleration presence, which is considered here at the level of 0.25 g. This paper is structured in mainly four sections. Section 2 presents the system concept of this work for fault detection. Section 3 is devoted to the model and evaluation of the electromagnetic harvester. Section 4 introduces the passive energy management solution and its evaluation. Section 5 describes the selected wireless sensor node and its performance. Finally, the feasibility of powering the wireless sensor node through the developed energy harvester system is discussed.

2. System Concept for Fault Detection

This paper proposes an autonomous WSN solution for predictive maintenance. Three main parts were developed, as shown in Figure 1. The first element expresses the power unit, including an energy harvester based on a vibration converter that can generate a relevant energy level under a weak vibration level from 0.1 g of acceleration. It consists of the electromagnetic converter detailed in Section 3 and is characterized by good robustness and high energy density with a large frequency bandwidth.

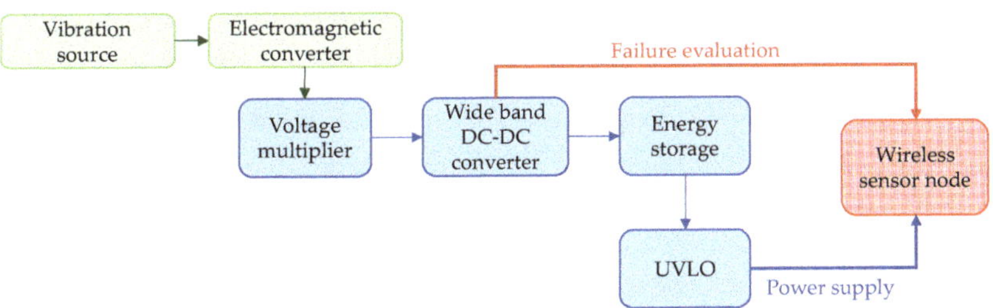

Figure 1. Schematic of the proposed concept.

The energy harvester is followed by the passive energy-management circuit described in Section 4, which is advantageous in terms of its charging time and ability to manage the load requirement in terms of energy. The complete power unit enables the continuous powering of the wireless sensor node, which is responsible for the communication of the acceleration measurement; fault detection is conducted by the WSN and is presented in Section 5.

The developed solution aimed to perform acceleration measurement and detect failures through the presence of high acceleration. Failure is detected based on the measurement of the converter voltage based on the ADC of the used WSN. When the measured acceleration exceeds 0.25 g, the converter voltage increases, and a message will be sent within 1 min to declare that the system has a failure. The treated scenario in terms of the acceleration profile and failure detection is presented in Figure 2.

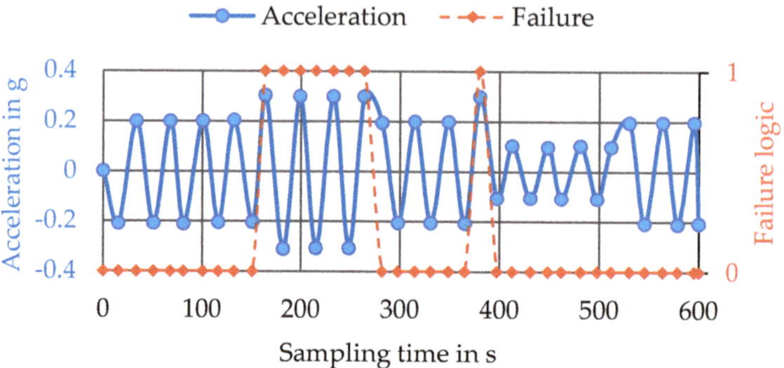

Figure 2. Scenario with the identification of possible failures from 0.25 g of acceleration.

3. Harvester Structure and Concept

The aim is to design an energy harvester able to deliver enough energy, especially for low acceleration starting from 0.1 g, to provide continuous energy for the WSN. Further, in this section, the focus is on the design and realization of a robust harvester that is compact and works in a relevant bandwidth frequency.

3.1. Mechanical Structure of the Energy Harvester

The harvester consists of an electromagnetic harvester based on planar spring architecture. In particular, it consists of a planar spring where a magnet is attached, which is the moving part of the harvester. The coil is placed surrounding its housing, which is placed at the level of the moving magnet. The proposed harvester is shown in Figure 3. The design is quite challenging and has several parameters that can affect the harvester's behavior. These especially include the planar spring design, which is decisive for the working frequency of the harvester.

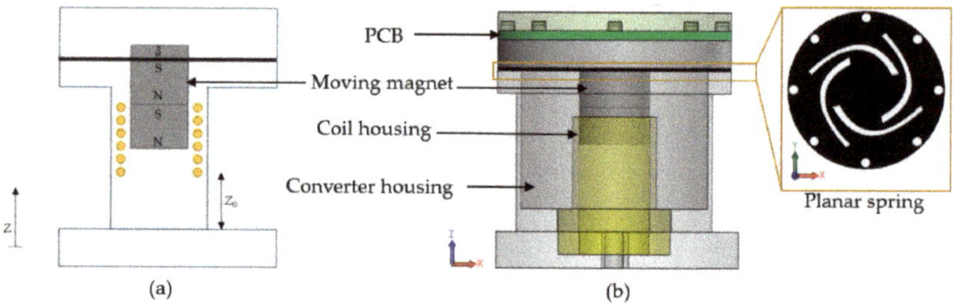

Figure 3. Designed energy harvester system: (**a**) schematic, (**b**) 3D illustration.

3.2. Design and Evaluation of the Planar Spring

In this work, the aim of using the planar spring is to ensure a more compact structure for the converter as well as more flexibility. This section is devoted to investigating the design of the planar spring, in particular its geometry and size, and their influence on the resonance frequency. The main evaluated parameter is the resonance frequency of the spring relative to the geometry. The planar spring has a diameter of 50 mm, and the spring geometry is presented in Figure 4.

The influences of the thickness, the number of beams, and the material of the spring on the resonance frequency are investigated based on finite element analysis. As the first evaluation, the effect of the material is evaluated for a spring thickness of 0.2 mm and using four beams. The results show that resonance frequencies of 12 Hz, 156 Hz, and 193 Hz can be reached for FR4, aluminum, and copper materials respectively (Figure 5a).

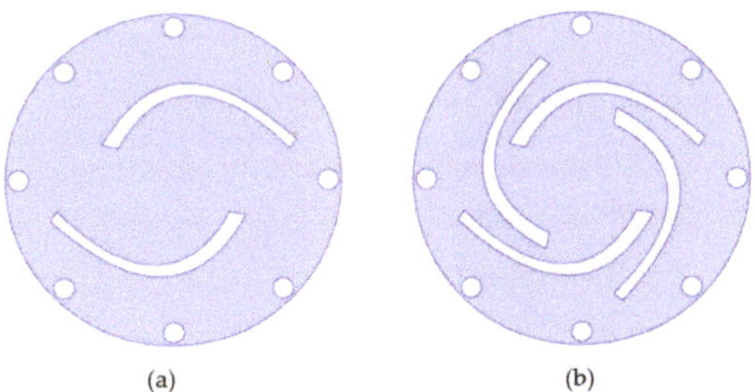

(a)   (b)

Figure 4. Evaluated spring geometries: (**a**) 2 beams, (**b**) 4 beams.

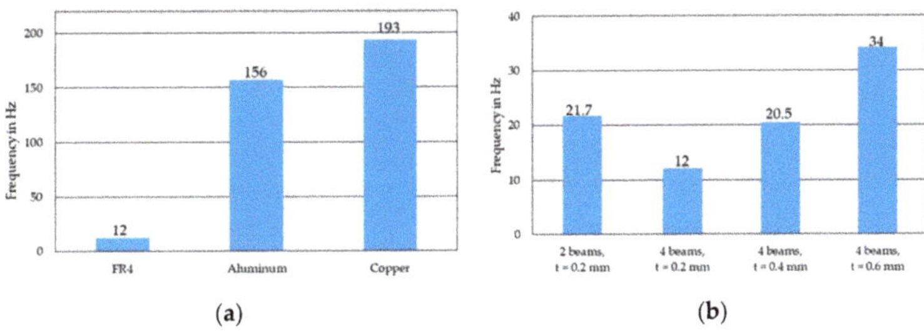

(a)   (b)

Figure 5. Resonance frequency of the planar spring depending on (**a**) the spring material, (**b**) the number of beams, and the spring thickness.

Further, an increase in the number of beams leads to a decrease in the resonance frequency. For example, for an FR4 spring with a thickness of 0.2 mm, the resonance frequency decreases from 21.7 Hz to 12 Hz for two beams and four beams, respectively (Figure 5b). This is due to the stiffness decrease when increasing the number of the beams, and hence the decrease in the resonance frequency. A part of that, i.e., the spring thickness effect while maintaining the same number of beams, was evaluated (Figure 5b). This showed that the increase in the spring thickness was proportional to the frequency increase due to the increase in the spring stiffness.

3.3. Experimental Setup

The converter parts were produced with a 3D printer using Clear V4 as material. The planar spring was produced using laser machining. To avoid the burning of the material, the cutting parameters were optimized in terms of power and speed. The converter had a total volume of ~35 cm^3. To test the converter, an electromagnetic shaker (VebRobotron Type 11077) was used to generate the mechanical vibration, which was controlled through a laser control system and an acceleration sensor. Figure 6 presents the experimental setup used for the test. The converter open circuit voltage was evaluated under a harmonic signal with an acceleration starting from 0.1 g to 0.2 g and a frequency range from 8 Hz to 30 Hz.

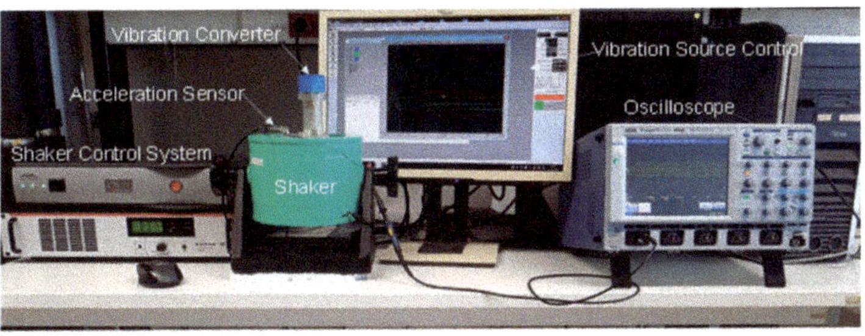

Figure 6. Schematic of the experimental setup.

Table 1 presents the different parameters for the converter design.

Table 1. Key parameters for the design of the converter.

Parameters	Value
Magnet radius	7.5 mm
Magnet height	8 mm
Magnet material	NdFeB 42
Magnet mass	23.3 g
Inner coil radius	8 mm
Outer coil radius	9 mm
Coil wire diameter	0.1 mm
Coil height	15 mm
Coil turns	1300
Coil resistance	175 Ω
Spring thickness	0.2 mm
Spring diameter	50 mm
Spring material	FR4

3.4. Experimental Evaluation of the Converter Output

The open-circuit voltage through the converter for acceleration from 0.1 g to 0.2 g for a frequency range from 8 Hz to 30 Hz with a step of 1 Hz was evaluated. The used coil for the experimental results was based on a coil wire thickness equal to 0.1 mm and has 1300 turns with a resistance equal to 175 Ω. The results show that the converter was able to generate a minimum output voltage of 0.42 V at an acceleration equal to 0.1 g and a frequency equal to 8 Hz, as illustrated in Figure 7. A maximum open-circuit output voltage of 3.98 V was reached for a frequency equal to 15 Hz and acceleration equal to 0.2 g. For a bandwidth frequency equal to 3 Hz and an acceleration of 0.1 g, the converter showed a minimum output voltage of 0.77 V. For an acceleration equal to 0.2 g, the frequency bandwidth reached 5 Hz, where the voltage was 2 V. This enables storing the energy through the proposed circuit in Section 4, not only at resonance, but also for bandwidth frequency.

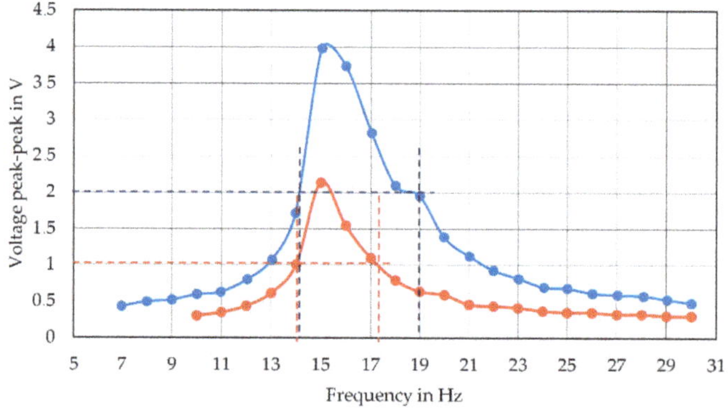

Figure 7. Open-circuit voltage output of the electromagnetic converter for an applied frequency between 7 Hz and 30 Hz for an acceleration of 0.1 g and 0.2 g.

To conclude the experimental evaluation of this section, the converter had a structure that was flexible to be adjusted to the required resonant frequency based on different parameters, principally the planar spring structure and the used mass. This makes the converter a good solution to be adjusted to the desired defect that needs to be detected based on its characteristic frequency and acceleration. Further, the converter showed a good performance, with a minimum output equal to 0.42 V, even under weak acceleration equal to 0.1 g, which is interesting for several applications.

4. Wideband Energy-Management Circuit

4.1. Energy Management for EM Converter

The energy-management circuit of the EM converter serves primarily to provide the necessary output voltage for WSN's operation. In particular, it produces the voltage needed to power the WSN at different vibration levels. Different energy-management architectures can be proposed with different main operational elements. One interesting architecture among all is the use of a rectifier stage. In this structure, a rectifier stage is used as Voltage Multiplier (VM) configuration connected directly to a super capacitor and a WSN [12]. However, in this case, the input signal can be lower or even higher than the optimal voltage level required for charging the super capacitor, which can cause the node to malfunction or even damage to the storage element. Consequently, a voltage regulator can be added to the previous architecture to protect the super capacitor from being over-charged [24]. However, the voltage regulator can dramatically reduce the system efficiency and requires a minimum input voltage to function properly, which is not always guaranteed. Alternatively, a DC-DC converter can be used instead of the voltage regulator to amplify or reduce the input voltage to a predefined super capacitor voltage level. This architecture allows reaching a wide working range, as well as the protection of the super capacitor in the case of high input voltage levels.

In contrast, the energy-management circuit can be based on discrete components [20], MEMS circuits [25,26], or commercial converters in a chip [27]. The circuit with discrete components was built using basic power electronic components, including diodes, switches, and comparators, which increase the circuit size and complexity. Integrated circuits, such as MEMS circuits, are based on transistors and require a high fabrication cost within a long period of investigation and fabrication for the design of a suitable solution. Due to their efficiency, ease of integration, small size, and low cost, commercial on-chip converters are becoming more popular for energy-harvesting systems. Besides, they can integrate sophisticated features, such as Maximum Power Point Tracking (MPPT), rectifier stages, and cold-start circuits [28].

4.2. Proposed Energy-Management Circuit

The structure of the adopted energy-management circuit for an electromagnetic converter is shown in Figure 8. It consists of a VM associated with a wideband DC-DC converter that charges the super capacitor. Besides, an Under-Voltage Lock-Out (UVLO) circuit is associated with the energy-management circuit to activate the WSN when the super capacitor is full of charge.

The rectification stage aims to transform the AC signal of the EM to a DC signal. The rectification stage reduces the output voltage according to the properties and numbers of the used diode. Generally, the rectification stage is associated with a DC-DC converter to ensure a constant output voltage to the super capacitor. The used EM converter generates a maximum open voltage of around 4 V peak-to-peak, with an acceleration of 0.2 g and a frequency of 13 Hz (Figure 7). The main requirement of the proposed energy-management circuit is to charge the super capacitor for different input voltages with a range between 2 V peak-to-peak to 4 V peak-to-peak, where three main possible approaches can be suggested. In the first approach, the use of a bridge rectifier is promoted to reduce the DC-DC converter input voltage of $2V_{drop}$ related to the used diodes. In this case, the DC-DC converter should be able to provide a low input voltage, where their efficiency and time to start-up are critical. In the second approach, wide-band converters are used, which are composed of a combination of dual DC-DC converters with specific control circuits, resulting in high-cost and complex circuits. Besides, the use of switches with a higher inner resistance causes energy losses. Another approach that relies on this is to use VM circuits in conjunction with AC signals to achieve higher DC-DC converter input voltages. Through this approach, a DC-DC converter based on buck architecture can reduce the cold-start phase and charge the super capacitor within a wide band of acceleration.

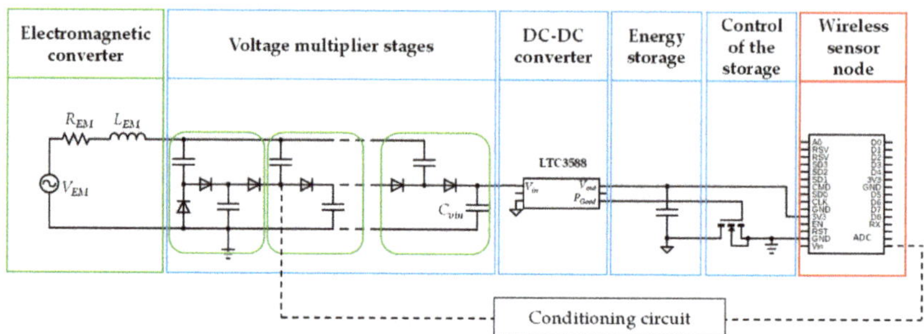

Figure 8. Schematic of the proposed energy-management circuit for the electromagnetic converter.

Various commercial DC-DC converters have been developed and used for energy harvesting with vibration converters; LTC3588-1 was selected, with an efficiency of 82% [28] for energy storage in the super capacitor. It presents a wideband hysteresis voltage supervisor like that of buck converters. As the super capacitor is being charged, the LTC3588-1 maintains a high impedance state to minimize leakage current during the energy-conversion phase. The output voltage of the used DC-DC converter is defined as 3.3 V, which presents the maximum possible supply voltage of the WSN.

Various VM architectures have been presented and compared in the literature [28], including Dickson and Cockcroft-Walton architectures. Both architectures can reach the main required output voltage. However, the Dickson architecture is more recommended for low-power applications due to its small capacitor stage. The Dickson VM is composed of dual diodes and capacitors on each stage, as shown in Figure 8. Typically, with an input voltage (V_{in}) of 0.65 V, the VM output voltage ($V_{multiplier}$) based on three stages becomes around 4 V. However, as shown in Figure 9, in this case with real implementation, the output voltage of the three stages of the voltage multiplier is around 2.1 V. In fact, in Figure 9, for an acceleration of 0.15 g, the input signal presents a maximal amplitude of 0.65 V, expressed by the red color, where stages 1, 2, and 3 of the VM are expressed by blue, green, and purple colors, respectively. The VM voltage amplitudes are 0.75 V, 1.3 V, and 2.1 V for stages 1, 2, and 3, respectively. This voltage drop ($\Delta(n)$) is related to various parameters, as shown in Equations (1) and (2), such as the losses due to the used capacitors ($V_{dcap,L}$) and diodes ($V_{diode,L}$), where the error related to the working frequency as well as the number of VM stages (n).

$$V_{multiplier} = 2nV_{in} - \Delta(n), \tag{1}$$

$$\Delta(n) = V_{dcap,L}(2n) + V_{diode,L}(2n) + \delta_o(n), \tag{2}$$

$$V_{ripple} = \frac{I_o\left(n^2 + \frac{n}{2}\right)}{8fC_{VM}} \tag{3}$$

As shown in Figure 9, the output voltage of the voltage multiplier presents some ripples in the first stages. These ripples are related mainly to the selected VM capacitors (C_{VM}). Typically, the VM capacitors should be selected according to the electromagnetic converter output current (I_o), the number of the VM stages, as well as the working frequency (f), as shown in Equation (3). By increasing the VM stages, the VM capacitors' values are required to be larger in terms of capacity (Equation (3)). In reality, increasing the value of the VM capacitors does not only reduce the voltage ripple (V_{ripple}) in the output voltage, but also increases their time of charge and the super capacitor. For this, in this paper, smaller capacitor values have been adopted in the first two VM stages of about 1 µF, while stage numbers 3, 4, and 5 use capacitors with 10 µF of capacity. The last voltage multiplier stage uses a capacitor of 20 µF, and the capacitor connected to the DC-DC input (C_{Vin}) has 1000 µF to generate a current pick to charge the super capacitor for a longer time (Figure 8). With these values, the activation of the WSN in a proper way is guaranteed.

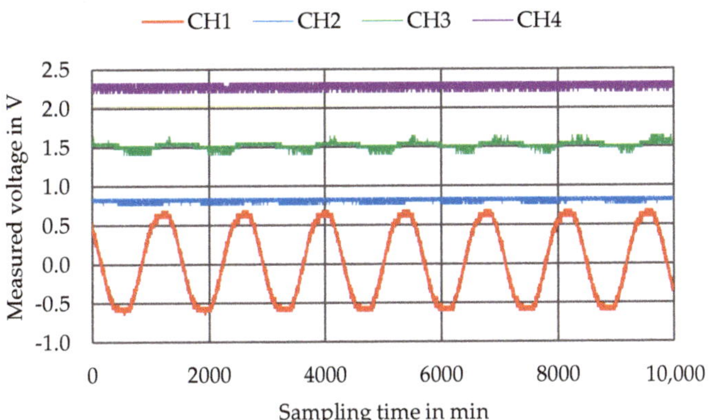

Figure 9. Output voltage at different stages: red, EM converter signal; blue, first stage of the VM; green, second stage of the VM; and purple, third stage of the VM outcomes.

When the WSN is connected to the super capacitor, some current losses flow to supply the node properly, which reduce the time to charge of the super capacitor. For this, disconnecting the load from the super capacitor during the charging process is required for fast charging. This can be achieved through the utilization of Under-Voltage Lock-Out (UVLO) circuits [29,30] designed to use hysteresis comparator-based circuits and MOSFET-based circuits [31]. By using active components, such as comparators, switches, or even detectors, they increase the circuit cost and consumption, which makes the circuit with a singular MOSFET more recommended. Figure 8 shows the proposed circuit to connect the WSN and the super capacitor when their voltage is between 3 V and 3.3 V. The circuit is composed of an N-channel MOSFET connected between the supply circuit ground and the wireless sensor node ground. The proposed architecture does not cause a voltage drop of the super capacitor voltage that supplies the WSN. The MOSFET becomes conductive by a gate voltage higher than 1 V delivered by the P_{Good} pin of the used DC-DC converter. However, the activation time of the WSN by the P_{Good} signal becomes high when the super capacitor voltage is about 3 V, which is sometimes not sufficient to transmit the message by the WSN. For this, a higher capacitor in the final stage of the VM circuit is connected to the input of the DC-DC converter that injects a higher current to supply the WSN.

One of the major aspects to follow is the sizing of the circuit capacitors, which should not be too big to reduce the time to charge and not too small that they are not able to properly supply the WSN. The size of the super capacitor ($C_{super-Cap}$) can be defined analytically based on the explored energy of the used WSN ($E_{explored}$), as well as the minimal (V_t) and the maximal (V_i) required input voltage of the WSN, as presented in Equation (4). The explored energy can be defined based on the power consumption (P_{consum}), the time of activation, and the associated DC-DC converter energy-management circuit power consumption ($P_{EMcircuit}$). Besides, it considers the power dissipation due to the super capacitor discharge $(P_{self-dis})$ when the WSN is not activated. For this, the accurate super capacitor size can be explored, as shown in Equation (5).

$$C_{super-cap} = 2 \cdot \frac{E_{explored}}{\left(V_i^2 - V_t^2\right)} \tag{4}$$

$$C_{super-Cap} = 2 \cdot \Delta t \cdot \frac{P_{consum} + P_{self-dis} + P_{EMcircuit}}{\left(V_i^2 - V_t^2\right)} \tag{5}$$

The proposed circuit uses the WSNs themselves to detect failure. This can be established by the use of their internal ADC that measures the peak voltage of the electromagnetic converter. The peak voltage is measured by the outcome of the first stage of the voltage multiplier circuit and with the help of a conditioning circuit. The conditioning circuit uses an additional resistor of a few kΩ connected as the voltage divider and helps to limit the current flow to the WSN. Furthermore, an additional capacitor of 1 μF is connected to the ground to improve the filtering aspect and reduces the ripples for accurate voltage measurement by the WSN. Besides, a Zener diode of 3 V is associated with the circuit to protect from over-voltage that may occur in the case of damage. The measured voltage in normal working conditions is about 2 V peak-to-peak when the acceleration is 0.2 g.

5. Wireless Sensor Node Design

The EM converter provides a sustainable power source to satisfy the energy requirements for the wireless sensor node. In practice, the electronic specifications of the wireless module, such as the communication module, the sampling rate, and the resolution of the ADC, influence its overall energy consumption [3]. The choice of the microcontroller (MCU) is crucial in the operation of the wireless sensor node and represents the main energy-consuming component [31,32]. Therefore, the MCU must enable a high level of performance while maintaining a low level of energy consumption. In particular, the choice of the MCU is based on the applications' requirements, such as the computational capabilities of the node, power consumption, and size. Considering our application's requirements, the system-on-chip microcontroller CC430F5137 [33] integrated with an RF transceiver core CC1001 [34] is selected to collect the vibration information and forward it to the user. The CC430F5137 MCU works at a frequency of 868 MHz and consumes a minimum current of 0.5 μA in the deep sleep mode with a 3 V supply voltage. In particular, CC430F5137 MCU enables three main working modes: active, sleep, and idle. It consumes around 13 mA and 36 mA for the receiving and the sending of a data packet, respectively. During the runtime, the wireless node collects the vibration information from the converter and then transmits it through the RF transceiver to the base station node, which is connected to the computer. The main characteristics of the wireless node are presented in Table 2. Figure 10 presents the node hardware used in this work.

Table 2. Hardware specification of the wireless sensor node.

Parameters	Values
MCU	CC430F5137 (MSP430 core + CC1101 radio SOC)
Operating input voltage	3–3.3 V
Operating frequency	868 MHz
Current drain	Sleep: 1–2 μA Tx: 36 mA (max) Rx: 18 mA (max)
Sending power	+12 dBm (max)
ADC	12 bits

Figure 10. Wireless sensor node used for failure detection based on system-on-chip microcontroller CC430F5137 integrated with an RF transceiver core CC1001.

In our application, the wireless sensor node was powered using the harvested energy of the EM converter. For this, given the difference between the EH and communication link budgets, investigating the energy consumption and ideal efficient current for the node's operation was required. To do so, a small wireless network was created using two wireless nodes as a proof of concept. One node is identified as the transmitter node, which is connected to the harvester output, responsible for following the change in the voltage output. The collected voltage output is later sent to the base station node, which is interfaced with the computer. Periodic voltage sensor data are collected from the connected pin and transferred to the base station. The energy consumption of the node during its different working modes was evaluated to extract the necessary tuning parameters for the energy-management circuit. To characterize the needs of the wireless sensor node in terms of power consumption, the wireless module was investigated over its different working activity modes. The energy consumption during the run time of the wireless node, the Agilent Keysight E5270b, was used to characterize the current-voltage consumption [35]. In practice, the four-wire, resistance measurement method was used to quantify the energy consumption of the wireless module during its different working modes (active and sleep). Continuous data measurement and transmission were considered to identify the complete working cycle of the node in a normal scenario. To do so, an accurate energy measurement is necessary to characterize the wireless module, which can be generalized according to the consumed current in each state, which is defined in Equation (6).

$$E_i = I_i \times V \times \Delta_{ti} \tag{6}$$

where E_i, I_i, and Δ_{ti} are the consumed energy in J, consumed current in A, and activity time in sec, receptively, for a defined activity i, which can be transmission, reception, or idle/sleep.

Figure 11 illustrates the measured current drains of the wireless node during a complete measurement cycle. For measuring the voltage output of the harvester, the wireless node consumes around 5 mA and requires approximately 20 mA to send the measured voltage to the base station.

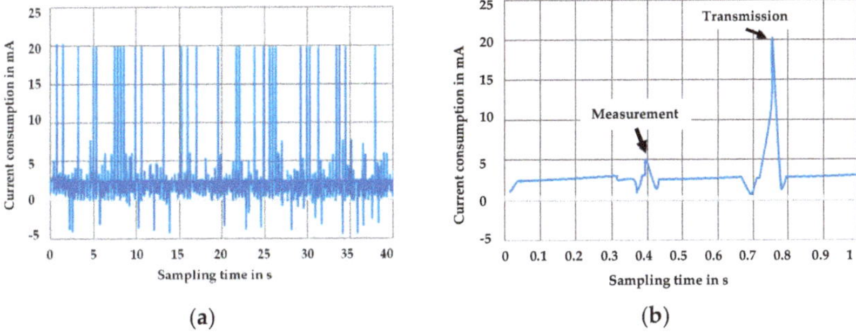

Figure 11. Current drain during continuous data transmission measured with a sampling rate of 1 ms: (**a**) all measurement samples and (**b**) one measurement and one transmission cycle.

6. Energy Autonomous Failure Detection System

Failure detection was carried out based on changes in the measured acceleration of the EM converter. In this case, when the measured acceleration exceeds ± 0.25 g, the system declares the current case as a failure and sends alert messages to the control unit. In practice, an increase in the acceleration refers to an increase in the generated voltage level by the electromagnetic converter to reach a voltage above the voltage threshold noted as V_{thr}. The designed EM converter presents a voltage threshold of 2 V. The threshold is defined by a priori vibration measurements on the machine during operation. The complete execution scenario of the proposed failure detection is illustrated in Figure 12. The wireless sensor node was connected to the voltage output of the EM converter. The node continuously measures the voltage and compares the obtained results to the defined V_{thr}. In the beginning, the "status" register of the WSN is initiated as "0" to indicate the normal working mode of the harvester. In case the obtained voltage exceeds the V_{thr}, the WSN updates its register's status to "1". The voltage data and the status information are transmitted to the base station for processing and failure identification. Considering the characteristic behavior of the vibration converter, the obtained voltages are converted to acceleration values, whereas the status information is used to print out the existence of a failure or not.

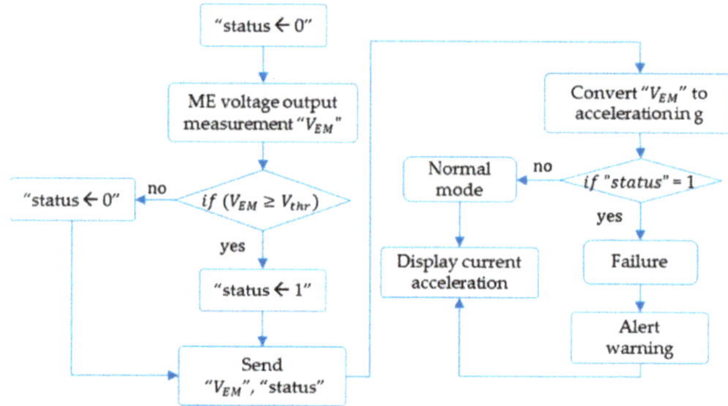

Figure 12. Flow-chart of the failure detection system based on the voltage output of the EM converter.

The proposed system works in two major modes, where the first is starting with initially a fully charged super capacitor and the second mode is the recharge of the super capacitor after each established communication by the WSN that consumes the energy stored. The wireless sensor node requires a supply voltage between 3 V and 3.3 V, with a total energy consumption of 25 mA on each measurement and communication cycle. For three consecutive messages, the power consumption of the node is about 225 mW. For this, the super capacitor is calculated based on Equation (5), and it should be higher than 0.6 F. The selected super capacitor has a size of 0.69 F and is composed of dual parallel capacitors with capacities of 0.49 F and 0.2 F. Figure 13 shows the charging of the super capacitor process for an acceleration of 0.3 g. The super capacitor is fully charged in around 14 min, where the recharging time is estimated to be 0.41 min. The voltage of the LTC3588 P_{Good} pin can be activated during the charging process when a high current injection by the fully charged voltage multiplier capacitors causes an increase in the super capacitor voltage to 3 V for a short time.

When the voltage of P_{Good} reaches 3.3 V, the UVLO circuit connects the charged super capacitor to the WSN that can send a message to the base station. When communication is established, the super capacitor voltage decreases and the voltage of P_{Good} reduces until the super capacitor gets charged again (Figure 13b). The time of the recharging of the super capacitor after the communications is related to the input voltage level that is generated by the electromagnetic converter. The time is measured experimentally for different accelerations from 0.195 g to 0.3 g (Figure 14). For a low acceleration level of 0.2 g, the startup time of the system is about 45 min, where the recharging time is around 28 s, whereas, for a high acceleration of 0.25 g, the recharging time is about 24 s. Typically, the system can deliver the current situation information within a few seconds.

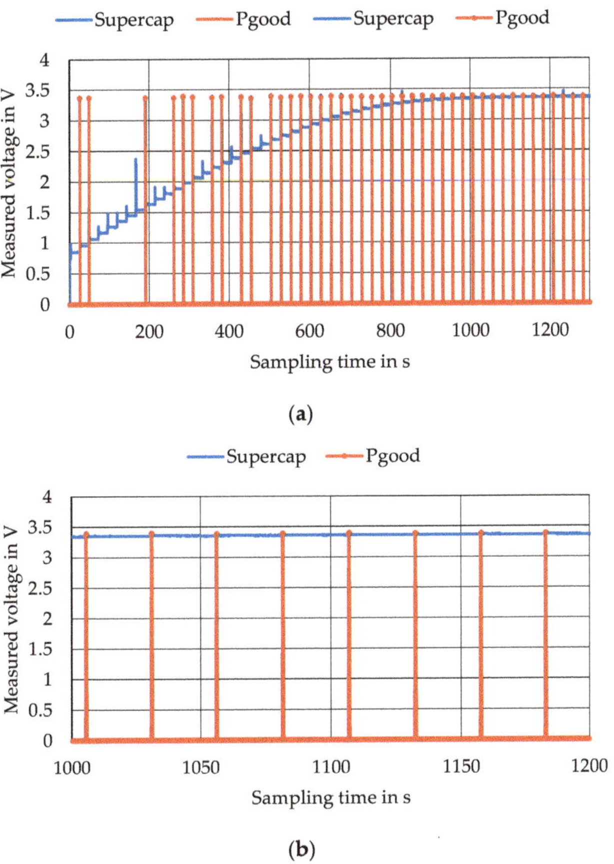

(a)

(b)

Figure 13. (**a**) Charging of the super capacitor and (**b**) re-charging of the super capacitor after communication.

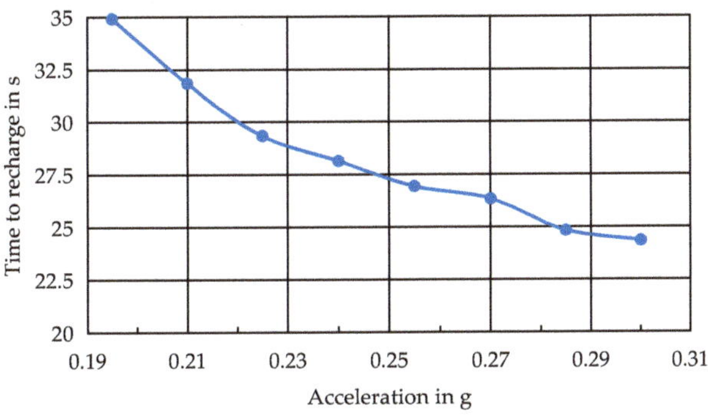

Figure 14. Time to recharge of the super capacitor for different acceleration levels.

The selection of the number of VM stages influences the ability to charge the super capacitor under different conditions. The increase in the number of VM stages influences the system in two different ways. The first is by increasing the system losses by the association of diodes that may drop the input voltage [28]. Secondly, it influences the DC-DC converter efficiency, which is related to the input voltage [28]. The procedure to reach fast-charging was investigated experimentally by testing the time to recharge the super capacitor after communication with the lower acceleration of 0.2 g. The results show that the system can work properly within VM stage numbers higher than five. Besides, for five stages of VM, the recharge time is about 2 min and 6 s, whereas, for six VM, the required time for re-charging the super capacitor is around 31.9 s (Figure 13). This is related mainly to the used DC-DC converter, which works with higher efficiency in the case of higher input voltage. The system efficiency directly affects the required time to charge the super capacitor in the initial phase and the recharging time.

The proposed failure detection system was evaluated under two different working conditions. In the first case, a gradual acceleration change was realized to follow the system's response to progressive changes. The second case involved triggering an abrupt change between two acceleration levels of the vibration converter to study the response time and adaptability to changes in the system. Specifically, we aimed to study the response time and output voltage of the complete proposed solution in both cases.

In the first scenario, the applied acceleration varied from 0.1 g to 0.25 g with a small step of approximately 0.015 g. During the increase in the applied acceleration, the measured acceleration at the harvester output was collected along with the probability of failure, as shown in Figure 15. The experimental acceleration presents some noise in comparison with the expected acceleration variation, which can result from the associated variations from the devices in proximity. In this case, an unexpected change in the acceleration is not regarded as a failure; since the failure needs to occur for a certain time other than this, it is regarded as erroneous acceleration due to a real working environment. Therefore, the transmitted message of failure is sent only when the real failure occurs with acceleration above 0.25 g after 10 min, as shown in Figure 15a. At the same time, as the acceleration increases, the recharging time decreases, which is reduced due to the increase in the input voltage of the energy-management circuit. In Figure 15b, the ADC input of the energy-management circuit is presented. The ADC shows a gradual variation of the voltage, which follows the increase of the applied acceleration. This proves that the EM converter can provide a continuous and sustainable power supply for the WSN.

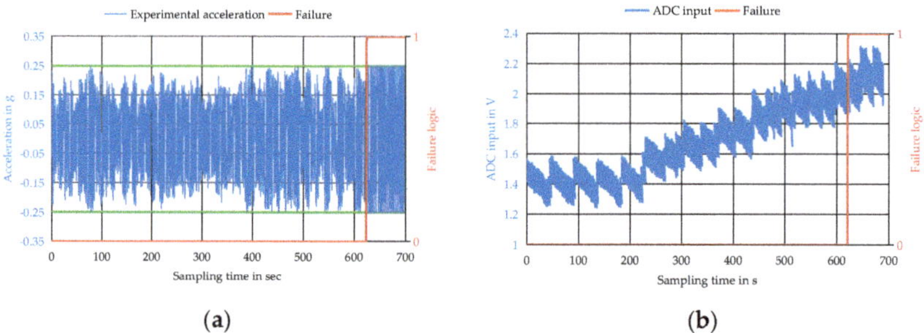

Figure 15. Failure detection for different (**a**) acceleration levels and (**b**) ADC input voltages.

The response time of the proposed failure detection system was determined based on the comparison between the real failure existence and the declaration of the failure at the base station. In Figure 16, a representation of the real acquisition data between the real acceleration changes and the identification of failure is plotted. The failure occurred after 644 s from the initialization of the system, which was declared at the base station after approximately 669 s from the initial state. Hence, failure detection was realized after 25 s from its actual occurrence.

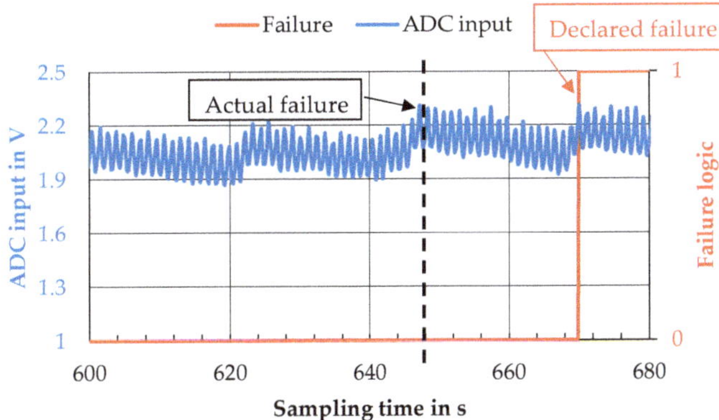

Figure 16. Identification of the time delay for the failure detection system.

Besides, some errors can typically occur when the usual communication happens simultaneously with the error caused by the acceleration. This effect can be shown by alternating between failure and normal conditions several times. In the second scenario, an abrupt change of the acceleration between 0.2 and 0.25 was realized to study the fast response of the system. Figure 17 shows an alternation between 0.2 g and 0.25 g acceleration. When the acceleration was less than 0.2 to 0.23, the ADC input voltage increased from about 1.3 V, where a message that indicated that there was no failure was shown (Figure 17a). However, when the increase was more than 0.25 g, the ADC input voltage increased to more than 2 V, and failure messages appeared, where few errors can be present due to fluctuations in the applied acceleration, as shown in Figure 17b. This fluctuation in terms of acceleration can be usual in real system operations, and the proposed system shows good sensitivity to the reorganization of the dysfunctionality of the system behavior.

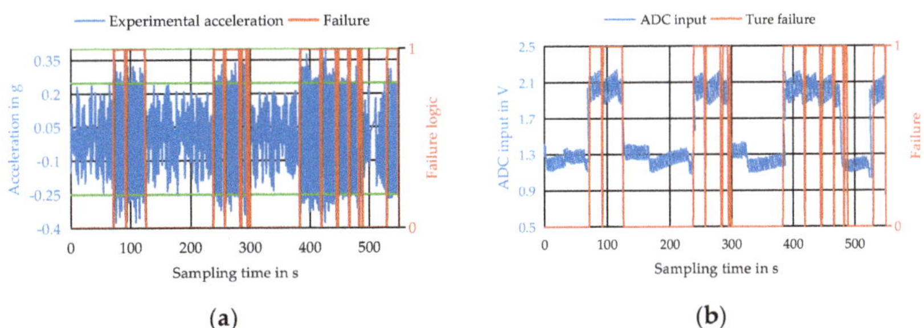

Figure 17. Failure detection in case of alternating between 0.2 g and 0.25 g acceleration: (**a**) acceleration level and (**b**) ADC input voltages.

7. Conclusions

This paper proposes an autonomous WSN powered through an electromagnetic vibration converter with a passive energy management solution for predictive maintenance. The system is able to detect failure without the use of additional sensors, such as accelerometers or temperature sensors, which makes it more compact and have lower energy consumption. The proposed solution enables the detection of failure at a low constant frequency of 13 Hz and low acceleration levels, which enables early-stage failure detection. The proposed system consists of an electromagnetic converter-based planar spring that generated a maximum output voltage for the converter of up to 4 V peak-to-peak voltage for a coil resistance equal to 175 Ω with an applied vibration of 13 Hz and 0.2 g in terms of frequency and acceleration, respectively. The reached output is stored in a super capacitor through an energy-management circuit, which consists of a six-stage voltage multiplier circuit connected to a wide-band DC-DC converter, as well as an under-voltage lock-out (UVLO) circuit to connect to the storage device to the WSN. Under the failure condition with a frequency of 13 Hz and an acceleration of 0.3 g, the charging time of the super capacitor when defined without any initial energy is about 14 min, where the recharging time is estimated to be 0.41 min. For a low acceleration level of 0.2 g, the startup time of the system is about 45 min and the recharging time is around 31 s. The last and crucial part, which is the sensor node, has good performance in terms of response time and data reliability. The wireless node operates at a frequency of 868 MHz, which helps to reduce energy consumption. It has a sensitivity of −116 dBm, which ensures a reliable and high data transmission rate. Once the solution starts working, only a few seconds up to 35 s is required to transmit the information in normal cases and up to 24 s in case of failure detection. Further, the proposed solution enables the detection of failure at low frequencies and acceleration, which enables early-stage failure detection. As a future perspective, the implementation of the solution for specific fault types and machine learning model will be conducted.

Author Contributions: S.B., G.B. and D.E.H. contributed equally to the conceptualization, investigation, methodology, data analysis, validation, and the original draft preparation and writing, O.K. contributed to the conceptualization, the original draft preparation and writing, reviewing, and editing. All authors have read and agreed to the published version of the manuscript.

Funding: This research was funded by the Bundesministeriums für Wirtschaft und Energie (BMWi), within the project "Entwicklung einer KI-gestützten, miniaturisierten energieautarken Multisensorplattform als universelle IoT-Lösung (KI-NO)" (grant number 16KN087924).

Institutional Review Board Statement: Not applicable.

Informed Consent Statement: Not applicable.

Conflicts of Interest: The authors declare no conflict of interest. The funders had no role in the design of the study; in the collection, analyses, or interpretation of data; in the writing of the manuscript, or in the decision to publish the results.

References

1. Pozo, B.; Araujo, J.Á.; Zessin, H.; Mateu, L.; Garate, J.I.; Spies, P. Mini Wind Harvester and a Low Power Three-Phase AC/DC Converter to Power IoT Devices: Analysis, Simulation, Test and Design. *Appl. Sci.* **2020**, *10*, 6347. [CrossRef]
2. Ahmad, I.; Hee, L.M.; Abdelrhman, A.M.; Imam, S.A.; Leong, M.S. Scopes, challenges and approaches of energy harvesting for wireless sensor nodes in machine condition monitoring systems: A. review. *Measurement* **2021**, *183*, 109856. [CrossRef]
3. Aftabuzzaman, M.; Sarker, S.; Lu, C.; Kim, H.K. In-depth understanding of the energy loss and efficiency limit of dye-sensitized solar cells under outdoor and indoor conditions. *J. Mater. Chem. A* **2021**, *44*, 24830–24848. [CrossRef]
4. Kanoun, O.; Bradai, S.; Khriji, S.; Bouattour, G.; El Houssaini, D.; Ben Ammar, M.; Naifar, S.; Bouhamed, A.; Derbel, F.; Viehweger, C. Energy-aware system design for autonomous wireless sensor nodes: A comprehensive review. *Sensors* **2021**, *21*, 548. [CrossRef]
5. Kanoun, O.; Keutel, T.; Viehweger, C.; Zhao, X.; Bradai, S.; Naifar, S.; Trigona, C.; Kallel, B.; Chaour, I.; Bouattour, G.; et al. Next generation wireless energy aware sensors for internet of things: A review. In Proceedings of the 2018 15th International Multi-Conference on Systems, Signals & Devices (SSD), Yasmine Hammamet, Tunisia, 19–22 March 2018. [CrossRef]
6. Hosangadi, P.S.; Sunil, M.; Udaya, K.B.; Dipti, G. Triboelectric effect based self-powered compact vibration sensor for predictive maintenance of industrial machineries. *Meas. Sci. Technol.* **2021**, *32*, 095119. [CrossRef]
7. Dragunov, V.P.; Ostertak, D.I.; Pelmenev, K.G.; Sinitskiy, R.E.; Dragunova, E.V. Electrostatic vibrational energy converter with two variable capacitors. *Sens. Actuators A Phys.* **2021**, *318*, 112501. [CrossRef]
8. Valery, P.D.; Dmitriy, I.O.; Dmitry, E.K.; Evgeniya, V. Dragunova; Impact-enhanced electrostatic vibration energy harvester. *J. Appl. Comput. Mech.* **2021**, *8*, 671–683. [CrossRef]
9. Al-Yafeai, D.; Darabseh, T.; Abdel-Hamid, I. Mourad. A state-of-the-art review of car suspension-based piezoelectric energy harvesting systems. *Energies* **2020**, *13*, 2336. [CrossRef]
10. Zhemin, W.; Yu, D.; Tianrun, L.; Zhimiao, Y.; Ting, T. A flute-inspired broadband piezoelectric vibration energy harvesting device with mechanical intelligent design. *Appl. Energy* **2021**, *303*, 117577. [CrossRef]
11. Zhijie, F.; Han, P.; Yong, C. A dual resonance electromagnetic vibration energy harvester for wide harvested frequency range with enhanced output power. *Energies* **2021**, *14*, 7675. [CrossRef]

12. Bradai, S.; Bouattour, G.; Naifar, S.; Kanoun, O. Electromagnetic energy harvester for battery-free IoT solutions. In Proceedings of the 2020 IEEE World Forum on Internet of Things, WF-IoT 2020-Symposium Proceedings, New Orleans, LA, USA, 2–16 June 2020. [CrossRef]
13. Bradai, S.; Naifar, S.; Trigona, C.; Baglio, S.; Kanoun, O. An electromagnetic/magnetoelectric transducer based on nonlinear RMSHI circuit for energy harvesting and sensing. *Meas. J. Int. Meas. Confed.* **2021**, *177*, 109307. [CrossRef]
14. Ju, S.; Chae, S.H.; Choi, Y.; Jun, S.; Park, S.M.; Lee, S.; Lee, H.W.; Ji, C.H. Frequency up-converted low frequency vibration energy harvester using trampoline effect. *J. Phys. Conf. Ser.* **2013**, *476*, 012089. [CrossRef]
15. Li, M.; Deng, H.; Zhang, Y.; Li, K.; Huang, S.; Liu, X. Ultra-Low Frequency Eccentric Pendulum-Based Electromagnetic Vibrational Energy Harvester. *Micromachines* **2020**, *11*, 1009. [CrossRef] [PubMed]
16. Ashraf, K.; Md Khir, M.H.; Dennis, J.O.; Baharudin, Z. A wideband, frequency up-converting bounded vibration energy harvester for a low-frequency environment. *Smart Mater. Struct.* **2013**, *22*, 049601. [CrossRef]
17. Kuang, Y.; Hide, R.; Zhu, M. Broadband energy harvesting by nonlinear magnetic rolling pendulum with subharmonic resonance. *Appl. Energy* **2019**, *255*, 113822. [CrossRef]
18. Kangqi, F.; Hengheng, Q.; Yipeng, W.; Tao, W.; Fei, W. Design and development of a rotational energy harvester for ultralow frequency vibrations and irregular human motions. *Renew. Energy* **2020**, *156*, 1028–1039. [CrossRef]
19. Yulong, Z.; Anxin, L.; Yifan, W.; Xiangtian, D.; Yan, L.; Fei, W. Rotational electromagnetic energy harvester for human motion application at low frequency. *Appl. Phys. Lett.* **2020**, *116*, 05390. [CrossRef]
20. Xia, C.; Zhang, D.; Pedrycz, W.; Fan, K.; Guo, Y. Human body heat based thermoelectric harvester with ultra-low input power management system for wireless sensors powering. *Energies* **2019**, *12*, 3942. [CrossRef]
21. Szarka, G.D.; Burrow, S.G.; Stark, B.H. Ultralow power, fully autonomous boost rectifier for electromagnetic energy harvesters. *IEEE Trans. Power Electron.* **2012**, *28*, 3353–3362. [CrossRef]
22. Ullrich, M.; Wolf, M.; Rudolph, M.; Diller, W.; Root, J. Sensornetzwerke zur Kugellagerdiagnose am rotierenden Innenring. *Z. Für Wirtsch. Fabr.* **2021**, *116*, 603–607. [CrossRef]
23. Zanelli, F.; Castelli-Dezza, F.; Tarsitano, D.; Mauri, M.; Bacci, M.L.; Diana, G. Design and field validation of a low power wireless sensor node for structural health monitoring. *Sensors* **2021**, *21*, 1050. [CrossRef] [PubMed]
24. Dos Santos, A.D.; de Brito, S.C.; Martins, A.V.; Silva, F.F.; Morais, F. Thermoelectric energy harvesting on rotation machines for wireless sensor network in industry 4.0. In Proceedings of the 2021 14th IEEE International Conference on Industry Applications (INDUSCON), São Paulo, Brazil, 15–17 August 2021; pp. 694–697. [CrossRef]
25. Hehn, T.; Bleitner, A.; Goeppert, J.; Hoffmann, D.; Schillinger, D.; Sanchez, D.A.; Manoli, Y. Energy-harvesting applications and efficient power processing. In *NANO-CHIPS 2030. The Frontiers Collection*; Murmann, B., Hoefflinger, B., Eds.; Springer: Cham, Switzerland, 2020. [CrossRef]
26. Tan, Y.; Dong, Y.; Wang, X. Review of MEMS electromagnetic vibration energy harvester. *J. Microelectromech. Syst.* **2017**, *26*, 1–16. [CrossRef]
27. Ma, Y.; Ji, Q.; Chen, S.; Song, G. An experimental study of ultra-low power wireless sensor-based autonomous energy harvesting system. *J. Renew. Sustain. Energy* **2017**, *9*, 054702. [CrossRef]
28. Available online: https://www.analog.com/en/products/ltc3588-1.html (accessed on 25 November 2021).
29. Chaour, I.; Fakhfakh, A.; Kanoun, O. Enhanced passive RF-DC converter circuit efficiency for low RF energy harvesting. *Sensors* **2017**, *17*, 546. [CrossRef]
30. Boitier, V.; Estèbe, P.D.; Monthéard, R.; Bafleur, M.; Dilhac, J.M. Under voltage lock-out design rules for proper start-up of energy autonomous systems powered by supercapacitors. *J. Phys. Conf. Ser.* **2013**, *476*, 012121. [CrossRef]
31. Le, T.N.; Nguyen, T.H.N.; Vo, T.P.; Phan-Dinh, T.D.; Pham, H.A. Penalty shutdown mitigation in wireless sensor networks powered by ambient energy. In *Computational Data and Social Networks. CSoNet 2018. Lecture Notes in Computer Science*; Chen, X., Sen, A., Li, W., Thai, M., Eds.; Springer: Cham, Switzerland, 2018; Volume 11280. [CrossRef]
32. Khriji, S.; Chéour, R.; Goetz, M.; El Houssaini, D.; Kammoun, I.; Kanoun, O. Measuring energy consumption of a wireless sensor node during transmission: Panstamp. In Proceedings of the 2018 IEEE 32nd International Conference on Advanced Information Networking and Applications (AINA), Krakow, Poland, 16–18 May 2018; pp. 274–280. [CrossRef]
33. Texas Instruments, CC430F5137. Available online: https://www.ti.com/product/CC430F5137 (accessed on 25 November 2021).
34. Texas Instruments, CC1101. Available online: https://www.ti.com/lit/ds/symlink/cc1101.pdf?ts=1637823334345 (accessed on 25 November 2021).
35. Götz, M.; Khriji, S.; Chéour, R.; Arief, W.; Kanoun, O. Benchmarking-Based Investigation on Energy Efficiency of Low-Power Microcontrollers. *IEEE Trans. Instrum. Meas.* **2020**, *69*, 7505–7512. [CrossRef]

Article

A Dual Resonance Electromagnetic Vibration Energy Harvester for Wide Harvested Frequency Range with Enhanced Output Power

Zhijie Feng, Han Peng * and Yong Chen

School of Electrical and Electronic Engineering, Huazhong University of Science and Technology, Wuhan 430074, China; feng_z_j@hust.edu.cn (Z.F.); m201971640@hust.edu.cn (Y.C.)
* Correspondence: pengh@hust.edu.cn

Abstract: A dual resonance vibration electromagnetic energy harvester (EMEH) is proposed in this paper to extend frequency range. Compared with the conventional dual resonance harvester, the proposed system realizes an enhanced "band-pass" harvesting characteristic by increasing the relative displacement between magnet and coil among two resonance frequencies with a significant improvement in the average harvested power. Furthermore, two resonant frequencies are decoupled in the proposed system, which leads to a more straightforward design. The proposed dual resonance EMEH is constructed with a tubular dual spring-mass structure. It is designed with a serpentine planar spring and the coil position is optimized for higher power density with an overall size of 53.9 cm^3 for the dual resonance EMEH. It realizes an output power of 11 mW at the first resonant frequency of 58 Hz, 14.9 mW at the second resonant frequency of 74.5 Hz, and 0.52 mW at 65 Hz, which is in the middle of the two resonance frequencies. The frequency range of output power above 0.5 mW is from 55.8 Hz to 79.1 Hz. The maximum normalized power density (NPD) reaches up to $2.77 \text{ mW}/(\text{cm}^3{\cdot}\text{g}^2)$. Compared with a single resonance harvester design under the same topology and outer dimension at a resonant frequency of 74.5 Hz, the frequency range in the proposed EMEH achieves more than a $2\times$ times extension. The proposed dual resonance EMEH also has more than 2 times wider frequency range than other state-of-art wideband EMEHs. Therefore, the proposed dual resonance EMEH is demonstrated in this paper for a high maximum NPD and higher NPD over a wide frequency range.

Keywords: dual resonance frequencies; vibration electromagnetic energy harvester; wide harvested frequency range; enhanced "band-pass" harvested power; independent resonant frequencies

Citation: Feng, Z.; Peng, H.; Chen, Y. A Dual Resonance Electromagnetic Vibration Energy Harvester for Wide Harvested Frequency Range with Enhanced Output Power. *Energies* **2021**, *14*, 7675. https://doi.org/10.3390/en14227675

Academic Editors: Dibin Zhu and Paolo Visconti

Received: 29 August 2021
Accepted: 8 November 2021
Published: 16 November 2021

1. Introduction

Vibration energy harvesting (EH) converts ambient vibration energy into electrical energy, which is regarded as a premium technology to replace batteries to power wireless sensors [1–3]. An electromagnetic energy harvester (EMEH) extracts electricity through the induced electromotive force generated by the relative displacement between the magnet and the coil [4–6]. Traditional EMEHs usually have a single resonant frequency, through which the maximum displacement can be achieved, and a maximum harvested power is realized. However, when vibration frequency deviates from the resonant frequency, the harvested power drops dramatically [7–9], as shown in Figure 1. Actual vibrations in the environment always have wide dynamic range. For example, the main vibration frequencies of a pile head excited by a train passing through a tunnel vary between 50 and 80 Hz [10]. It is hard to keep the single resonance EH system constantly operating at the optimal point [11]. To maintain sufficient harvested power, extending the frequency coverage for EMEH is vital.

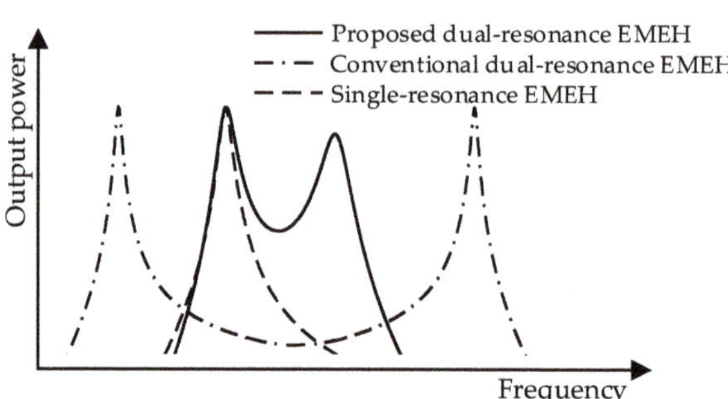

Figure 1. Comparison of the output power of a single resonance EMEH, conventional dual resonance EMEH, and the proposed dual resonance EMEH.

Several techniques have been explored to realize wide frequency range for EMEHs. EH arrays were adopted in [12,13] by combining multi-narrow-band harvesters into a board-band system at the cost of large volume and weight. Sufficient space must be guaranteed among different subsystems to avoid mutual influences, which further reduces overall power density. Resonant frequency tuning technology in [14] employed an additional unit to adjust the stiffness of the spring to match with excitation frequency, which brought in extra power losses and increased volume. Nonlinear characteristics in mechanical–electrical conversion utilized a non-linear system force to produce non-linear dynamic responses for harvested frequency extension. For example, [15,16] used nonlinear springs to extend the frequency range, but it was relatively narrow. The bi-stable harvesters in [17,18] had two stable equilibrium positions and would generate higher power when translators oscillated between the two potential wells. However, such architectures are either complicated for actual practices or with still limited frequency range. The up-conversion harvesters in [19,20] converted low-frequency vibration into high-frequency vibration through internal collisions. Due to the inevitable energy loss in the collision process, it was not clear whether the output power can be effectively increased. Continuous collisions will also reduce system reliability. Hybrid harvesters, such as piezoelectric–electromagnetic mechanisms, are also effective in both improving energy conversion efficiency and expanding bandwidth [21–23], which are usually realized through nonlinear effects [24] or multi-resonance structures [25,26].

Dual resonance technology combines two resonators to form two resonant frequencies [27–30], as depicted as the dashed line in Figure 1. Although this approach is straightforward, it contains significant limitations as the two resonant frequencies were cross-affected and were quite apart from each other. Hence, the harvested power was quite low in the region in between the two resonant frequencies and it is unrealistic to construct a high power "band-pass" harvester.

In this paper, an EMEH with dual uncoupled resonant frequencies is proposed to form a "band-pass" type of wide harvesting frequency range. The structure of the paper is organized as follows: Section 2 describes the basic topology of the proposed system and its frequency domain characteristics; Section 3 compares the design of dual resonant frequencies with a conventional dual resonance EMEH; the prototype of the proposed EMEH and measurement results are given in Section 4; and, finally, a conclusion is drawn in Section 5.

2. Basic Theorem and Frequency Range Extension of the Proposed Dual Resonance EMEH

The basic architecture of the proposed energy harvester is presented in Figure 2a. Mass m_1 and m_2 are composed by a nonmagnetic mass and a permanent magnet, respectively. The translators are connected to two springs and are designed for different resonant frequencies. When one translator oscillates relatively to the other, induced current is generated in the coil. According to Newton's second law of motion, the system's dynamic mechanical domain equations are established as:

$$\ddot{z}_1(t) + \frac{c_1}{m_1}\dot{z}_1(t) + \frac{k_1}{m_1}z_1(t) = -\ddot{y}(t) - \frac{K_e i_{load}(t)}{m_1} \tag{1}$$

$$\ddot{z}_2(t) + \frac{c_2}{m_2}\dot{z}_2(t) + \frac{k_2}{m_2}z_2(t) = -\ddot{y}(t) + \frac{K_e i_{load}(t)}{m_2} \tag{2}$$

where $z_1(t)$ and $z_2(t)$ are the relative displacement of the two translators; c_1 and c_2 are the mechanical damping of the two translators; k_1 and k_2 are the spring stiffness coefficients; y is the displacement of excitation source; K_e is the electromechanical transduction constant; and $K_e i_{load}(t)$ is the electromagnetic force, equaling to F_{EM} [31]. Figure 2b shows the fundamental equivalent circuit of the proposed EMEH, where the induced electromagnetic voltage $v_{EMF}(t)$ can be derived as:

$$v_{EMF}(t) = K_e\left(\dot{z}_1 - \dot{z}_2\right) = L_{coil}\dot{i}_{load}(t) + R_t i_{load}(t) \tag{3}$$

where R_{coil} and L_{coil} are the coil resistance and inductance; i_{load} is the loop current; R_{load} is the equivalent input resistance of the interface circuit; and R_t represents the summation of R_{coil} and R_{load}. As $\omega_{vib}L_{coil}$ is much smaller than R_t, L_{coil} is neglected in the following analysis. The electromagnetic damping to the translator is derived as $c_e = K_e^2/R_t$ [1,7,8]. To understand the frequency responses of the system, the transfer functions of the relative displacements to the output are defined as: $H_1(s) = Z_1(s)/Y(s)$, $H_2(s) = Z_2(s)/Y(s)$, and $H(s) = [Z_1(s) - Z_2(s)]/Y(s)$.

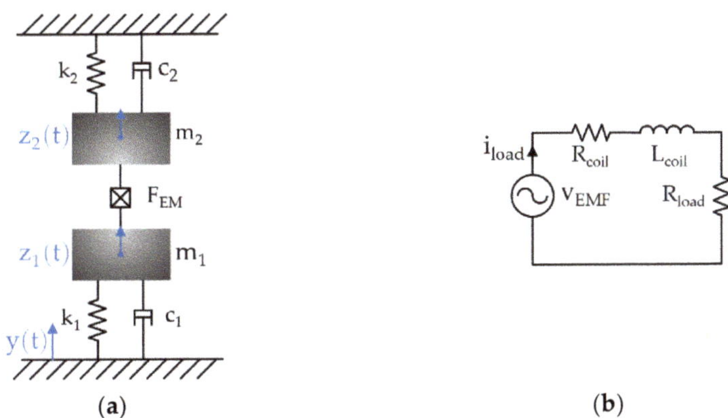

Figure 2. System topology of (**a**) the proposed dual resonance EMEH; and its (**b**) electrical equivalent circuit model.

With the assumption of a sinusoidal vibration source, i.e., $y(t) = Y_{max}\sin \omega_{vib}t$, and with a small displacement to ensure linearity in the spring, the displacement of two translators and overall displacements can be expressed as:

$$z_1(t) = |H_1(j\omega_{vib})|Y_{max}\sin(\omega_{vib}t + \arg(H_1(j\omega_{vib}))) \tag{4}$$

$$z_2(t) = |H_2(j\omega_{vib})|Y_{max}\sin(\omega_{vib}t + \arg(H_2(j\omega_{vib}))) \tag{5}$$

$$z(t) = z_1(t) - z_2(t) = |H(j\omega_{vib})|Y_{max}\sin(\omega_{vib}t + \arg(H(j\omega_{vib}))) \tag{6}$$

For convenient comparisons, transfer functions are transformed into the frequency domain and normalized by $\overline{\omega}_{vib} = \omega_{vib}/\omega_1$, as expressed in (7)–(9). ω_1 and ω_2 are the natural angular frequencies of resonators 1 and 2, which can be expressed as $\sqrt{k_1/m_1}$

and $\sqrt{k_2/m_2}$, respectively. ξ_1 and ξ_2 are the mechanical damping ratios of resonators 1 and 2 as $c_1/2m_1\omega_1$ and $c_2/2m_2\omega_2$, correspondingly. ξ_{e1} and ξ_{e2} are the electromagnetic damping ratios of resonators 1 and 2 as $c_e/2m_1\omega_1$ and $c_e/2m_2\omega_2$, respectively. α is the second natural angular frequency over the first natural angular frequency, as ω_2/ω_1. The frequency response of the dual resonance EMEH is studied with $\xi_1 = 0.004$, $\xi_2 = 0.0051$, $\xi_{e1} = 0.0173$, $\xi_{e2} = 0.0134$, and $\alpha = 1.28$, where the parameters are selected as follows: $c_1 = c_2 = 0.22\,\text{N}\cdot\text{s/m}$, $m_1 = 76$ g, $m_2 = 46$ g, $c_e = 0.74\,\text{N}\cdot\text{s/m}$, $\omega_1 = 364$ rad/s, and $\omega_2 = 465$ rad/s. The above parameters were chosen based upon the actual prototype designs that will be further discussed in Section 4; c_1 and c_2 were measured by Panasonic's laser sensor HL-C235. The frequency-domain characteristics of the transfer functions are plotted in Figure 3a. When one resonator works under resonance frequency, the other resonator is also affected by coupled electromagnetic force and produces a small output spike, as circled in Figure 3a. In the region of $1 < \overline{\omega}_{vib} < 1.28$, the magnitude of $H(j\overline{\omega}_{vib})$ equals to the summation of $|H_1(j\overline{\omega}_{vib})|$ and $|H_2(j\overline{\omega}_{vib})|$. This is because the angular difference between $H_1(j\overline{\omega}_{vib})$ and $H_2(j\overline{\omega}_{vib})$ is almost 180°. The high harvested power in the harvested frequency band is one main advantage of this proposed dual resonance EMEH. A single resonance EMEH is also analyzed under the same parameters and with only one translator of normalized resonant angular frequency 1.28, as depicted in Figure 3b. The normalized frequency range of magnitude above 8 in the dual resonance EMEH is from 0.953 to 1.336 and is from 1.207 to 1.36 for the single resonance EMEH. The frequency coverage range in the dual resonance EMEH is more than 2.5 times of that in the single resonance EMEH, and therefore more output power can be harvested with extended frequency ranges. The model of the proposed EMEH and the above analysis were carried out under the assumptions of the small displacement and linear movements of the spring. If a large deformation happens, the spring will generate non-linear effects that vary the frequency responses.

$$H_1(j\overline{\omega}_{vib}) = \frac{-(j\overline{\omega}_{vib})^2 \cdot \left[-\overline{\omega}_{vib}^2 + 2(\xi_2\alpha + \xi_{e2}\alpha + \xi_{e1})j\overline{\omega}_{vib} + \alpha^2\right]}{\left[-\overline{\omega}_{vib}^2 + 2(\xi_1 + \xi_{e1})j\overline{\omega}_{vib} + 1\right]\left[-\overline{\omega}_{vib}^2 + 2\alpha(\xi_2 + \xi_{e2})j\overline{\omega}_{vib} + \alpha^2\right] + 4\xi_{e1}\xi_{e2}\alpha\overline{\omega}_{vib}^2} \tag{7}$$

$$H_2j\overline{\omega}_{vib} = \frac{-(j\overline{\omega}_{vib})^2 \cdot \left[-\overline{\omega}_{vib}^2 + 2(\xi_1 + \xi_{e1} + \xi_{e2}\alpha)j\overline{\omega}_{vib} + 1\right]}{\left[-\overline{\omega}_{vib}^2 + 2(\xi_1 + \xi_{e1})j\overline{\omega}_{vib} + 1\right]\left[-\overline{\omega}_{vib}^2 + 2\alpha(\xi_2 + \xi_{e2})j\overline{\omega}_{vib} + \alpha^2\right] + 4\xi_{e1}\xi_{e2}\alpha\overline{\omega}_{vib}^2} \tag{8}$$

$$H(j\overline{\omega}_{vib}) = \frac{-(j\overline{\omega}_{vib})^2 \cdot \left[2(\xi_2\alpha - \xi_1)j\overline{\omega}_{vib} + \alpha_1^2 - 1\right]}{\left[-\overline{\omega}_{vib}^2 + 2(\xi_1 + \xi_{e1})j\overline{\omega}_{vib} + 1\right]\left[-\overline{\omega}_{vib}^2 + 2\alpha(\xi_2 + \xi_{e2})j\overline{\omega}_{vib} + \alpha^2\right] + 4\xi_{e1}\xi_{e2}\alpha\overline{\omega}_{vib}^2} \tag{9}$$

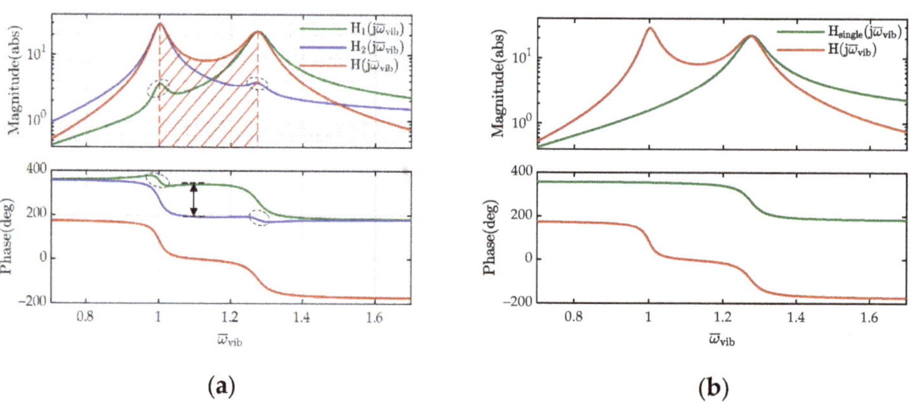

Figure 3. (**a**) Frequency–domain characteristics of the proposed dual resonance EMEH; (**b**) Frequency-domain characteristics comparison of the single resonance EMEH and the proposed dual resonance EMEH.

3. Resonant Angular Frequency of the Dual Resonance Electromagnetic EH

The resonant angular frequencies of the proposed system are defined as ω_{n1} and ω_{n2}, and further normalized as $\overline{\omega}_{n1}\omega_{n1}/\omega_1$ and $\overline{\omega}_{n2}\omega_{n2}/\omega_1$, respectively. Equation (10) can be derived by ignoring the electromagnetic and mechanical damping ratios in Equation (9), and the normalized resonance angular frequencies of the proposed dual resonance EMEH can be extracted as $\overline{\omega}_{n1} = 1$ and $\overline{\omega}_{n1} = 2 = \alpha = \omega_2/\omega_1$:

$$\left(-\overline{\omega}_{vib}^2+1\right)\left(-\overline{\omega}_{vib}^2+\alpha^2\right)= 0 \tag{10}$$

Figure 4 shows the magnitude frequency characteristic of $H(j\overline{\omega}_{vib})$ under different α. The higher α is, the further apart from each other two resonant frequencies are. Therefore, the actual resonance frequencies of the proposed EMEH can be adjusted by the natural angular frequencies of the resonators through spring and mass designs, which follow $\omega_{n1}= \omega_1 = \sqrt{k_1/m_1}$ and $\omega_{n2}= \omega_2 = \sqrt{k_2/m_2}$. The two resonant frequencies of the proposed EMEH are decoupled and have independent freedom to adjust.

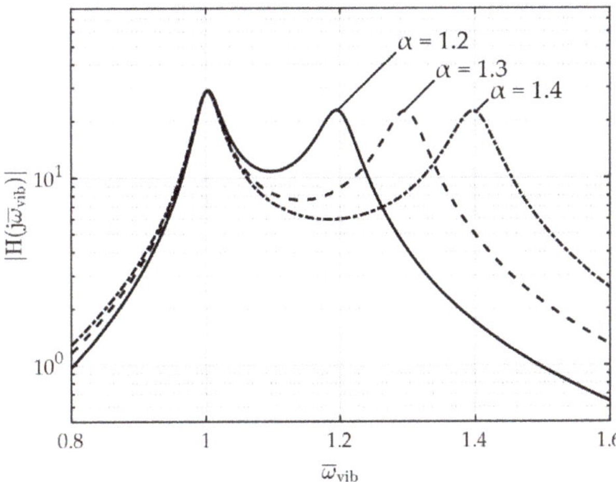

Figure 4. The magnitude characteristic of the dimensionless transfer function for the proposed dual resonance EMEH under different α.

To further demonstrate this feature, the resonant frequencies for a conventional dual resonant EMEH were studied and compared with the proposed system. As depicted in Figure 5a, the accessory resonator in a conventional dual resonant EMEH is connected to the main resonator through the spring. The transducer can be placed in 2 positions: between mass 1 and the frame, as configuration A, and between mass 1 and mass 2, as configuration B [32]. The dynamic mechanical domain equations are established as shown in Equations (11) and (12) [32,33], where $z_1'(t)$ and $z_2'(t)$ are the relative displacement; c_1' and c_2' are the mechanical damping; k_1' and k_2' are the spring stiffness coefficients; y is the displacement of the excitation source; F_{EM}' is the electromagnetic force; and m_1' and m_2' are the masses of the two translators in the conventional system, respectively. The transfer functions for the conventional dual resonant EMEH are expressed as $H_A(s)= Z_1'(s)/Y(s)$, for configuration A, and $H_B(s) = [Z_1'(s)-Z_2'(s)]/Y(s)$, for configuration B. The frequency domain transfer functions become $H_A(j\overline{\omega}_{vib}')$ and $H_B(j\overline{\omega}_{vib}')$, normalized by $\overline{\omega}_{vib}' = \omega_{vib}/\omega_1'$, as listed in Equations (13) and (14). The cross-coupling parameter μ is defined as m_2'/m_1'. Under undamped conditions, the resonant angular frequencies of the conventional dual resonant EMEH are derived as shown in Equation (15). The resonant frequencies of the conventional dual resonant EMEH are reliant on the quotient μ of the masses and α'. Figure 5b displays that magnitude frequency characteristics of $H_A\left(j\overline{\omega}'_{vib}\right)$ and $H_B\left(j\overline{\omega}'_{vib}\right)$ at $\alpha' = 1.2$ under different μ. It shows that μ has a positive impact on pushing two resonant frequencies apart from each other.

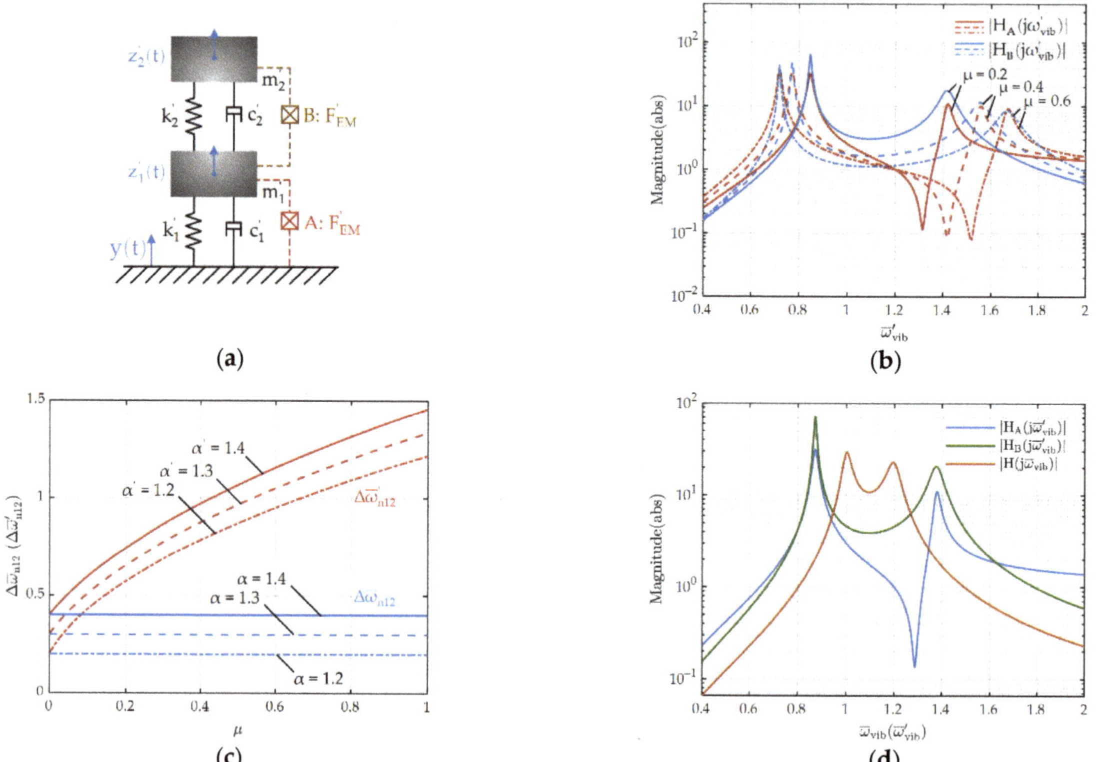

Figure 5. (**a**) The conventional dual resonance EMEH model. (**b**) The magnitude frequency characteristic of the normalized transfer functions for the conventional dual resonance EMEH under different μ at $\alpha' = 1.2$. (**c**) The differences between the two resonant angular frequencies for the proposed and conventional dual resonance EMEHs vs. μ at different α and α'. (**d**) The magnitude characteristic of the normalized transfer functions for the proposed and conventional dual resonance EMEHs at $\alpha = \alpha' = 1.2$ and $\mu = 0.2$.

The differences of the two resonance angular frequencies are defined as $\Delta\overline{\omega}'_{n12} = \overline{\omega}'_{n1} - \overline{\omega}'_{n2}$, for the conventional dual resonance EMEH, and $\Delta\overline{\omega}_{n12} = \overline{\omega}_{n1} - \overline{\omega}_{n2}$, for the proposed system. Figure 5c exhibits the resonance differences under different μ, α, and α'. The resonant frequencies of the proposed system are completely unrelated with μ, only depended on the natural angular frequencies of the two resonators. The resonant resonances of the conventional system are highly nonlinearly related to μ. Although the two resonant frequencies of the conventional system are relatively close when μ is small, it is difficult to realize in actual practice. A small μ corresponds to a small m'_2 compared with m'_1, resulting in a low k'_2. A spring with low stiffness usually takes up more volume. Figure 5d shows the magnitude frequency characteristic of the transfer functions for the proposed and conventional dual resonance EMEHs at $\alpha = \alpha' = 1.2$ and $\mu = 0.2$. Compared with the conventional system, the proposed system has two closer resonances and higher magnitudes between resonances.

$$\ddot{z}_1'(t) = -\frac{k_2'}{m_1'}\left[z_1'(t) - z_2'(t)\right] - \frac{c_2'}{m_1'}\left[\dot{z}_1'(t) - \dot{z}_2'(t)\right] - \frac{k_1'}{m_1'}z_1'(t) - \frac{c_1'}{m_1'}\dot{z}_1'(t) - \ddot{y}(t) - \frac{K_e i_{load}(t)}{m_1'} \tag{11}$$

$$\ddot{z}_2'(t) = -\frac{k_2'}{m_2'}\left[z_2'(t) - z_1'(t)\right] - \frac{c_2'}{m_2'}\left[\dot{z}_2'(t) - \dot{z}_1'(t)\right] - \ddot{y}(t) \tag{12}$$

$$H_A\left(j\overline{\omega}_{vib}'\right) \frac{-\left(j\overline{\omega}_{vib}'\right)^2 \cdot \left[-j\overline{\omega}_{vib}'^2 + 2(1+\mu)\xi_2'\alpha'j\overline{\omega}_{vib}' + (1+\mu)\alpha'^2\right]}{\left[-\overline{\omega}_{vib}'^2 + 2\left(\xi_1' + \xi_{e1}' + \mu\xi_2'\alpha'\right)j\overline{\omega}_{vib}' + 1 + \mu\alpha'^2\right]\left[-\overline{\omega}_{vib}'^2 + 2\xi_2'\alpha'j\overline{\omega}_{vib}' + \alpha'^2\right] - \mu\left(2\xi_2'\alpha'j\overline{\omega}_{vib}' + \alpha'^2\right)^2} \tag{13}$$

$$H_B\left(j\overline{\omega}_{vib}'\right) = \frac{-\left(j\overline{\omega}_{vib}'\right)^2 \cdot \left(2\xi_1'j\overline{\omega}_{vib}' + 1\right)}{\left[-\overline{\omega}_{vib}'^2 + \left(2\xi_1' + 2\mu\left(\xi_2' + \xi_{e2}'\right)\alpha'\right)j\overline{\omega}_{vib}' + 1 + \mu\alpha'^2\right]\left[-\overline{\omega}_{vib}'^2 + \left(2\left(\xi_2' + \xi_{e2}'\right)\alpha'\right)j\overline{\omega}_{vib}' + \alpha'^2\right] - \mu\left(2\left(\xi_2' + \xi_{e2}'\right)\alpha'j\overline{\omega}_{vib}' + \alpha'^2\right)^2} \tag{14}$$

$$\overline{\omega}_{n1,2}' = 0.707\sqrt{\left(1 + (1+\mu)\alpha'^2\right) \pm \sqrt{1 + (1+\mu)^2\alpha'^2 + 2(\mu-1)\alpha'^2}} \tag{15}$$

4. Prototype Design and Experiment Verification

4.1. Prototype Design

The proposed EMEH is composed by two resonators, one of which resonator is composed by a nonmagnetic mass and a coil, and the other resonator is composed by a permanent magnet. Two masses are connected to the frame though planar springs. The planar spring is composed of multiple serpentine sectors, as plotted in Figure 6a. The serpentine structure allows for lower stiffness and reduces the occupied area. This structure also overcomes the buckling problem caused by the residual stress [34] and further prolongs the reliability of the EMEH. The material of the spring is copper. Finite element analysis (FEA) is employed to determine the stiffness of the spring. When a ramp displacement of 2 mm is applied to the center of the spring, the force reaction is observed and the stiffness is derived as 11,730 N/m, as plotted in Figure 6b. Two identical planar springs were used in the proposed harvester, so that $k_1 = k_2 = 11,730$ N/m. As discussed in Section 3, the resonant frequencies of the proposed harvester is equal to the natural frequency of the resonator, which can be calculated as $f_{n1} = 1/(2\pi) \cdot \sqrt{k_1/m_1} = 62.5$ Hz and $f_{n2} = 1/(2\pi) \cdot \sqrt{k_2/m_2} = 80.4$ Hz. The key parameters of the dual resonators are listed in Table 1.

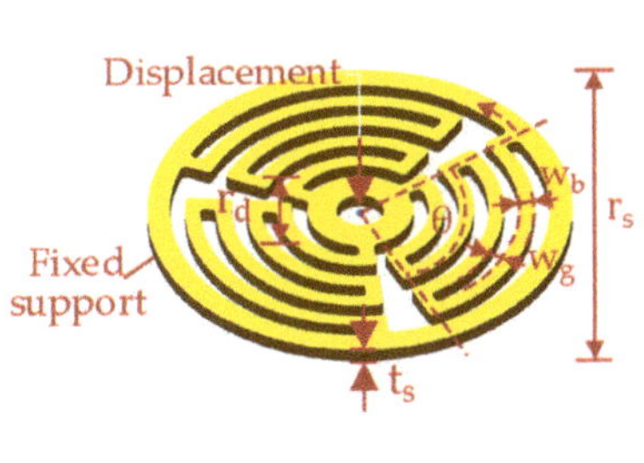

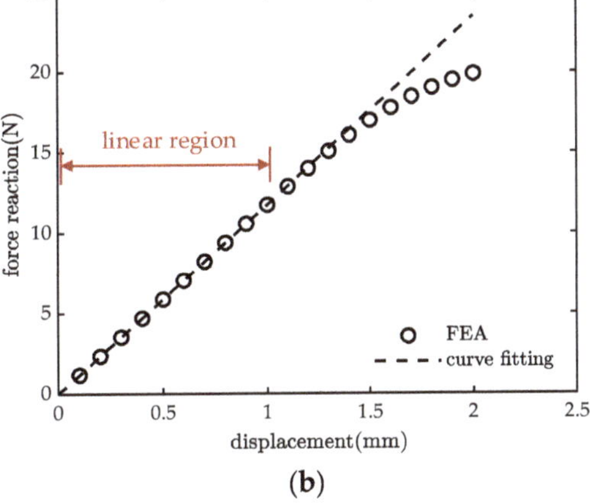

(a)   **(b)**

Figure 6. (**a**) Designed planar spring drawing. (**b**) Force reaction vs. displacement of the designed planar spring.

Table 1. Key Design Parameters of the Proposed EMEH.

Parameter	Description	Value
m_1, m_2	Mass 1, Mass 2	76 g, 46 g
Y_{max}	Vibration source amplitude	20, 40, 60 μm
R_{coil}	Internal resistance of coil	273 Ω
L_{coil}	Internal inductance of coil	77.8 mH
r_m	Magnet radius and height	9.9 mm
h_m	Magnet height	19.5 mm
B_r	Remanent magnetic flux density of magnet	1.23 T
h_{coil}	Ring coil height	5 mm
r_{coili}	Ring coil inner radius	5.2 mm
r_{coilo}	Ring coil outer radius	11.5 mm
N_{coil}	Ring coil turns	2350
θ	Spring sector angle	100°
t_s	Spring thickness	0.9 mm
w_b	Spring beam width	1 mm
w_g	Spring beam gap width	1 mm
r_d	Spring central disc radius	3.5 mm
r_s	Spring outer radius	14 mm

To induce high EMF voltage and power, coils should be placed in the position where the radial magnetic flux density is large to achieve high K_e. The expression of K_e for a single-turn coil is listed in Equation (16) [35], where rm and hm are magnet radius and height, respectively; r_{coil} and z_{coil} are the single-turn coil radius and the vertical position, respectively, as depicted in Figure 7a. The flux gradient of a single-turn coil with a radius of 12 mm at different positions is plotted in Figure 7b. A higher K_e is obtained near the upper and lower surfaces of the permanent magnet. Therefore, a ring coil is located above the magnet with the key parameters listed in Table 1. The total flux gradient of the designed coil is 18.99 Wb/m.

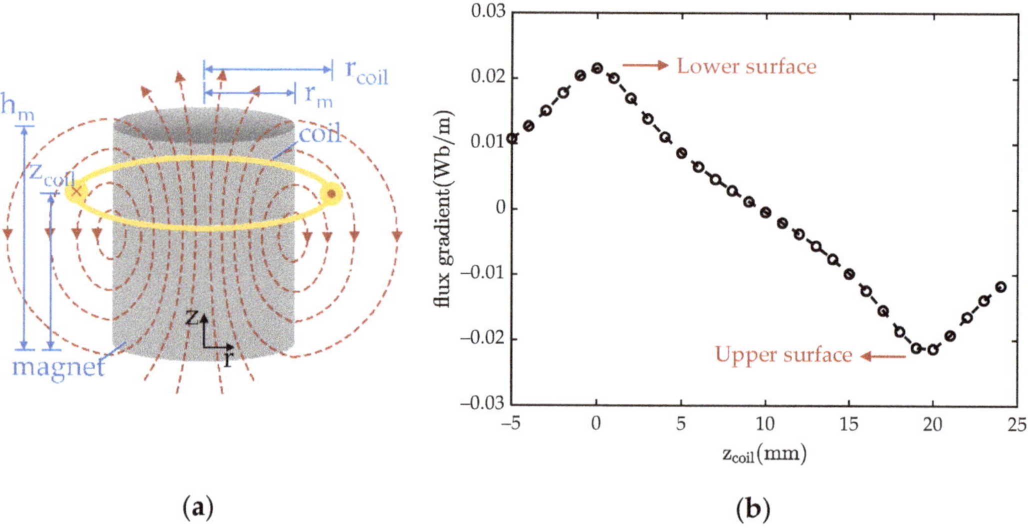

(a)

(b)

Figure 7. (a) Magnetic field distribution of the axial-magnetized cylindrical magnet. (b) Flux gradient of the single-turn coil vs. vertical position.

Figure 8a exhibits the three-dimensional assembly drawing of the designed dual resonance EMEH. The planar springs were processed by laser and the frames were processed by 3D printing. m_1 was selected as a copper mass at 76 g. A NdFeB magnet of 46 g was adopted for m_2. A tubular dual resonance EMEH was fabricated with a diameter of 31.2 mm and height of 70.5 mm, as shown in Figure 8b. It is worth pointing out that the copper mass 1 was selected just for manufacturing convenience. To further reduce the eddy current loss induced in the copper mass of m_1, a non-magnetic and non-conductive material can be employed for future designs.

$$K_e = 0.5 B_r r_m r_{coil} \cdot \int_0^{2\pi} \left[\frac{\cos\theta}{\sqrt{r_m^2 + r_{coil}^2 - 2r_m r_{coil}\cos\theta + (h_m - z_{coil})^2}} - \frac{\cos\theta}{\sqrt{r_m^2 + r_{coil}^2 - 2r_m r_{coil}\cos\theta + z_{coil}^2}} \right] d\theta. \quad (16)$$

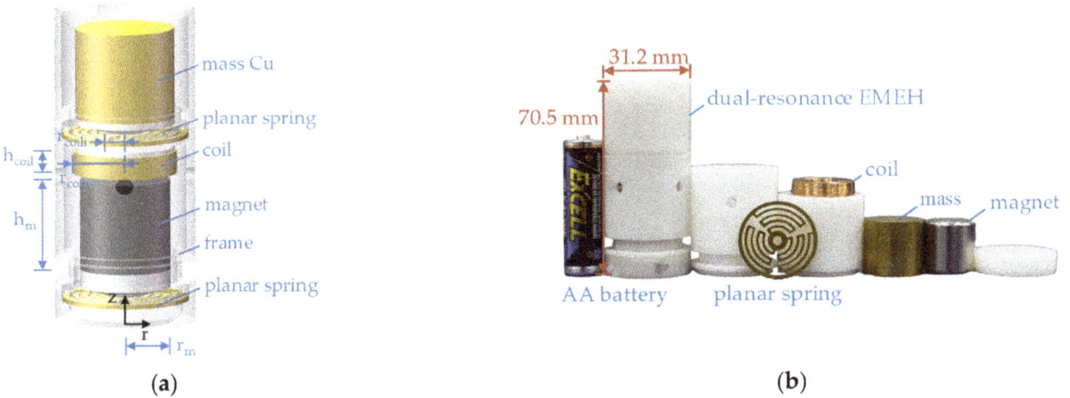

Figure 8. (**a**) 3D assembly drawing of the designed EMEH. (**b**) Prototype of the designed EMEH.

4.2. Experimental Results and Discussion

Figure 9 shows the electric vibration generation system ES-3-150 manufactured by Donglingtech (Suzhou, China), including a host computer, a power amplifier, and a vibration generator. A variable resistance box was used as the load of the EMEH, and the voltage signal was measured by a digital oscilloscope WaveRunner 604Zi manufactured by Teledyne LeCroy (New York, NY, USA). Panasonic's (Osaka, Japan) laser position sensor HL-C235 was employed to measure the displacement of masses under no-load conditions to further extract the mechanical damping [36,37]. Under no load conditions, the EMEH can be treated as a traditional mass-spring-damper model and the transfer function $|H_1(j\overline{\omega}_{vib})|$ at the resonant frequency of ω_{n1} would be simplified as $|H_1(j\overline{\omega}_{vib})| = 1/2\xi_1$. By obtaining the maximum displacement through the laser position sensor, the mechanical damping of translator 1 is further calculated accordingly, as 0.22 N/(m/s) in this design, which is also the same as the mechanical damping of translator 2.

Figure 10a,b present output voltage and power of the designed dual resonance EMEH under different excitation amplitudes with a load resistor of 300 Ω. At the first resonance of 58 Hz, the output power under the excitation amplitudes of 20 μm, 40 μm, and 60 μm are 10.96 mW, 19.4 mW, and 24.6 mW, respectively. At the second resonance of 74.5 Hz, the output power under three excitation amplitudes is 14.9 mW, 32.4 mW, and 37.2 mW, respectively. At the excitation frequency of 65 Hz, the proposed EMEH has the lowest output power of 0.52 mW, 2.17 mW, and 4.99 mW under three excitation amplitudes, respectively. It is well known that output power is proportional to the square of the excitation amplitude at each excitation frequency [1,7,8]. However, this rule does not hold for dual resonance EMEH because that output power is saturated at two resonance frequencies, which is limited by the translators' displacements. The parasitic damping is increased when the planar springs suffer from large deformations at two resonances [38]. When the excitation amplitude increases, two resonant frequencies of the EMEH are reduced slightly, especially for the second resonance, as shown in Figure 10a,b. Such phenomenon is caused by the nonlinear restoring force of the spring [15,16].

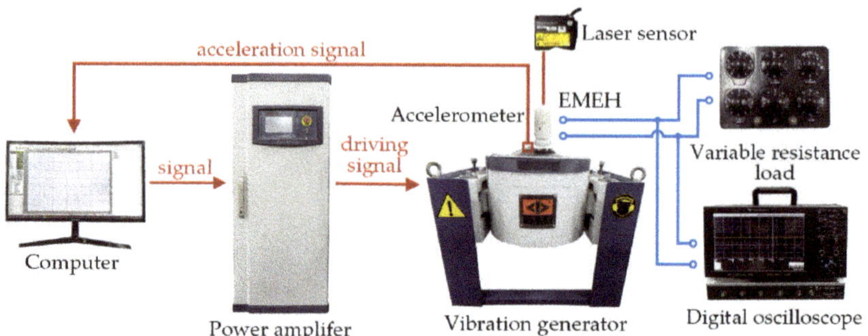

Figure 9. Testing bench of the vibration generation system.

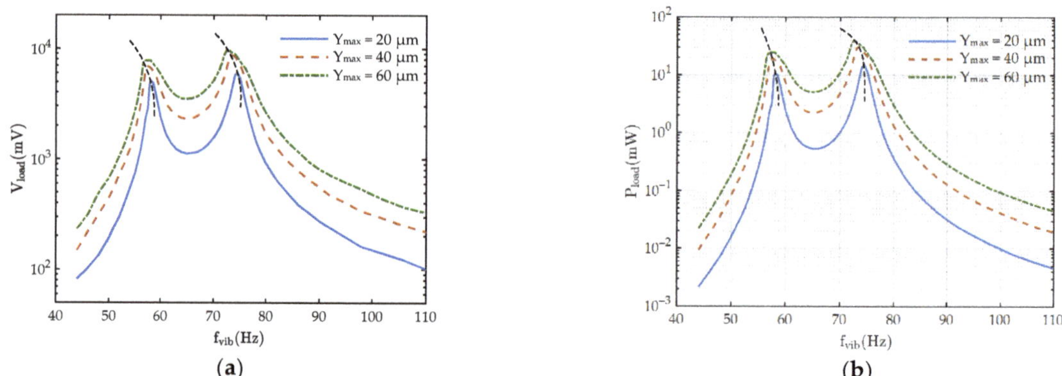

Figure 10. (**a**) Output peal-to-peak voltage vs. vibration frequency under different sinusoidal excitations at $R_{load} = 300\ \Omega$. (**b**) Output power vs. vibration frequency under different excitation amplitudes at $R_{load} = 300\ \Omega$.

Figure 11a shows the output peak-to-peak voltage of the designed dual resonance EMEH under different loads with an excitation amplitude of 20 μm. Output voltage increases as the load increases. One reason for this is that the translators have larger displacements with the decrease of electromagnetic damping. The other reason is that larger load resistor has a larger output voltage due to the law of KVL. Figure 11b illustrates the relationship between output power and excitation frequency under different loads. The output power of the EMEH is lowest at the excitation frequency of 65 Hz, which is right in the middle of two resonances. Figure 12a shows the relationship between output power and loads resistors under the excitation frequency of 65 Hz. Hence, to improve output power between two resonant frequencies, a 300 Ω load is preferred for the proposed EMEH. The steady-state output voltage waveform is presented in Figure 12b with a vibration frequency and amplitude of 74.5 Hz and 20 μm, and a load resistance of 300 Ω. The output voltage is sinusoidal and the peak-to-peak voltage reaches 6.161 V. Acceleration follows a linear relationship with vibration amplitude, as $A_{max} = 4\pi^2 f_{vib}^2 Y_{max}$. Figure 13 further provides the harvested output voltage and power at different vibration frequencies, but with a constant maximum acceleration (A_{max}). The output characteristics follow the same trend as Figure 11.

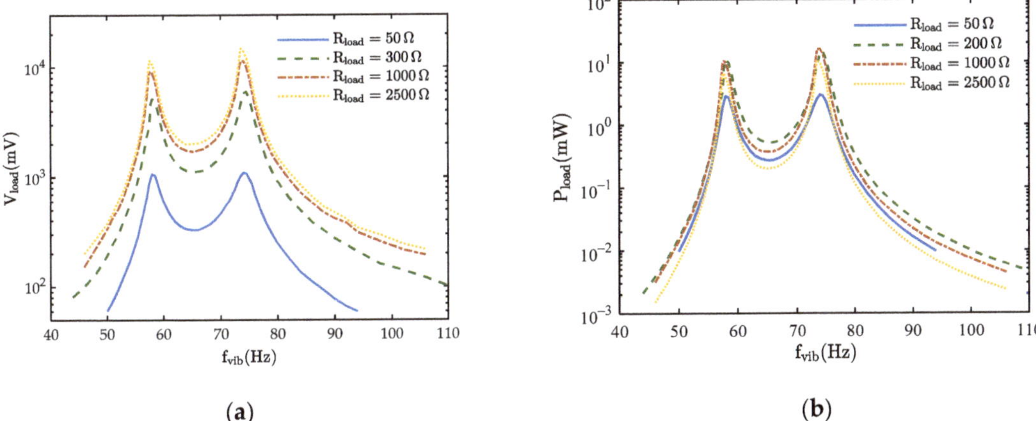

Figure 11. (**a**) Output peal-to-peak voltage vs. vibration frequency under different loads at $Y_{max} = 20$ µm. (**b**) Output peal-to-peak power vs. vibration frequency under different loads at $Y_{max} = 20$ µm.

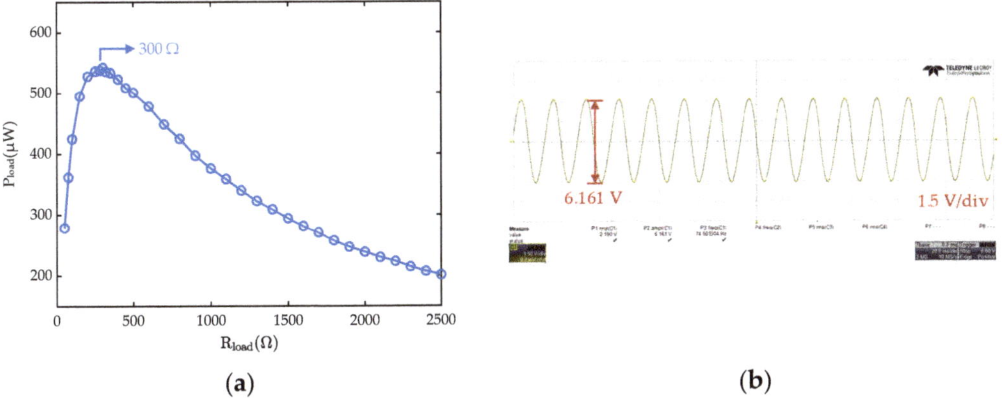

Figure 12. (**a**) Output power vs. load under the excitation frequency of 65 Hz at $Y_{max} = 20$ µm. (**b**) Steady waveform of output voltage at $f_{vib} = 74.5$ Hz, $Y_{max} = 20$ µm, and $R_{load} = 300$ Ω.

Two single resonance harvesters were designed and fabricated with the same design parameter at the resonant frequencies of 58 Hz and 74.5 Hz, respectively. Figure 14a shows output power of two single resonance EMEHs and the dual resonance EMEH under different excitation frequencies. The load is set as 300 Ω and the excitation amplitude is 20 µm. At the excitation frequency of 65 Hz, the dual resonance harvester reaches an output power of 523.8 µW. Two single resonance harvesters reach an output power of 72.9 µW and 200.5 µW, respectively. The dual resonance harvester achieves a higher output power than the summation of the two single resonance devices. In the dual resonance EMEH, the frequency range of the output power above 0.5 mW is from 55.8 Hz to 79.1 Hz. While for the single resonance EMEH with a resonance frequency 74.5 Hz, the frequency range of the output power above 0.5 mW is limited from 70.1 Hz to 80.1 Hz. The frequency range in the proposed dual resonance harvester is more than twice of that in single resonance harvester. Experiment results demonstrate the advantage of the dual resonance harvester of extending the frequency range and higher harvested power.

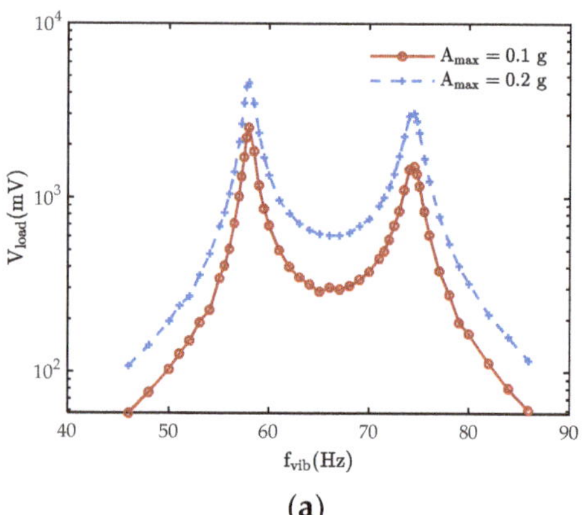

(a)

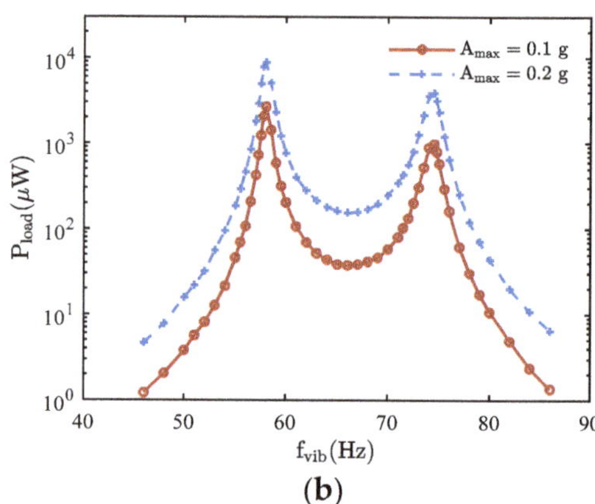

(b)

Figure 13. (**a**) Output peal-to-peak voltage vs. vibration frequency under two fixed excitation accelerations (A_{max}) at R_{load} = 300 Ω. (**b**) Output power vs. vibration frequency under two fixed excitation accelerations (A_{max}) at R_{load} = 300 Ω.

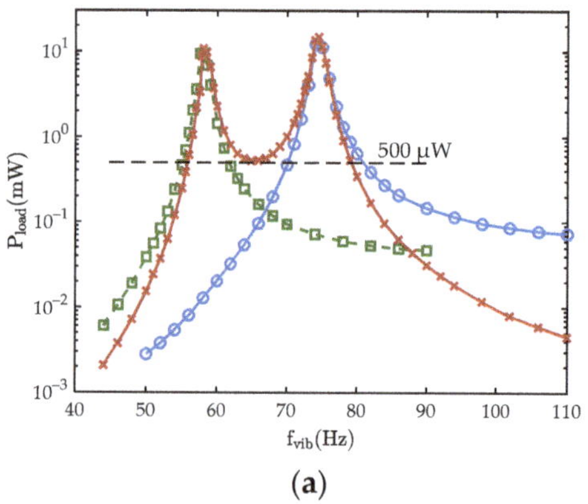

(a)

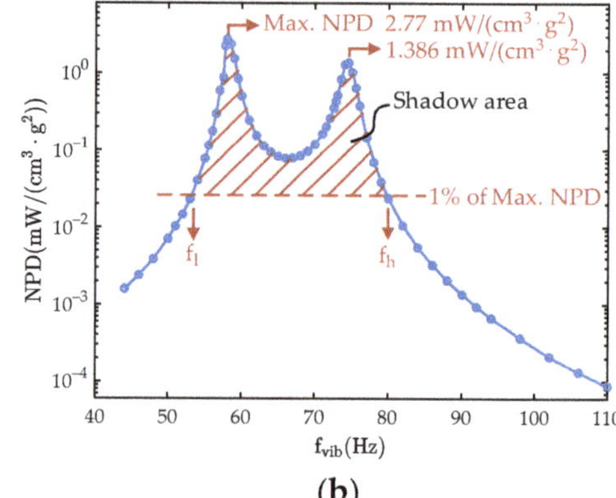

(b)

Figure 14. (**a**) Output power of two single resonance harvesters and the dual resonance EMEH vs. vibration frequency at R_{load} = 300 Ω, Y_{max} = 20 μm. (**b**) Normalized power density of the proposed dual resonance EMEH vs. vibration frequency at R_{load} = 300 Ω, Y_{max} = 20 μm.

Since wideband EMEHs in the literature were designed under significant different specifications, the normalized power density (NPD) is proposed for a fair comparison [39], which is expressed in Equation (17). To further evaluate the covered frequency bandwidths among state-of-art apparatuses, index S, shown in Equation (18), is defined as the integral of the NPD higher than 1%, which represents the coverage area of the NPD higher than 1% of the maximum NPD in the frequency domain.

The designed EMEH has a maximum NPD of 2.77 mW/(cm^3·g^2) at the excitation frequency of 58 Hz and S of 10.53 mW/(s·cm^3·g^2) from frequency of 53.4 Hz to 79.7 Hz. Compared to the state-of-art EMEHs with wide frequency ranges in Table 2, the maximum NPD is high due to the adoption of a planar spring and the optimization of the coil position. Furthermore, the S parameter is more than 2× that of the state-of-art apparatuses. Hence, the proposed EMEH is demonstrated to have a high maximum NPD and high output power over wide harvesting frequencies.

$$NPD = \frac{P_{load}}{\left(4\pi^2 f_{vib}^2 Y_{max}/g\right)^2 \cdot Volume} \tag{17}$$

$$S = \int_{f_l}^{f_h} \text{NPD}(f_{vib}) \cdot df_{vib} \tag{18}$$

Table 2. Performance Comparisons Among State-of-art Apparatuses.

Ref	Transduction Type and Employed Technology	Spring Type	Resonant Frequencies (Hz)	Max. Output Power (mW)	Volume (cm^3)	Max. NPD (mW/(cm^3·g^2))	S (mW/(s·cm^3·g^2))
[40]	Electromagnetic; dual resonance, nonlinear	Magnetic spring	7.5, 18.5	2.58 @ 7.5 Hz, 0.5 g	9.73	1.06	5.2247
[27]	Electromagnetic; dual resonance	MEMS planar spring	326, 391	9.6×10^{-7} @ 391 Hz, 0.12 g	0.29	2.3×10^{-4}	null
[12]	Electromagnetic; EH array	Magnetic spring	7, 8, 9, 10	2.09 @ 8.5 Hz, 0.5 g	40.18	0.208 @ 8.5 Hz, 0.5 g	0.681
[18]	Electromagnetic; bi-stable EH	FR4 spring	36	0.0193 @ 36 Hz, 1.5 g	~1.171	7.325×10^{-3}	null
[19]	Electromagnetic; up-conversion	Helical spring	null	11.89 @ 5.17 Hz, 2.06 g	6.47	0.07847	null
[16]	Electromagnetic; nonlinear	Magnetic and helical springs	9	1.15 @ 9 Hz, 0.8 g	12.2	0.1473	0.7352
This Work	Electromagnetic; dual-resonance	Planar spring	58, 74.5	39.8 @74.5 Hz, 0.45 g	53.9	2.77 @ 58 Hz, 0.27 g	10.53 @ 53.4–79.7 Hz

5. Conclusions

An electromagnetic energy harvester (EMEH) with dual resonant frequencies is proposed in this paper, which is constructed by two spring-mass resonators. Due to the angular differences between the two resonators, the relative displacement is increased within the frequency bandwidth to achieve enhanced output power. Compared with single resonance harvester with the same designed parameters, the proposed dual resonance EMEH achieves a 2.5 times larger frequency range. Furthermore, the two resonators are decoupled in the proposed configuration, so that the resonant frequencies of the proposed EMEH can be determined independently and are not cross-affected. Therefore, a high power "band-pass" EMEH can be achieved. A tubular dual spring-mass EHEM is designed with a diameter of 31.2 mm and height of 70.5 mm. Two resonant frequencies are designed at 58 Hz and 74.5 Hz. An output power of 14.9 mW is achieved at 74.5 Hz with a load of 300 Ω and an excitation amplitude of 20 μm. The frequency range of the output power above 0.5 mW is from 55.8 Hz to 79.1 Hz, reaching the width of 23 Hz. The maximum normalized power density (NPD) reaches to 2.77 mW/(cm^3·g^2), which is more than 2 times larger than state-of-art wideband EMEHs. The proposed dual resonance EMEH is verified to provide a descent high output power over "band-pass" frequency range. The proposed technology can be a promising technology when vibrations have wide frequency coverage.

Author Contributions: H.P. was the corresponding author and proposed the conceptual design of the device and supervised Z.F. and Y.C. for this work. Z.F. is responsible for the main technical work. Y.C. provided support for the design of the planar spring. All authors have read and agreed to the published version of the manuscript.

Funding: This research received no external funding.

Institutional Review Board Statement: Not applicable.

Informed Consent Statement: Informed consent was obtained from all subjects involved in the study.

Data Availability Statement: The data presented in this study are available on request from the corresponding author.

Conflicts of Interest: The authors declare no conflict of interests. The authors declare that they have no known competing financial interests or personal relationships that could have appeared to influence the work reported in this paper.

References

1. Priya, S.; Inman, D.J. *Energy Harvesting Technologies*, 1st ed.; Springer US: Boston, MA, USA, 2008.
2. Citroni, R.; Di Paolo, F.; Livreri, P. A Novel Energy Harvester for Powering Small UAVs: Performance Analysis, Model Validation and Flight Results. *Sensors* **2019**, *19*, 1771. [CrossRef]
3. Citroni, R.; Di Paolo, F.; Livreri, P. Evaluation of an optical energy harvester for SHM application. *AEU Int. J. Electron. Commun.* **2019**, *111*, 152918. [CrossRef]
4. Gu, Y.; Liu, W.; Zhao, C.; Wang, P. A goblet-like non-linear electromagnetic generator for planar multi-directional vibration energy harvesting. *Appl. Energy* **2020**, *266*, 114846. [CrossRef]
5. Zhang, J.; Su, Y. Design and analysis of a non-resonant rotational electromagnetic harvester with alternating magnet sequence. *J. Magn. Magn. Mater.* **2021**, *540*, 168393. [CrossRef]
6. Zhou, N.; Hou, Z.; Zhang, Y.; Cao, J.; Bowen, C.R. Enhanced swing electromagnetic energy harvesting from human motion. *Energy* **2021**, *228*, 120591. [CrossRef]
7. Williams, C.B.; Shearwood, C.; Harradine, M.A.; Mellor, P.H.; Birch, T.S.; Yates, R.B. Development of an electromagnetic micro-generator. *IEE P-Circ. Dev. Syst.* **2001**, *148*, 337–342. [CrossRef]
8. Stephen, N.G. On energy harvesting from ambient vibration. *J. Sound Vib.* **2006**, *293*, 409–425. [CrossRef]
9. Chen, Y.; Peng, H.; Cheng, Z.; Tong, Q.; Kang, Y. A Planar PCB Based Energy Harvester with Voltage Multiplier. In Proceedings of the 2020 IEEE Energy Conversion Congress and Exposition (ECCE), Detroit, MI, USA, 11–15 October 2020; pp. 975–980.
10. Makovička, D. Response Analysis of Building Loaded by Groundborne Transient Vibration. In Proceedings of the III European Conference on Computational Mechanics, Lisbon, Portugal, 5–8 June 2006; p. 748.
11. Tang, L.; Yang, Y.; Soh, C.K. Toward Broadband Vibration-based Energy Harvesting. *J. Intel. Mat. Syst. Str.* **2010**, *21*, 1867–1897. [CrossRef]
12. Foisal, A.R.M.; Hong, C.; Chung, G. Multi-frequency electromagnetic energy harvester using a magnetic spring cantilever. *Sens. Actuators A Phys.* **2012**, *182*, 106–113. [CrossRef]
13. Ferrari, M.; Ferrari, V.; Guizzetti, M.; Marioli, D.; Taroni, A. Piezoelectric multifrequency energy converter for power harvesting in autonomous microsystems. *Sens. Actuators A Phys.* **2008**, *142*, 329–335. [CrossRef]
14. Challa, V.R.; Prasad, M.G.; Fisher, F.T. Towards an autonomous self-tuning vibration energy harvesting device for wireless sensor network applications. *Smart Mater. Struct.* **2011**, *20*, 0250042. [CrossRef]
15. Mann, B.P.; Sims, N.D. Energy harvesting from the nonlinear oscillations of magnetic levitation. *J. Sound Vib.* **2009**, *319*, 515–530. [CrossRef]
16. Fan, K.; Cai, M.; Liu, H.; Zhang, Y. Capturing energy from ultra-low frequency vibrations and human motion through a monostable electromagnetic energy harvester. *Energy* **2019**, *169*, 356–368. [CrossRef]
17. Palagummi, S.; Yuan, F.G. A bi-stable horizontal diamagnetic levitation based low frequency vibration energy harvester. *Sens. Actuators A Phys.* **2018**, *279*, 743–752. [CrossRef]
18. Podder, P.; Amann, A.; Roy, S. Combined Effect of Bistability and Mechanical Impact on the Performance of a Nonlinear Electromagnetic Vibration Energy Harvester. *IEEE-Asme T Mech.* **2016**, *21*, 727–739. [CrossRef]
19. Halim, M.A.; Cho, H.; Park, J.Y. Design and experiment of a human-limb driven, frequency up-converted electromagnetic energy harvester. *Energy Convers. Manag.* **2015**, *106*, 393–404. [CrossRef]
20. Zorlu, O.; Topal, E.T.; Kulah, H. A Vibration-Based Electromagnetic Energy Harvester Using Mechanical Frequency Up-Conversion Method. *IEEE Sens. J.* **2011**, *11*, 481–488. [CrossRef]
21. Wu, Z.; Cao, Z.; Ding, R.; Wang, S.; Chu, Y.; Ye, X. An electrostatic-electromagnetic hybrid generator with largely enhanced energy conversion efficiency. *Nano Energy* **2021**, *89*, 106425. [CrossRef]
22. Wu, Y.; Zeng, Q.; Tang, Q.; Liu, W.; Liu, G.; Zhang, Y.; Wu, J.; Hu, C.; Wang, X. A teeterboard-like hybrid nanogenerator for efficient harvesting of low-frequency ocean wave energy. *Nano Energy* **2020**, *67*, 104205. [CrossRef]
23. Li, P.; Gao, S.; Cai, H. Modeling and analysis of hybrid piezoelectric and electromagnetic energy harvesting from random vibrations. *Microsyst. Technol.* **2015**, *21*, 401–414. [CrossRef]
24. Mahmoudi, S.; Kacem, N.; Bouhaddi, N. Enhancement of the performance of a hybrid nonlinear vibration energy harvester based on piezoelectric and electromagnetic transductions. *Smart Mater. Struct.* **2014**, *23*, 075024. [CrossRef]
25. Iqbal, M.; Khan, F.U. Hybrid vibration and wind energy harvesting using combined piezoelectric and electromagnetic conversion for bridge health monitoring applications. *Energy Convers. Manag.* **2018**, *172*, 611–618. [CrossRef]

26. Toyabur, R.M.; Salauddin, M.; Cho, H.; Park, J.Y. A multimodal hybrid energy harvester based on piezoelectric-electromagnetic mechanisms for low-frequency ambient vibrations. *Energy Convers. Manag.* **2018**, *168*, 454–466. [CrossRef]

27. Tao, K.; Wu, J.; Tang, L.; Xia, X.; Lye, S.W.; Miao, J.; Hu, X. A novel two-degree-of-freedom MEMS electromagnetic vibration energy harvester. *J. Micromech. Microeng.* **2016**, *26*, 0350203. [CrossRef]

28. Hu, G.; Tang, L.; Das, R.; Marzocca, P. A two-degree-of-freedom piezoelectric energy harvester with stoppers for achieving enhanced performance. *Int. J. Mech. Sci.* **2018**, *149*, 500–507. [CrossRef]

29. Wang, H.; Tang, L. Modeling and experiment of bistable two-degree-of-freedom energy harvester with magnetic coupling. *Mech. Syst. Signal. Process.* **2017**, *86*, 29–39. [CrossRef]

30. Wang, Z.; Ding, H.; Chen, L. Nonlinear oscillations of a two-degree-of-freedom energy harvester of magnetic levitation. *J. Vib. Shock* **2016**, *35*, 55–58.

31. Cammarano, A.; Burrow, S.G.; Barton, D.A.W.; Carrella, A.; Clare, L.R. Tuning a resonant energy harvester using a generalized electrical load. *Smart Mater. Struct.* **2010**, *19*, 0550035. [CrossRef]

32. Tang, L.; Yang, Y. A multiple-degree-of-freedom piezoelectric energy harvesting model. *J. Intell. Mat. Syst. Struct.* **2012**, *23*, 1631–1647. [CrossRef]

33. Xiao, H.; Wang, X.; John, S. A dimensionless analysis of a 2DOF piezoelectric vibration energy harvester. *Mech. Syst. Signal. Process.* **2015**, *58–59*, 355–375. [CrossRef]

34. Barillaro, G.; Molfese, A.; Nannini, A.; Pieri, F. Analysis, simulation and relative performances of two kinds of serpentine springs. *J. Micromech. Microeng.* **2005**, *15*, 736–746. [CrossRef]

35. Cepnik, C.; Radler, O.; Rosenbaum, S.; Stroehla, T.; Wallrabe, U. Effective optimization of electromagnetic energy harvesters through direct computation of the electromagnetic coupling. *Sens. Actuators A Phys.* **2011**, *167*, 416–421. [CrossRef]

36. Spreemann, D.; Manoli, Y. *Electromagnetic Vibration Energy Harvesting Devices: Architectures, Design, Modeling and Optimization*; Springer Science & Business Media: Dordrecht, The Netherlands, 2012; Volume 35.

37. Dayal, R.; Dwari, S.; Parsa, L. A New Design for Vibration-Based Electromagnetic Energy Harvesting Systems Using Coil Inductance of Microgenerator. *IEEE T Ind. Appl.* **2011**, *47*, 820–830. [CrossRef]

38. Wang, P. Study on the Micro Electromagnetic Vibration Energy Harvester Based on MEMS Technology. Ph.D. Thesis, Shanghai Jiao Tong University, Shanghai, China, 2010.

39. Arnold, D.P. Review of microscale magnetic power generation. *IEEE T Magn.* **2007**, *43*, 3940–3951. [CrossRef]

40. Fan, K.; Zhang, Y.; Liu, H.; Cai, M.; Tan, Q. A nonlinear two-degree-of-freedom electromagnetic energy harvester for ultra-low frequency vibrations and human body motions. *Renew. Energy* **2019**, *138*, 292–302. [CrossRef]

Article

A Fully Featured Thermal Energy Harvesting Tracker for Wildlife

Eiko Bäumker *, Luca Conrad, Laura Maria Comella and Peter Woias *

Laboratory for the Design of Microsystems, Department of Microsystems Engineering—IMTEK, University of Freiburg, 79106 Freiburg im Breisgau, Germany; luca.conrad@imtek.uni-freiburg.de (L.C.); laura.comella@imtek.uni-freiburg.de (L.M.C.)
* Correspondence: eiko.baeumker@imtek.uni-freiburg.de (E.B.); woias@imtek.de (P.W.)

Abstract: In this paper, we describe a novel animal-tracking-system, solely powered by thermal energy harvesting. The tracker achieves an outstanding $100\,\mu W$ of electrical power harvested over an area of only 2 times $20.5\,cm^2$, using the temperature difference between the animal's fur and the environment, with a total weight of $286\,g$. The steps to enhance the power income are presented and validated in a field-test, using a system that fulfills common tracking-tasks, including GPS with a fix every $1.1\,h$ to $1.5\,h$, activity and temperature measurements, all data wirelessly transmitted via LoRaWAN at a period of $14\,min$. Furthermore, we describe our ultra low power design that achieves an overall sleep power consumption of only $8\,\mu W$ and is able to work down to temperature differences of $0.9\,K$ applied to the TEGs.

Keywords: TEG; thermoelectricity; thermal energy harvesting; tracker; wildlife; animal; ultra low power

Citation: Bäumker, E.; Conrad, L.; Comella, L.M.; Woias, P. A Fully Featured Thermal Energy Harvesting Tracker for Wildlife. *Energies* **2021**, *14*, 6363. https://doi.org/10.3390/en14196363

Academic Editor: Dibin Zhu

Received: 26 August 2021
Accepted: 25 September 2021
Published: 5 October 2021

Publisher's Note: MDPI stays neutral with regard to jurisdictional claims in published maps and institutional affiliations.

1. Introduction

Wildlife populations are under increasing pressure. Climate change, habitat loss and invasive species are only some examples of numerous threats to resident and migratory wildlife throughout the world. Additionally, in some cases, wildlife populations can negatively impact economy and agriculture. The result is a human–wildlife conflict that has negative consequences for both worlds [1]. In order to reestablish balance and harmonize the opposite requirements between human society and the need to preserve biodiversity, effective wildlife management solutions must be developed. Such development requires a deep understanding of wildlife behavior, possible through high-tech tracking devices able to gather valuable data from animals in their natural environment, in the best case in a non-disturbing and interaction-free manner. Recorded positions and activity levels of wildlife can be used to derive migratory paths, reactions induced by stress and the impact of human-made habitat modifications or losses. Given the long duration of those processes, a continuous and long-time monitoring is a mandatory requirement.

Despite recent advances in the development of tracking devices, most of them are still powered from batteries. Depending on the number and frequency of tasks required, the limited energy of batteries may shorten the possible monitoring time. An increase of power capacities and run-times with this technology typically comes with an increased battery size, hence an increased size and weight, which may have a negative influence on the animals to be monitored. A weight below 5% of the body weight is, e.g., considered acceptable as a general rule of thumb for a collar applied to a mammal, but even then, an impact on the animal can not be excluded [2]. Moreover, the limited energy available requires a severe restriction of features of the tracking devices. To prolong the device lifetime, in fact, a compromise must be found between reducing the set of features or carefully budgeting the energy by reducing the frequency of actions. An exchange of batteries may circumvent these problems; however, this is not a practicable solution. The animal should be approached and recaptured causing stress and undesired impact on its behavior.

Energy harvesting (EH) techniques, i.e., the gathering energy from the direct environment of the animal, may be an alternative solution to power an animal tracker. Several possibilities exists—one may use kinetic energy such as the movement of animals or vibrations via piezoelectric, triboelectric or electromagnetic harvesters [3]. In [4], a tracker is presented that uses electromagnetism to harvest energy from an elephant's motions. Other sources of energy can be the body of the animal itself, either by using electrochemical energy via biofuels or thermal energy in form of its body heat [5]. Other studies use environmental available energy drawn from radio frequency [6], which is typically radiated by another devices or the radiation of the sun.

With regard to commercially available systems, solar energy harvesting is one of the most used concepts to power wildlife-tracking devices. The *LifeTag*™ *(Cellular Tracking Technologies)* is an example of a small tracker running exclusively by harvesting solar power. The system periodically transmits a beacon signal that may be intercepted to get the animal location [7]; however, it does not have any additional function. Another solar-powered but feature-rich tracker, equipped with temperature sensors and global positioning system (GPS), is presented in [8]. A similar device in [9] uses solar cells as an additional energy source to prolong its battery lifetime. Although solar power represents a good alternative or assistance to battery power for wildlife-tracking-applications, it has a significant drawback: The energy income is unsteady and unpredictable. First, energy can only be harvested for approximately a half of a day during sunlight, given that the tracked animal is not living north or south of the respective polar circles. Second, the incoming amount of energy is highly dependent on environmental factors such as cloud cover, shadowing, seasons, or even the current angle of the solar cell to the sun [10]. If a continuous operation is desired, a secondary energy storage is needed in any case that can compensate the lack or reduction of energy income for at least a couple of days.

In comparison with solar energy harvesting, several advantages are obtained by using thermal energy harvesting. In an application at warm-blooded animals, the temperature gradient between animal and environment can be used to harvest electrical energy. The use of thermoelectric generators (TEGs) as a power source makes the system very robust and durable, as no moving or fragile parts are required for its operation. Harvesting from the near-constant body temperature of warm-blooded animals promises a more predictable energy income that might not be as dependent on environmental factors, except the ambient temperature, when compared to the usage of solar cells. Despite the opportunities that such a system could provide in tracking applications, there is few literature about thermal harvesting at animals. We have shown already in [11–13] that a temperature difference at a reasonably sized TEG between 2.5 K to 3.5 K can be used to supply a simple collar tracker, regularly transmitting a beacon signal for localization. In this paper, we will present our more thoroughly designed feature-rich tracking device optimized for an application at mammals. The developed tracking collar is able to record position, temperature, animal activity and transmit these data wirelessly over long distances. Its performance has been analyzed in a long running experiment. The developed system is used in this case as a measurement device to characterize the whole scenario and its boundary conditions. Valuable data about the energy income, occurring temperatures, and other performance parameters are gathered and analyzed. The paper is organized as follows: In Section 2, our design focus is explained, followed by a description of the collar's design and its field test. A detailed description of the system design and its effects on the overall performance is discussed in Section 3. The structure of our field test and its results are then presented in Section 4 and finally addressed in Section 5.

2. Aims and Scope

Our proposed tracking device and experimental setup follows two purposes. The first purpose is the development of a full-featured tracking device that is solely driven by thermal energy harvested from the temperature difference between animal and environment. With regard to the limited energy income, we investigate how to reduce power consumption in the tracking sub-system while maximizing power input from the harvesting sub-system. Power income and consumption are being recorded in a field test, together with information about how and when each feature has been activated to verify the device's functionality. The feature-set of the wildlife tracker consists of the following functions:

The animal's activity is tracked and recorded as a single value, indicating how much the animal has moved over the last period. This measurement is implemented following the common approach of using an accelerometer that records events exceeding a certain threshold. More sophisticated approaches may gain more information about the type of movements and detect the type of moving or current animal position. These, however need significantly more energy due to higher sampling rates and processing power, for example, via a frequency analysis of accelerometer values in the tracker itself [14]. For our power-optimized tracker that may use the animal activity to schedule tasks, implementing a simple step counter is sufficient.

The position in our device should be gathered via GPS as this is the most commonly used technology and offers worldwide coverage. It did become an essential tool for studying migratory routes, home ranges or the migratory behavior of animals [15]. As other global navigation satellite system (GNSS) systems such as Galileo or Glonass work with a similar principle, GPS is a good representative for this category of location technology. Depending on the intention of the tracking device, positions are recorded at very short time intervals down to a record every few minutes. Long-term studies typically gather a position every hour or with intervals up to 40 h [15,16]. For our device, that aims on long-running observations, one location per hour is an adequate requirement and compromise as higher frequencies are usually not required and are likely to not work due to the high power consumption of GPS.

Wireless transfer of the data is needed to gather the recorded data remotely. Technologies such as the transmission via satellite or the usage of cellular networks are common to achieve worldwide coverage. However, both require high amounts of energy, making them unsuitable for our purpose. An alternative with low power consumption is a custom-built world-spanning receiver network such as the Motus wildlife tracking system that is capable of receiving very high frequency (VHF) beacons [17]. It is used to locate an animal worldwide, but only works in areas where a receiver is deployed. The reception of a location history or of sensor data is not possible with that system and thus would negate the advantage of having GPS on-board. Our approach is to use a world-spanning network similar to *Motus*, but one that offers transmission of custom data at a minimum of energy and over a long range. We found the LongRangeWideAreaNetwork (LoRaWAN) to be a promising candidate as it is already widely used and can still be deployed manually. While the transmission of the recorded history is possible with that system, we have only implemented a message of the last record as all data-points will be stored and later be retrievable from the collar's internal memory.

The second purpose of our developed collar is the collection of environmental and performance-relevant data helping to analyze power input and thermal conditions at the animal. Valuable information are the thermal resistances of the device and the fur-device connections. To get an approximate value, sensors in the housing record the temperature gradient over the TEG and the ambient temperature. Due to the measurement method, we expect relatively high uncertainties of actual values, introduced by varying conditions such as the collar's varying looseness to the animal or environmental factors such as wind and rain. Still, these results are valuable as average, real-life data for improving the power outcome and weight of future devices.

Furthermore, internal data concerning the power income and usage are stored together with the device's state. It is desired to verify the power consumption for each task executed by the tracker and the performance of the implemented power management. With our implementation, a correlation between power income and ambient temperature, and its possible daily period can be derived.

3. System Design

The design of our tracker is inspired by available collars such as trackers from the companies *Vertex* or *ATS*. These collars consist typically of one or more enclosures. In our case, the collar holds three enclosures, as shown in Figure 1. Two of them contain the thermal energy harvesting part of the system, the third contains the electronics for power management, tracking and wireless communication. More details about the different collar parts will follow in the next sections.

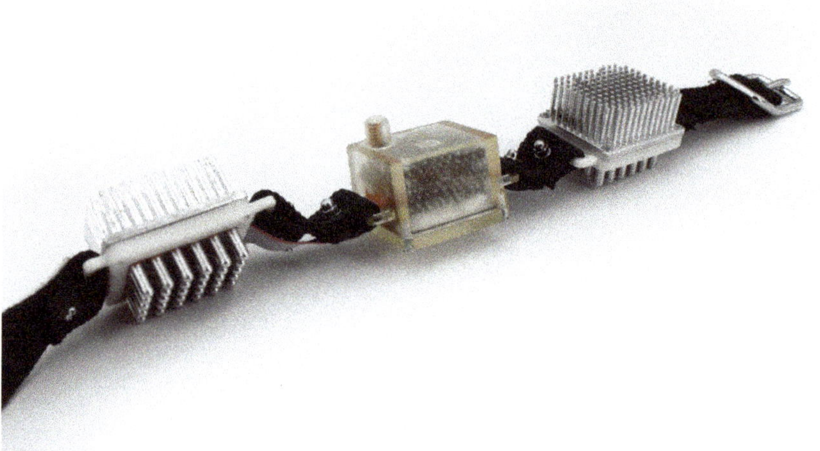

Figure 1. The tracker is implemented as a collar that can be attached to the desired animal. Two of the three housings are dedicated to thermal energy harvesting. The remaining one contains the main electronics.

3.1. Energy Harvesting

The thermal design of the harvester is already discussed in [13]. It consists of a TEG embedded between an aluminum heat connector at the animal side and a heat sink facing the environment, i.e., ambient air. Beside the TEG, two digital temperature sensors are placed on a PCB, which is embedded between the heat sink and the heat connector. The inside of the energy harvesting housing with its components is shown in Figure 2.

The heat sink is a commercially available from *Fischer Elektronik*. Its fins are cut down and modified to be screwed to the housing. The initial thermal resistance of the chosen heatsink is in the range of 0.5 K/W to 2.7 K/W, depending on the ambient wind speed [18]. We decreased the fin's height to 10 mm to achieve a less bulkier device and reduce risks of injuries. We expect a slight increase of the thermal resistance of the heat sink to a maximum of 10 K/W in a worst case condition. Compared to the heat connector's much higher thermal resistance this should be negligible.

The heat connectors' design is similar to that of the heat sink, as it ensures the closest contact of the heat connector and the animal skin. It is equipped with fins that penetrate through the fur of the animal to circumvent its high thermal resistance. Further details on the heat connector design are available in [13]. The length of the heat connectors' fins is a critical parameter to achieve a good performance, i.e., a high temperature gradient at the TEG. Too short fins can not overcome the high thermal resistance of the fur, whereas too long fins leave room between the housing and the fur where the ambient air would cool down the device and lead to heat losses. In a test with different fin lengths attached to the desired animal, here a cashmere goat, a height of 10 mm was identified as the most suitable.

The two temperature sensors for monitoring the hot and cold side temperature of the TEG are embedded into a small cutout which was milled into heat sink and heat connector (see Figure 2). In the desired temperature range between −20 °C to 50 °C, the sensors offer an accuracy of about 0.1 K [19]. Additionally they are equipped with a digital interface, which allows data transmission between the electronics and harvesting housing with a smaller number of wires in comparison to analog sensing solutions. Moreover, they are already calibrated and do not need an analog-to-digital-converter (ADC) at the main electronics board.

The housing at the animal's side and the outlets for the cables are sealed with PDMS. The heat sink is screwed to the housing against a rubber O-ring that protects the inside against water and moisture. In the housing, besides the structure holding the TEG in place, there is no additional isolation applied.

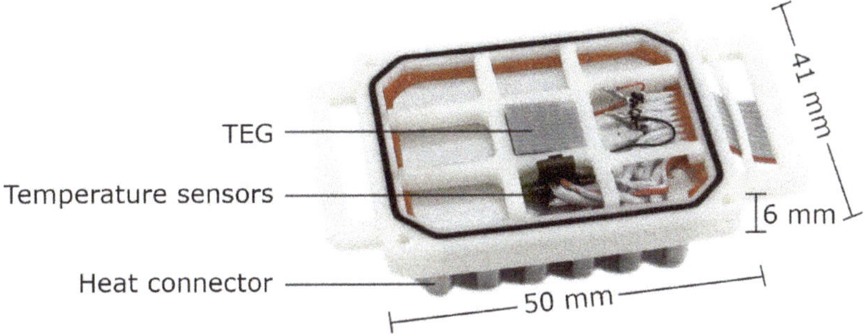

Figure 2. Housing of the harvesting part: TEG with two temperature sensors between two thermal coupling parts. The top-side heat sink is detached.

3.1.1. Output Power Considerations for TEG

To achieve a high output power of the TEG, a low thermal resistance, hence a good connection to the animal is required. Whereas this can be easily achieved by harvesting from big areas, this option is limited by the collar weight and size an animal can carry. Additionally, a careful adaptation of the TEG to the ambient thermal condition is needed. However, a high temperature gradient at the TEG and a high thermal flux resulting in high electrical output power are inherently contradictory demands. Our approach on the compromises to be taken, especially concerning the choice of a TEG, is described in the following.

In our case, we are harvesting at relatively low temperatures and temperature gradients. The temperature of the animal skin is in the range of 30 °C to 33 °C [20], while we expect the ambient temperature to be in the range of −20 °C to 30 °C. It is well known that TEGs made of bismuth telluride delivers the highest performance in this temperature region and achieve a figure-of-merit (ZT) of about 1.0 [21]. However, due to the thermal resistances of fur and ambient air, it can not be expected that the potentially high temperature differences between animal skin and ambient air can be used at the TEG. A former experiment found temperature gradients of only about 2.5 K to 3.5 K at the TEG [11]. Therefore, this situation is fundamentally different from thermoelectric harvesting devices embedded between highly conducting heat sources and heat sinks, operating at temperatures far beyond 100 °C and with much higher temperature differences ΔT. Consequently, the maximum output power and efficiency of such a device will be low due to the small temperature differences obtained: As an example, the theoretical power output limit when the thermal harvesting device is considered as an optimal Carnot-engine is at only 250 mW [22]. Even then, this output power is only achieved at very generous conditions, with a thermal harvesting device with a resistance of 10 K/W and an ambient temperature of −20 °C. For the more realistic scenario of a 3.5 K gradient, the limit drops down to 1 mW.

In the scenario when a TEG is used as a thermal harvesting device, the maximum possible output power is only a fraction of the Carnot-limit. The TEG's output power is not only dependent on the given temperature difference ΔT_{TEG}, but also on its Seebeck-coefficient α and its internal electrical resistance R_{eff}, all together described in Equation (1). For the design process of our tracker, Equation (1) states that a high α and ΔT_{TEG} are preferred over a low R_{eff}. However, if the resistance of electrical load attached to the TEG does not match R_{eff}, the possible output power given in Equation (1) will not be reached.

Another impact on output power is that the available ΔT_{TEG} at the TEG is further reduced when it is embedded between two thermal resistances and a ΔT is applied to the system. Here, ΔT reduces by a factor of K_{TEG}/K_{tot}, with a given K_{tot} as the sum of all three thermal resistances of the TEG itself and its two thermal interfaces (Equation (2)). In the scenario at the animal, we expect that our specific design of the thermal connection to the animal has a thermal resistance in the range of 35 K/W while the thermal resistance of the heat sink to the ambient air is in worst case in the range of 10 K/W, assuming calm wind [13]. The TEG's thermal resistance, however, is often below 10 K/W and further intensifies the mismatch of Equation (2), as a too-low thermal resistance of the TEG will virtually short-circuit the temperature difference across the TEG. Furthermore the surrounding thermal resistances decrease the available output power when a load is attached to the TEG and electrical power is drawn. This effect is modeled by introducing an additional term into the effective electrical resistance R_{eff} as stated in Equation (3). Both Equations (2) and (3) are put into Equation (1) to compare the achievable output powers for the following examined TEGs.

$$P_{out_{max}} = \frac{(\alpha \Delta T_{TEG})^2}{4 R_{eff}} \tag{1}$$

$$\Delta T_{TEG} = (\Delta T)\frac{K_{TEG}}{K_{tot}} \tag{2}$$

$$R_{eff} = R_0 + T_c \alpha^2 (K_c + K_h)\frac{K_{TEG}}{K_{tot}} \tag{3}$$

3.1.2. Optimal Commercially Available TEGs

For this project, we restricted our selection of TEGs exclusively to commercially available devices. Although a wide variety of TEGs exists with different thermal and electrical resistance, α_{TEG}, and size, for the purposes of this work it is sufficient to compare the TEG exclusively on a per thermocouple-leg basis. The optimal TEG with a maximum output power for a given thermal connection is then determined by choosing the corresponding number of thermocouples with a specific dimension. This is possible as, first, α is constant over all different sets of TEGs as all use the same thermoelectric materials, here bismuth telluride. Second, for the same dimension and material of a thermocouple, its thermal and electrical resistance are fixed. The overall α_{TEG} of the TEG and its thermal/electrical resistance is then determined by adding up the thermocouple parameters. It must, however, be considered that this approach is only valid as long as other effects such as additional resistances in the electrical connection of the thermocouples or parasitic thermal resistances are minor.

To compare the TEGs on the basis of their achievable maximum output power $P_{out_{max}}$ and output voltage V_{out}, the following assumptions are made: An ambient temperature of 25 °C and an animal skin temperature of 30 °C, giving a temperature difference of 5 K. The thermal resistances of the connections are fixed with K_h = 35 K/W and K_c = 10 K/W.

Here, we focus on the TEG series *1MC06-XXX_xx_TEG* of the company *TEC Microsystems*, which only differ in their thermocouple height and quantity. Other sets of this manufacturer only vary in the spacing of the thermocouples and thus have the same properties with a smaller overall size. To compare the TEGs at the thermocouple level, their parameters, including α, thermal resistance K_g, and electrical resistance R_0, are derived from the given datasheet of the whole TEG. We have verified that the calculated α per couple is the same for all variants with negligible differences. Each thermocouple's electrical and thermal resistance is, as expected, nearly constant (maximum deviation 3% and, 0.4 %, respectively) within each set of TEGs with the same height.

In total, we have compared five sets that differ in their thermocouple height from 5 mm to 16 mm. For each height variant and number of thermocouples, the corresponding output power and voltage related to our scenario are plotted in Figure 3. It can be seen that harvesting at low voltages is beneficial to get high output power. Thus, a low number of thermocouples with a small height is more promising in general. The configuration with the highest output power is a TEG with eight thermocouples and a height of 8 mm (curve ω in Figure 3), offering a promising power output of 70 µW, but has a low output voltage of less than 10 mV.

This low output voltage is a problem, as most CMOIS electronics, and also in our system, require a minimum supply voltage of about 1.8 V, which is several orders of magnitude higher. Therefore, an up-conversion of the TEG voltage is needed, which is accompanied by additional losses. Popular integrated circuits (ICs) for this task, such as the *LTC3108* from *Linear Technology* or other devices from *Texas Instruments* do not work at 10 mV or achieve a very low efficiency of only 40% for input voltages around 20 mV [23]. Therefore, it is wise to take the efficiency of the up-conversion at the corresponding voltage into account to maximize the overall usable energy. The use of commercial ICs would not allow us to harvest at very low input voltages. However, as this implies the drop of the maximum achievable power output of the TEG, for our setup a custom-built electronics for the up-conversion was used. It works with an efficiency of about 60% at 20 mV, quickly dropping for voltages lower than 15 mV. Given that, the *1MC06-048_15_TEG* was selected, as it delivers the highest output power at output voltages around 20 mV. To further increase the input power for the up-conversion, two TEGs are used in a parallel electrical connection. As all calculations are done for a worst-case scenario, the maximum power output and the voltage of the TEG will likely be higher.

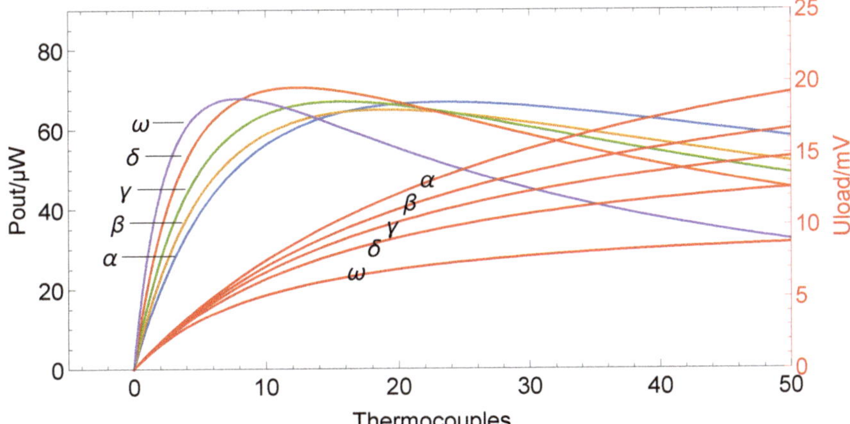

Figure 3. Power output and corresponding open-circuit voltage of TEG for a fin-length of $\alpha = 15$ mm, $\beta = 12$ mm, $\gamma = 10$ mm, $\delta = 8$ mm, $\omega = 5$ mm at a total temperature difference of 5 K and resistances of the thermal connections of $K_h = 35$ K/W, $K_c = 10$ K/W.

3.2. Power Management

The power management is based on having an intermediate storage that catches enough energy to fulfill the demanded tasks once. This is necessary due to the low output voltage and output power of the TEG that are not high enough to drive any electronics directly. For example, on the one hand, GPS and a wireless transmission module typically work at a minimum voltage of 1.8 V and a supply current of 30 mA, respectively, up to 100 mA. On the other hand, the TEGs' output power will be in the range of 24 µW to 220 µW, at presumed temperature gradients of 1 K to 3 K, here with the used TEG *1MC06-048_15_TEG*. At the same time, the TEG output voltage will be at 20 mV to 60 mV, which reduces to a range from 10 mV to 30 mV if an optimal load at maximum output power is connected.

Therefore, the power management in our system consists of a voltage up-conversion, a capacitor as intermediate storage, and a second voltage stabilization stage. Figure 4 shows the entire chain up to the power-consuming part: First, the output voltage of the TEG (10 mV to 30 mV) is up-converted and it charges a capacitor up to 5.5 V. A cold-start circuit monitors the capacitor voltage and turns on the rest of the system as soon as the voltage reaches a well-defined threshold. In the next step, the capacitor voltage is stepped down to a constant and regulated voltage for the main system. At this point, the main system takes over the power management, and only in the event of a very low energy state, the cold-start circuit will cut off the main system from the capacitor.

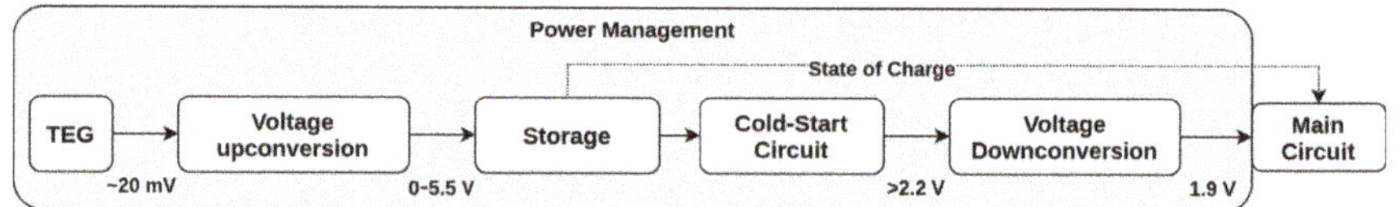

Figure 4. Concept of the power management circuit: The output voltage of the TEG is first upconverted and stored in a capacitor. A cold start circuit disconnects the rest of the system if the voltage falls below a threshold. Otherwise, the capacitor voltage is stabilized to the constant output voltage needed to drive the attached electronics.

For the up-conversion step, a custom-designed low-voltage boost converter is used based on the design explained in [24]. It provides an unregulated output voltage, a cold start capability down to 10 mV, requires no auxiliary power and has an efficiency above 40% over a wide range of the expected input voltage, with a maximum efficiency of approximately 60% at an input voltage of 20 mV. Load matching is not implemented in the boost-converter. As the capacitor is directly attached the output of the boost converter, the overall power input is heavily dependent on the capacitor's state-of-charge. The corresponding usable powers are shown in Figure 5: At a low capacitor voltage, the system can only retrieve low power from the boost converter, whereas at the capacitor voltage of around 2.8 V, the step-up converter delivers its maximum power. Therefore, the start-up phase from zero power for a system with an empty capacitor, needs a long time. Afterwards, the main electronics will monitor the capacitor voltage and schedule its tasks in such a way that the storage capacitor is most of the time in the optimal voltage range at 2.8 V. Although the optimal power point varies with the TEG's current temperature difference, this is not monitored. It turned out that in the expected temperature range, the maximum in Figure 5 does not vary too much.

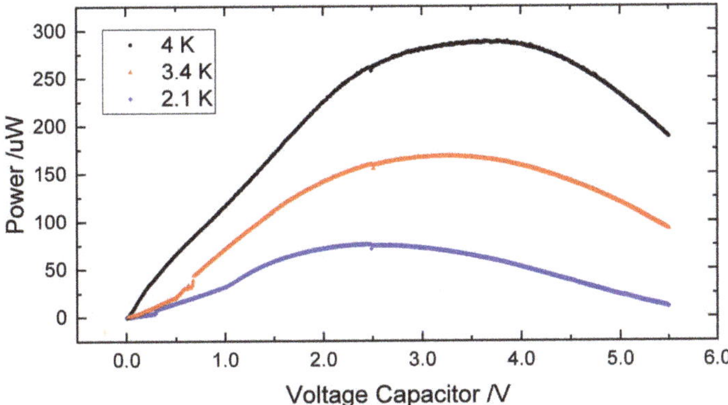

Figure 5. Measured charging power for different temperature differences directly applied to the two connected TEGs in the lab. Only the power-management circuitry (see Figure 4) is used. At 2.5 V, a slight power dip can be seen at which the step-down converter is turned on.

The energy storage capacity is chosen in such a way that it provides enough energy to complete the most power-hungry task, here the GPS, once. Long-term storage, using, e.g., a rechargeable battery, could be added; however, it is not foreseen in this concept, as we expect a continuous and constant energy income from the TEGs. In our case, the GPS module *CAM-M8 (u-blox)* needs around 26 mA in acquisition mode and can obtain a position within 7 s when the aided start functionality is used [25]. At a supply voltage of 1.8 V, the energy storage needs, therefore, to provide at least 328 mJ. However, since it is not guaranteed that a fix is always possible within 7 s a minimum stored energy of 500 mJ is desired. This is achievable by choosing the *BZ015B603Z_B (AVX)* as a storage capacitor that provides a capacitance of 60 mF. When charged up to its limit of 5.5 V and discharged down to 2.2 V, an energy budget of 760 mJ is available, which enables the GPS module to run approximately for 16 s.

The capacitor's leakage is an important parameter as it significantly impacts the overall power consumption of the system when in sleep mode for extended periods. The datasheet claims a maximum leakage current of 10 µA and a low equivalent series resistance (ESR) of 80 mΩ [26], suitable for the purposes of this work. A low ESR is further needed to prevent high voltage drops when a current up to 100 mA is needed during wireless data transmission. We experimentally verified that the leakage current is even lower with 5 µA near its maximum rated voltage at 5.5 V and around 1.6 µA at voltages below 3 V.

A constant regulated voltage supply for the microcontroller and periphery is realized by the *TPS62740 (Texas Instruments)* step-down converter connected to the capacitor (see Figure 4). It is suitable for handling input voltages up to 5.5 V, can provide a constant 1.9 V output at a reasonably high efficiency, and has a typical quiescent current of only 360 nA [27], which is advantageous to reduce the overall power consumption when the rest of the system is in sleep. Additional circuitry based on [28] is connected to the enable pin of the *TPS62740* as a voltage monitor and power switch and turns off the whole system when the capacitor voltage is below a certain threshold. This helps during the cold start phase when the energy storage is empty and prevents possible excessive current draw due to undefined behavior of the periphery in low voltage conditions. Detrimental effects of such a behavior are demonstrated in [29].

3.3. Main System and Periphery

The developed system can measure five physical properties, needed either for the wildlife-tracking purpose or to log the system performance: Time, voltage of the energy storage capacitor, activity, position of the tracker as well as the temperature difference over the TEGs in their separate housings. The gathered data is stored on the device itself, but eventually also transmitted wirelessly.

The overview in Figure 6 shows that the microcontroller unit (MCU) *MSP430FR5969 (Texas Instruments)* does centrally control all other components. It features an inbuilt real-time clock (RTC) that tracks time and wakes up the MCU periodically at a current draw of only 250 nA [30]. While the MCU is always connected to the power supply as long as the capacitor's voltage (see Figure 4) is above the turn-on threshold voltage, all other periphery can be individually disconnected from the power supply by the MCU with a low leakage switch [31]. This reduces the system's sleep current to an absolute minimum.

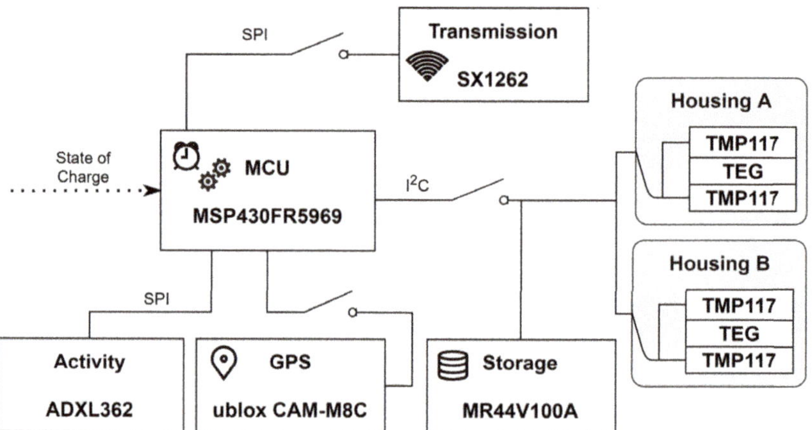

Figure 6. Part of the electrical system responsible for the data acquisition. The MCU controls via digital protocols all periphery and may completely cut each component from the power supply with an additional switch.

The temperature difference at the TEGs in each of the two housings is measured by two digital temperature sensors, *TMP117 (Texas Instruments)*, the accuracy of which is claimed to be 0.1 °C [19]. The four sensors are connected to the MCU via the same Inter-Integrated Circuit (I^2C)-bus. This avoids the use of additional external components and reduces the number of wires needed to connect the harvesting and the electronics housing.

Besides the built-in Ferroelectric Random Access Memory (FRAM) of the MCU, an additional one with 1 Mbit capacity (*MR44V064B (Lapis Semiconductor)*) is connected to the I^2C-bus [32]. It is used to store the aiding data for the GPS module and to save all measurement data. For the aiding data, 128 kB are reserved. The rest is used to store measurement data, each taking 30 B. For the experiment, a measurement was taken every 7 min, hence the FRAM's space lasts for about 18 days.

The *CAM-M8 (u-blox)*, the chosen GPS-receiver, has a built-in antenna and is placed at the very top in the electronics housing. At each activation of the module, the current time is checked by the MCU and the corresponding aiding-data from the external FRAM is transferred to the *CAM-M8* to speed-up the position calculation. The module's sleep mode is not used. Instead, it is completely disconnected from the energy storage to save the about 100 µA the module otherwise would draw.

Activity is detected via an *ADXL362 (Analog Devices)* accelerometer configured so that it wakes up the MCU when a threshold of motion is detected. A counter is then increased by one and is later reset after its content is saved in the FRAM with the other data.

The voltage of the storage capacitor is measured with the internal ADC of the MCU via an on/off switchable voltage-divider. It is first captured at the beginning of any wake-up-event and a second time right before the MCU goes back to the sleep state. Both values are later saved in the set of measurement records.

Data is transmitted at 868 MHz using the *SX1262 (Semtech)* chip embedded in a *Lambda62 (RF Solutions)* module. LongRange (LoRa) was selected for the physical layer and LoRaWAN to send out a package with a reduced set of data. A transmitted package has a payload of 11 B, respectively, 19 B when a GPS position is found. It includes a coarse voltage, the four temperature readings at the harvester-housings, activity, and time difference since the last wake-up. To this data, the LoRaWAN layer adds 9 B on additional information such as device address and a frame counter. As we are using a well documented and widely used technology that is not only compatible with self-deployed stations, already existing networks can receive and forward the data. With that, we connected our tracker to the world-spanning network TheThingsNetwork (TTN).

The main system with its periphery, excluding the temperature sensors in the TEG-housing, and power management is distributed over four custom-made PCBs as seen in Figure 7, stacked on top of each other in the electronics housing.

Figure 7. PCBs that are later stacked on top in the electronics housing. From left to right: Power management board with cold-start switch, step-down and step-up converter, but here without soldered transformer and capacitor, board for wireless transmission, board with activity sensor, MCU and FRAM, board with GPS-receiver.

3.4. Task Scheduler Algorithm

The scheduling of tasks in the main system has a critical impact on the overall available input power. As the TEG-step-up combination will deliver its maximum output power at a capacitor voltage around 2.3 V (see Figure 5) and as no other load-matching is integrated, it is the MCU's responsibility to track and adapt the load for optimal power. This is achievable by scheduling tasks according to their energy consumption and the current state of charge. High power-consuming tasks are only triggered when the capacitor voltage rises above the respective thresholds. Low-consuming tasks take place at a low state of charge and are done more often to the extend that energy is still saved for a later execution of more energy-hungry tasks. For example, a simple measurement recording happens more often than the dataset's transmission or a GPS-fix. It is worth mentioning that the system works best as long as there is always enough energy to continuously drive the main MCU with its internal RTC, allowing the collar to track and record time continuously. If the input power is too low and the capacitor voltage drops below 2.3 V, the main system will be shut off hard and will only be reconnected as soon as the voltage is in an acceptable range again. To stay on a point of maximum power and prevent a low-voltage condition, the algorithm chooses between the combination of three defined task sets: (1) the temperature readings of the hot side and cold side in the harvesting housing together with the animal activity, (2) the position calculation, and (3) the transmission of data. For the tasks, five energy levels are defined that correspond to specified capacitor voltages. In addition, a fixed time interval is given for most jobs. A task will be executed if both, the voltage level and the time interval match the settings defined for this task. If the time does not match, the task of the energy level below the current level will be executed. If the voltage level does not match, the task will be skipped and it will be triggered at its next interval. The intervals of the tasks are chosen so that its mean power consumption should be below the energy income. The tasks at levels four and five, acquiring the GPS position, are an exception. In those states, the GPS-receiver uses all available energy until it receives a position or until the capacitor voltage drops below 2.8 V. However, since the voltage in this state is measured once a second, it can also drop down to 2.4 V between two voltage readings. This means that the risk exists that no GPS signal is found and that at least 545 mJ are consumed. Table 1 summarizes the available tasks and the corresponding energy level.

Most of the time, the whole system is put into sleep with all components turned off and wakes up only either because of a RTC timeout or when an activity event is detected. In the latter case, the system only increments the activity value that is later sent together with the rest of the data before it is reset. The flowchart in Figure 8 shows the entire process for all events. To lower the energy overhead for the algorithm, the MCU wakes up only at a very low interval of seven minutes. This reduces energy needed for each wake-up and for capacitor voltage readings, but also restricts the time interval when tasks can be executed to this schedule. For our use case, the seven minute time grid is granular enough to prevent the missing of a capacitor overcharge. It also is relaxed enough to give enough time for charging: At an input power of around 50 µW, 21 mJ will be delivered to the capacitor per round. This is considerably higher than the energy consumption of the lower-level tasks that are done with the highest possible frequency, allowing us to save energy for the more energy-hungry tasks at higher voltage levels.

Table 1. The algorithm executes tasks according to a fixed time interval, but only when the capacitor's voltage is in the desired range or above. Otherwise, it is skipped.

Lvl	Voltage Range	Time Interval	Consumption	Task
0	0 V to 2.3 V	-	7 µW	Whole system is shut down by cold-start circuit
1	2.3 V to 2.5 V	-	8 µW	Only MCU and RTC is running, timetracking
2	2.5 V to 3 V	7 min	0.4 mJ	Measurements are done and stored internally
3	3 V to 5.1 V	14 min	8.3 mJ	Measurements are done and transmitted
4	5.1 V to 5.3 V	28 min	~500 mJ	Measurements are done, a GPS fix obtained and transmitted
5	>5.3 V	-	~500 mJ	Same as level 4

To conclude, the algorithm will take care that the voltage of the storage capacitor stays above 2.3 V as long as the power-income is at least 8 µW. Only the lower-level tasks are executed at low power incomes (between 8 µW to 9 µW), and all higher ones are skipped. The voltage of the capacitor will then oscillate between 2.3 V and 3 V. At higher power income, the capacitor will reach voltages up to 5.1 V. If this voltage is exceeded, the capacitor will be emptied down to 2.4 V by the high energy-consuming GPS-fix. Considering the efficiency of the energy-harvesting system, it follows that at a low power income situation, the system will be held in a state with relatively good efficiency according to Figure 5. The overall power could be improved by staying always in the high efficiency point, which was not possible in our case as the capacitor needed to be fully charged to provide enough energy for a GPS-fix. However, various concepts are conceivable to improve the situation, and will be followed in future studies.

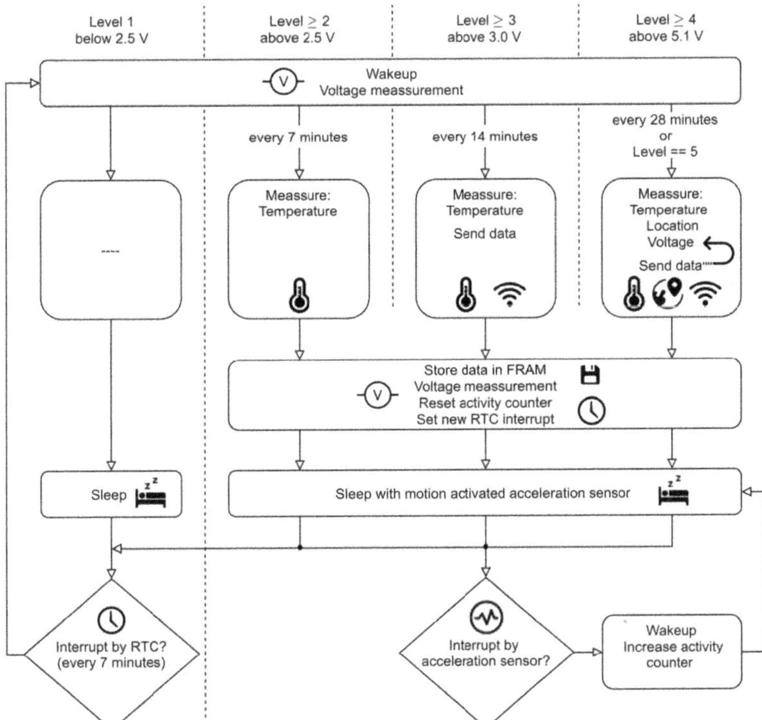

Figure 8. Flowchart of the algorithm. Most of the time, the tracker is put into sleep and only wakes up after a period of seven minutes, checks the voltage, and fulfills the task according to a defined time and voltage level.

4. Field Test and Validation

In a field test, three tracking collars were applied to cashmere goats. Those were able to graze freely on an area of about 15,000 m^2 at the zoo *Mundenhof* in Freiburg, Germany. The test took place from 26 May to 12 July 2020, with a duration of 17 days.

Two of the three collars were fully equipped as described in Section 3. The third collar, used as the reference, was equipped with a battery instead of the capacitor. In addition, the output voltage of the two TEGs was separately captured via an operational amplifier attached to the MCU's ADC instead of connecting them to the input of the power management circuit. In a small shelter for the animals, a LoRaWAN-receiver was placed that forwarded all transmitted packets of the collars to a central server via the online service TTN. This allowed us to check the current status of the collars in real-time and to demonstrate the successful functioning of the wireless transmission. The full data set for the analysis was recovered from the internal FRAM of all collars at the end of the field test.

The following data was gathered during the field test: The temperature at the hot/cold side of both TEG housings, the capacitor voltage at the very beginning when the MCU was woken up, and right before the system went again down to sleep, the time of the measurement points, the activity of the animal since the last measurement, status information about the GPS module and eventually the animal position. The ambient temperature, wind speed, and other environmental parameters were taken from the *meteorological city station Freiburg*, located around 5 km away from the animals shelter.

The average ambient temperature during the experiment was about 16.6 °C, with a maximum of 28 °C on 2 June 2020 and a minimum of 10.2 °C on 31 May 2020. On 6 June

2020, rain started with at least 1 mm/d until 10 June 2020, with a peak of 19 mm/d on 5 June 2020.

4.1. Results

Not all collars worked over the whole period. One collar was already heavily damaged after two days and was therefore removed. The second collar lost connection to one of the two energy-harvesting housings after five days and stopped collecting data after seven days. We also removed this collar the next day, as this was due to the same issue: The connection between belt and housing broke, and so the collar was held only by the electrical wiring between the housings (see Figure 9b). The reference collar, which was battery-driven, lasted fifteen days until one harvesting housing got loose in the same way. Although contact to the goat was from then on degraded, it was still working until the end, held only by the electrical wiring. Apart from the mechanical issues, all collars could harvest enough energy to constantly work without any interruption of the system, especially the RTC. During the 166 h run-time of the first tracker, 110 attempts for a GPS fix were made. The second tracker attempted 48 times for a fix during its total run-time of 54 h. In both cases, enough energy was harvested for a GPS-fix each 1.1 h to 1.5 h. Each GPS-fix used around 620 mJ. The scheduler started a major part of the GPS attempts because the capacitor voltage level reached the maximum allowed value (see Table 1). This happened mostly during night—at daytime, a high amount of scheduled fixes were omitted as the voltage was too low.

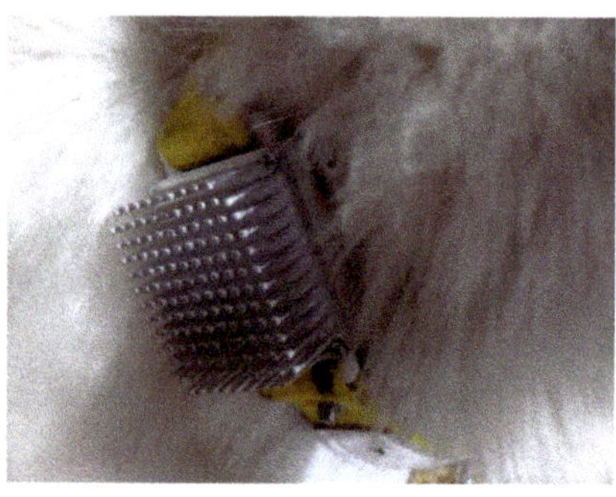

(a)   (b)

Figure 9. Side-by-side: Collar attached to the animal's fur in winter and late spring. As the goat lost its fur, the thermal connection is not optimal any more in the field test in spring. (**a**) Prototype of the thermal housing to determine the right fin length for an optimal penetration into the goat's fur—done in winter. (**b**) Goat with the tracker at the end of field-test in late spring. One housing is detached from the collar.

About 1429, respectively, 443 data points, were recorded in the FRAM by the energy-autonomous collars. The reference collar saved the maximum amount of 3252 datapoints and therefore started to overwrite the oldest data. Every collar was able to keep the intended seven minute record interval, thus having a high temporal resolution for the temperature and activity readings.

To compare the goats' activities, the number of detected activities are binned to each hour of the day. Figure 10 shows the median of each activity bin over all recorded days and goats. The number of days included in the figure is different for each goat. The median activity of goat two only includes two days, while for goat one and three, seven, respectively, fifteen days are taken into account. It can be concluded from the activity data that the goats are mainly inactive between 23 and 5 o'clock. A lower measured activity in the time from 8 to 9 o'clock correlates to the feeding time of the goats, demonstrating that the daily routine of the goats can be revealed by the activity sensor.

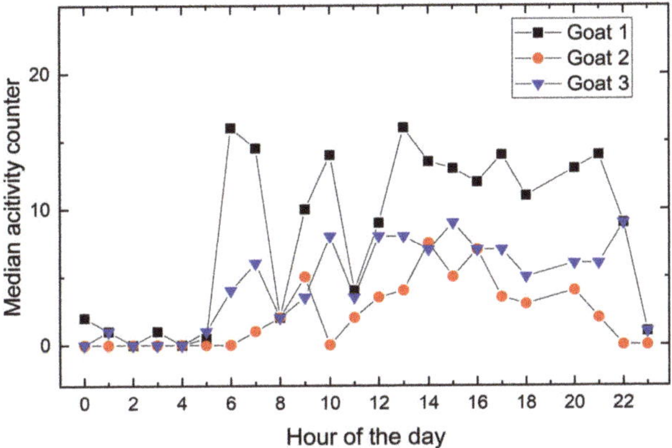

Figure 10. Median of activity events measured for each hour and goat during the trackers' lifetime. The daily routine such as sleeping and resting can be obtained.

4.1.1. Temperature Distribution

The average temperature-difference at the TEGs of each harvesting housing is (2.5 ± 1.0) K with maxima between -0.5 K to 6 K. Their distribution for each of the six harvesting housings is shown in Figure 11. Here, it can be seen that the temperature differences follow a normal distribution without a lot of extreme temperatures.

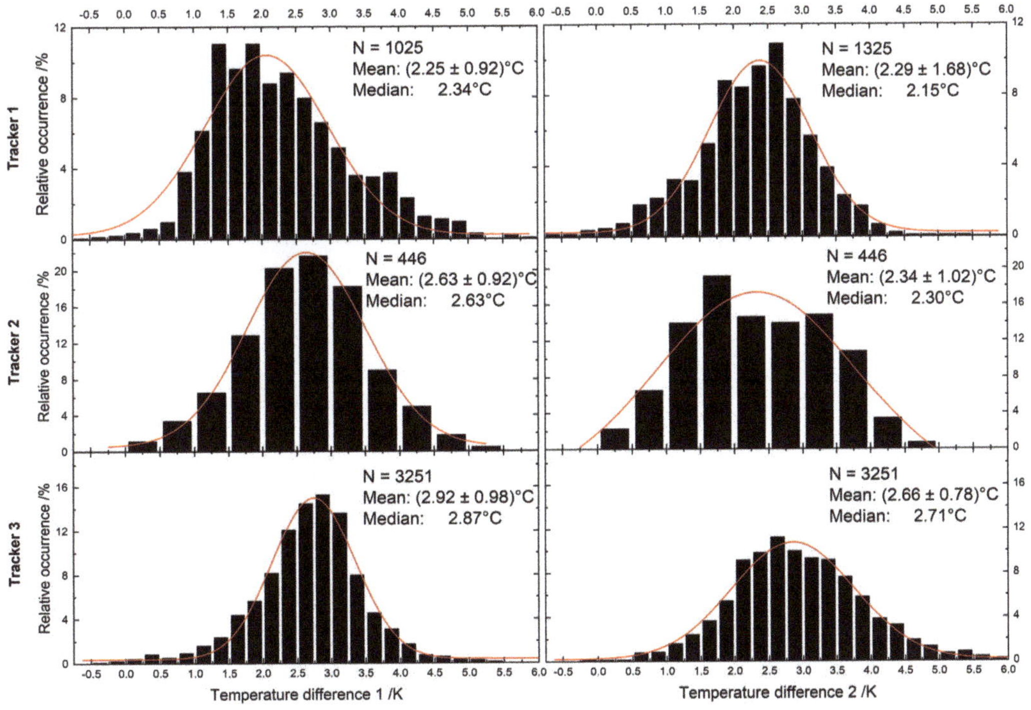

Figure 11. The distribution of temperature difference at the TEG in each of the trackers harvesting housings. Given: the total number of measurements for each tracker (N), mean and median for each housing, and a calculated normal distribution.

The temperature difference at the TEG (ΔT_{TEG}) against the ambient temperature (T_{amb}) is shown for the first goat in Figure 12a. Despite the large variety of ΔT_{TEG}, the trend reveals that lower ambient temperatures lead to higher ΔT_{TEG} and vice versa. To check for plausibility, a linear regression on T_{amb} and ΔT_{TEG} is done to derive the point of the x-axis intercept, at which ambient temperature equals the animal temperature. The linear relationship can be modeled by Equation (4) as long as constant thermal resistances are assumed and the ambient temperature deviation is small. In this case, the equation describes the thermal path from the animal's inner body as a heat source through the thermal resistances (K_c, K_h, K_{TEG}) to the environment according to [13].

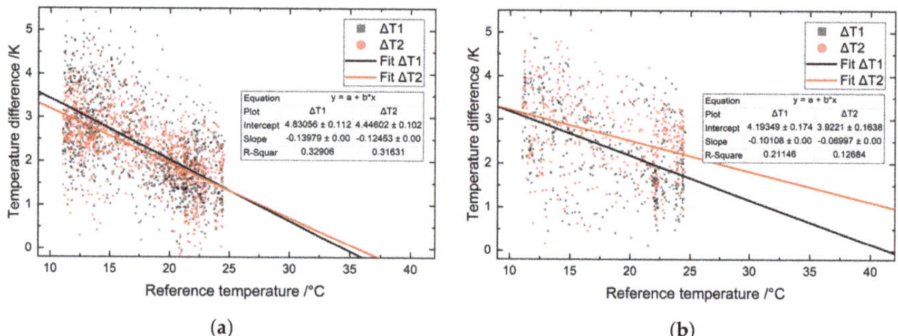

(a)	(b)

Figure 12. Temperature difference at the TEG versus ambient temperature measured by a weather station. A linear fit is shown to extrapolate the zero-crossing with the x-axis. At this point, the temperature difference at the TEG becomes zero and no thermal harvesting is possible. (**a**) First goat with both housings having a similar thermal connection. (**b**) Second goat with less data points.

The results should be close to 39 °C, which is the core body temperature for goats [33]. Indeed, the results are in the same range with 39.8 °C/37.2 °C for goat 3, and 34.6 °C/35.7 °C for goat 1. The second goat had the least number of measured points recorded, and its collar got loose in an early stage of the experiment. This might be the reason why the calculated body temperature of the animal results in 41.5 °C/56.1 °C. The large variety of obtained results was expected, as especially the thermal resistance of the heatsink varies due to wind speed or other environmental factors such as rain or sun. Above these temperatures, the ambient temperature is higher than that of the animal. The gradient and thus output voltage of the TEG would reverse and cannot be used by our device.

$$\Delta T_{TEG} = \frac{K_{TEG}}{K_c + K_w + K_{TEG}} T_{amb} + \frac{K_{TEG}}{K_c + K_w + K_{TEG}} T_{animal} \tag{4}$$

4.1.2. Charging Power

The usable input power of the system was always greater than the power consumption; thus, measurements were recorded at an interval of 7 min, as desired. To calculate the mean power income, the difference between the capacitor voltage right before the MCU goes to sleep and at the very beginning of the MCU's next wake-up is taken. Together with the known sleep-time and capacity, the power income is calculated. In the same manner, energy drawn by the system in its on-state is derived.

For the mean input power calculation, measurements right after a GPS fix are excluded as the capacitor shows a significant recovery effect: A test in the laboratory revealed that after a high and long current draw, here about 30 mA for 20 s when GPS is used, the voltage of the capacitor slowly recovers within several minutes. As the input power is derived from the voltage right after the high load and again after seven minutes, these points show unrealistic high charging powers above 300 µW. In addition, six measurement points of the first tracker and four of the second tracker were excluded as those were showing excessive power drain. We suspect that at those points, returning the full system into deep sleep, did not fully work. In total for goat 1 and 2, respectively, 1268 and 390 data points are used to calculate the mean charging power stated in Table 2.

Table 2. Mean charging power of the capacitor, including losses for the up-conversion, storage, and the attached system.

	Goat 1	**Goat 2**
Mean power income	$(103 \pm 91)\,\mu W$	$(136 \pm 100)\,\mu W$
Mean power income Night	$(157 \pm 96)\,\mu W$	$(217 \pm 96)\,\mu W$
Mean power income Day	$(65 \pm 68)\,\mu W$	$(78 \pm 80)\,\mu W$

There is a daily period for the power income: It is higher at night than at daytime. Two exemplary days, with their charging power, capacitor voltage and energy consumption are presented in Figure 13. The plot in the figure for the charging power is not filtered, and therefore the erroneous peaks right after a GPS are included. Each GPS event is clearly noticeable by peaks in all three graphs. As the MCU schedules GPS fixes according to the time and voltage of the capacitor (as stated in Table 1), more positions are obtained at high input power phases. Here, the capacitor is emptied to the absolute minimum. However, the maximum voltage is likely reached in a short period where GPS is turned on again.

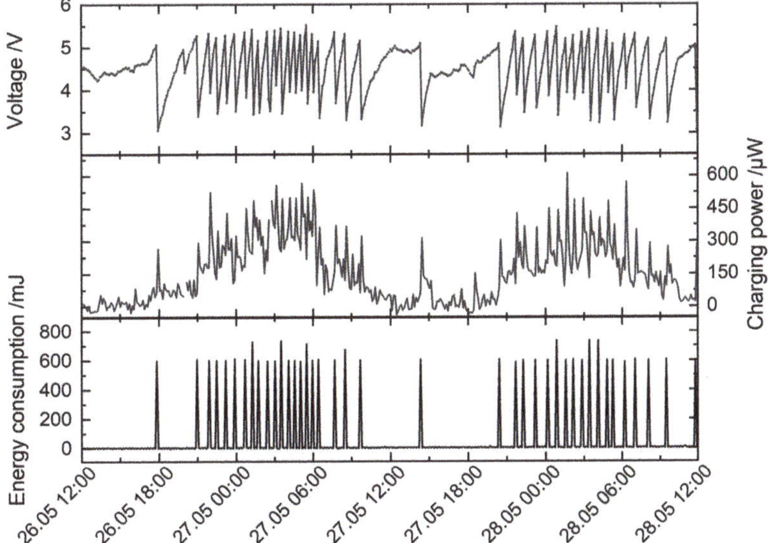

Figure 13. Voltage, charging power, and power consumption of tracker 1 for the first two days.

Charging power is directly dependent on the ΔT_{TEG} and thus on ambient temperature. We found that the system has a positive power income, including the power and conversion losses, down to a ΔT_{TEG} of (1.0 ± 0.3) K at the first goat and (0.8 ± 0.8) K at the second goat. It follows that the system installed on goat 1 and goat 2 can run on a maximum ambient temperature of (27.6 ± 1.5) °C and (38 ± 10) °C, respectively. These points are calculated via a 2nd-order polynomial fit on the recorded charging power and temperature at the TEGs. Again, all data points after a GPS fix are excluded. The measurements and their corresponding fit are shown in Figure 14 for the first goat. These values should be used with caution: The erratic influence of environmental influences such as wind-speed or local deviations from the recorded ambient temperature lead to non-constant thermal resistances, which was assumed for the calculation. Another impact is the change of thermal connection to the animal that happened, especially for the second collar that got loose during the field test. It is worth to mention that the charging power is also dependent on the state-of-charge of the capacitor at the time when the measurement was taken.

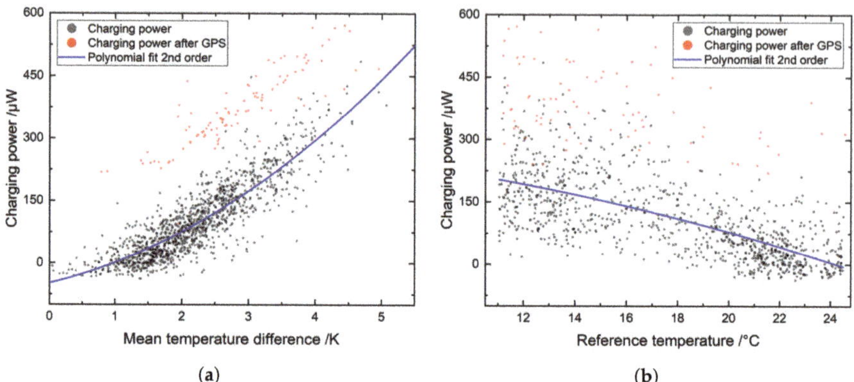

Figure 14. Charging power versus ambient temperature respectively the corresponding ΔT at the TEG from the tracker of goat 1. The marked data are excluded as these are outliers recorded after using high current periphery such as GPS. (**a**) Charging power versus ΔT_{TEG}. (**b**) Charging power versus the ambient temperature.

5. Discussion

The field test of our designed animal tracker proves that thermal harvesting systems can be used for tracking devices at warm-blooded animals. An average of $100\,\mu W$ is suitable for systems that spend most of their time in sleep. With a GPS-fix every 1.1 h to 1.5 h, our system fulfills the requirements of typical tracking devices, in which the position calculation is executed once per hour or even less. In our field experiment, nearly none of the GPS fixes were successful, this, however, is due to the poor antenna design and position: First, it turned out that a high damping of the satellite signals is introduced by the narrow housing, 3D-printed from photopolymer, around the antenna. A further issue was the size of the electronics ground-plane that did not meet the GPS-module requirements according to the datasheet. The same issues also degraded the range of the used LoRa-modules to less than one kilometer, although an output power of 15 dBm was set, which generally would have allowed remarkably higher ranges. The mentioned issues could be solved in a system that takes the antennas parameters into a better account and uses, for example, the animal's body as a ground reference.

Our setup uses a harvesting area at the animal of $2 \times 20.5\,\text{cm}^2$ and a total weight of 286 g. It has been proven that if loads carried by the animal are below 5 % of its weight, they have minimal impact on the animal's health and behavior; however, their influence may not be negligible [2]. From that, it follows that our system is already suitable for smaller warm-blooded animals such as the lynx. Higher input power can easily be achieved for larger and heavier animals with larger harvesting areas but higher system weight. The constant power income and robustness of a thermal-harvesting tracker are of great advantage for long-term studies. However, the capacity of today's batteries is already in a range that can achieve long run-time when used with our low-power system. For example, the *CRV3 (Panasonic)* with 3000 mA h at 3 V could deliver similar amounts of power over a period of 10 years, weighing only 39 g. Still, leakage and degradation over that long time are concerns for battery-driven systems. Nevertheless, using both approaches in conjunction would allow us to double the system functionality and would deliver a fall-back operation for the failure of one power supply.

Outstanding features of the tracker are both a high amount of usable power harvested at low temperature-gradients of only 2 K and a very low power but full-featured electronic system. The input power achieved with this design is doubled compared to our previous design in [12]. Substantial improvements are: First, a careful thermal adaption to the animal's fur as already discussed in [13]. Second, harvesting at very low TEG output voltages down to 15 mV, which we managed to up-convert with low losses. This increases the available energy as the TEGs work closer to their maximum output-power point. Third, the indirect load-matching via the task-scheduler increased the up-conversion efficiency by selecting the tasks in an order for the system to stay close to the optimal capacitor voltage. A narrower control with regard to the best voltage could further increase the usable power. Most of the time, the capacitor voltage was between 4 V to 5 V, while the optimal one is around 3 V (see Figure 5). This is a design choice, as the charging of the capacitor to higher voltages is necessary to provide enough energy for a GPS-fix. Larger capacitances or the usage of batteries would allow to stay near the optimal voltage of the intermediate storage. However, most alternatives have higher losses and leakage, which is an important factor, as the system is in deep sleep most of the time. The main electronics all together use less than 1 µA in that state, and thus our already low leakage capacitor (60 mF *BZ015B603ZLB (Bestcap)*) has the most significant impact on the overall power loss, with a leakage current of 10 µA. It is worth mentioning that sleep currents could be further lowered by a step-down converter that has even better efficiency than the one we used (*TPS26740 (Texas Instruments)*).

6. Summary

We have designed, built and tested a collar to track position, activity and temperature of a goat. It is the first collar that is equipped with power-hungry modules such as a GPS or a LoRa-transmitter and solely supplies itself via thermal energy-harvesting from the animal.

In a field test with cashmere goats, a high mean output power of around 100 µW is achieved with two TEGs that operate on the low temperature gradient between animal and environment. The usable mean power at night was higher with up to 211 µW compared to 85 µW during the day. Directly at the TEG, we found temperature differences ΔT_{TEG} of (2.5 ± 1.0) K. According to our calculations, our system can work down to a ΔT_{TEG} of (0.90 ± 0.33) K that corresponds to an ambient temperature of (27.6 ± 1.5) °C. This is achieved with a harvesting area of 2 times 20.5 cm^2 at a total system weight of 286 g.

The high input power is achieved by: First, a careful thermal adaption and the choice of TEGs with low output voltages but low internal electrical resistance. Second, a highly efficient up-conversion of TEG-output voltages down to 15 mV. Third, a power management that holds the energy storage close to the voltage with the highest input power. Fourth, very low energy consumption in deep sleep for the entire system, including losses, of 8 µW due to extra circuitry that disconnects most of the components from the energy storage. Most of the remaining sleep current is due to leakage of the storage capacitor.

With the harvested energy, the collar can try a position-fix every 1.1 h to 1.5 h, each fix using approximately 621 mJ. The wireless transmission of data happens every 14 min at an output power of 15 dBm. With that, our developed collar already fulfills the usual tasks of commercially available trackers. There are chances that such systems might become a robust and long-living alternative to batteries in today's trackers and enable long-term studies at the animal without needing to recapture them just for changing batteries.

Author Contributions: Conceptualization, E.B. and P.W.; methodology, E.B.; software, E.B.; validation, E.B., L.C. and L.M.C.; formal analysis, E.B. and L.C.; investigation, L.C. and E.B.; resources, P.W.; data curation, E.B. and L.C.; writing—original draft preparation, E.B.; writing—review and editing, L.M.C. and P.W.; visualization, E.B. and L.C.; supervision, E.B. and P.W.; project administration, P.W.; funding acquisition, E.B. All authors have read and agreed to the published version of the manuscript.

Funding: The article processing charge was funded by the Baden-Wuerttemberg Ministry of Science, Research and Art and the University of Freiburg in the funding program Open Access Publishing.

Institutional Review Board Statement: Ethical review and approval were waived for this study as animals chosen for this study are used to wear collars and regular contact to humans. No additional measures were taken compared to commonly used collars.

Informed Consent Statement: Not applicable.

Data Availability Statement: The data presented in this study are openly available in Zenodo at https://doi.org/10.5281/zenodo.5547295 (accessed on 24 September 2021).

Acknowledgments: The authors gracefully thank the Mundenhof Freiburg for their cooperation and their valuable contribution. They supported the experiment with their expertise in handling and taking care of the animals. We also thank the animal welfare officer of the Uniklinik Freiburg for her council and supervision of the experiment.

Conflicts of Interest: The authors declare no conflict of interest.

References

1. White, P.C.L.; Taylor, A.C.; Boutin, S.; Myers, C.; Krebs, C.J.; White, P.C.L.; Taylor, A.C.; Boutin, S.; Myers, C.; Krebs, C.J. Wildlife Research in a changing world. *Wildl. Res.* **2009**, *36*, 275–278. [CrossRef]
2. Brooks, C.; Bonyongo, C.; Harris, S. Effects of Global Positioning System Collar Weight on Zebra Behavior and Location Error. *J. Wildl. Manag.* **2008**, *72*, 527–534. Available online: https://wildlife.onlinelibrary.wiley.com/doi/pdf/10.2193/2007-061 (accessed on 28 May 2021). [CrossRef]
3. Adu-Manu, K.S.; Adam, N.; Tapparello, C.; Ayatollahi, H.; Heinzelman, W. Energy-Harvesting Wireless Sensor Networks (EH-WSNs): A Review. *Acm Trans. Sen. Netw.* **2018**, *14*, 1–50. [CrossRef]
4. Wijesundara, M.; Tapparello, C.; Gamage, A.; Gokulan, Y.; Gittelson, L.; Howard, T.; Heinzelman, W. Design of a Kinetic Energy Harvester for Elephant Mounted Wireless Sensor Nodes of JumboNet. In Proceedings of the 2016 IEEE Global Communications Conference (GLOBECOM), Washington, DC, USA, 4–8 December 2016; pp. 1–7. [CrossRef]
5. Dagdeviren, C.; Li, Z.; Wang, Z.L. Energy harvesting from the animal/human body for self-powered electronics. *Annu. Rev. Biomed. Eng.* **2017**, *19*, 85–108. [CrossRef] [PubMed]
6. Luo, Y.; Pu, L.; Zhao, Y. RF Energy Harvesting Sensor Networks for Healthcare of Animals: Opportunities and Challenges. *arXiv* **2018**, arXiv:abs/1803.00106.
7. *LifeTag*™; Cellular Tracking Technologies: Rio Grande, NJ, USA, 2019.
8. Loreti, P.; Catini, A.; De Luca, M.; Bracciale, L.; Gentile, G.; Di Natale, C. The Design of an Energy Harvesting Wireless Sensor Node for Tracking Pink Iguanas. *Sensors* **2019**, *19*, 985. [CrossRef] [PubMed]
9. Jain, V.R.; Bagree, R.; Kumar, A.; Ranjan, P. wildCENSE: GPS based animal tracking system. In Proceedings of the 2008 International Conference on Intelligent Sensors, Sensor Networks and Information Processing, Sydney, Australia, 15–18 December 2008; pp. 617–622. [CrossRef]
10. Knight, C.; Davidson, J.; Behrens, S. Energy Options for Wireless Sensor Nodes. *Sensors* **2008**, *8*, 8037–8066. [CrossRef] [PubMed]
11. Woias, P.; Schule, F.; Bäumke, E.; Mehne, P.; Kroener, M. Thermal Energy Harvesting from Wildlife. *J. Phys. Conf. Ser.* **2014**, *557*, 012084. [CrossRef]
12. Bäumker, E.; Schüle, F.; Woias, P. Development of a batteryless VHF-Beacon and tracker for mammals. *J. Phys. Conf. Ser.* **2018**, *1052*, 012005. [CrossRef]
13. Bäumker, E.; Beck, P.; Woias, P. Thermoelectric Harvesting Using Warm-Blooded Animals in Wildlife Tracking Applications. *Energies* **2020**, *13*, 2769. [CrossRef]
14. Terrasson, G.; Llaria, A.; Marra, A.; Voaden, S. Accelerometer based solution for precision livestock farming: Geolocation enhancement and animal activity identification. *Iop Conf. Ser. Mater. Sci. Eng.* **2016**, *138*, 012004. [CrossRef]
15. Latham, A.D.M.; Latham, M.C.; Anderson, D.P.; Cruz, J.; Herries, D.; Hebblewhite, M. The GPS craze: Six questions to address before deciding to deploy GPS technology on wildlife. *N. Z. J. Ecol.* **2015**, *39*, 143–152.
16. Harrity, E.J.; Conway, C.J. Satellite transmitters reveal previously unknown migratory behavior and wintering locations of Yuma Ridgway's Rails. *J. Field Ornithol.* **2020**, *91*, 300–312. [CrossRef]
17. Taylor, P.; Crewe, T.; Mackenzie, S.; Lepage, D.; Aubry, Y.; Crysler, Z.; Finney, G.; Francis, C.; Guglielmo, C.; Hamilton, D. The Motus Wildlife Tracking System: A collaborative research network to enhance the understanding of wildlife movement. *Avian Conserv. Ecol.* **2017**, *12*, 8. [CrossRef]
18. Fischer Elektronik GmbH & Co. KG. Datenblatt Produkt ICK S 50 × 50 × 20. *Datasheet* 2020. Available online: https://www.fischerelektronik.de/web_fischer/de_DE/$catalogue/fischerData/PR/ICKS50x50x20_/datasheet.xhtml?branch=Kühlkörper (accessed on 26 May 2021) .
19. Texas Instruments Incorporated. TMP117 High-Accuracy, Low-Power, Digital Temperature Sensor. *Datasheet* 2018. Available online: https://www.ti.com/lit/ds/symlink/tmp117.pdf?ts=1633404466025&ref_url=https%253A%252F%252Fwww.google.com.sg%252F (accessed on 26 May 2021).
20. Cena, K.; Clark, J. Heat balance and thermal resistances of sheep's fleece. *Phys. Med. Biol.* **1974**, *19*, 51. [CrossRef] [PubMed]
21. Rowe, D.M. *Thermoelectrics Handbook: Macro to Nano*; CRC Press: Boca Raton, FL, USA, 2018.

22. Penfield, P. Available Power from a Nonideal Thermal Source. *J. Appl. Phys.* **1961**, *32*, 1793–1794. [CrossRef]
23. Analog Devices, Inc. Ultralow Voltage Step-Up Converter and Power Manager. *Datasheet.* 2019. Available online: https://www.analog.com/media/en/technical-documentation/data-sheets/3108fc.pdf (accessed on 1 April 2021).
24. Woias, P.; Islam, M.; Heller, S.; Roth, R. A low-voltage boost converter using a forward converter with integrated Meissner oscillator. *J. Phys. Conf. Ser.* **2013**, *476*, 012081. [CrossRef]
25. u-blox. u-blox M8 Concurrent GNSS Antenna Modules. *Datasheet.* 2019. Available online: https://www.u-blox.com/sites/default/files/CAM-M8-FW3_DataSheet_%28UBX-15031574%29.pdf (accessed on 3 May 2021).
26. AVX Corporation. BestCap® Ultra-low ESR High Power Pulse Supercapacitors. *Datasheet.* 2017. *BZ015B603ZAB.* Available online online: http://catalogs.avx.com/BestCap.pdf (accessed on 14 March 2016).
27. Texas Instruments Incorporated. TPS6274x 360nA IQ Step Down Converter For Low Power Applications. *Datasheet.* 2013. Available online: https://www.ti.com/document-viewer/TPS62740/datasheet (accessed on 17 May 2021).
28. Biancuzzi, G.; Wischke, M.; Peter, P.D.I.W. Elektrische Schaltung. DE DE102011014512A1, 20 September 2012.
29. Woias, P.; Wischke, M.; Eichhorn, C.; Fuchs, B. An energy-autonomous wireless temperature monitoring system powered by piezoelectric energy harvesting. In Proceedings of the PowerMEMS 2009, Washington, DC, USA, 1–4 December 2009; pp. 209–212.
30. Texas Instruments Incorporated. MSP430FR596x, MSP430FR594x Mixed-Signal Microcontrollers. *Datasheet.* 2014. Available online: https://www.mouser.de/pdfdocs/TI_MSP430FR59xx_Datasheet.PDF (accessed on 26 February 2021).
31. Vishay Intertechnology, Inc. SiP32431, SiP32432 10 pA, Ultra Low Leakage and Quiescent Current, Load Switch with Reverse Blocking. *Datasheet.* 2017. Available online: https://www.vishay.com/docs/66597/sip32431.pdf (accessed on 18 May 2021).
32. LAPIS Technology Co., Ltd. 64k(8,192-Word × 8-Bit) FeRAM. *Datasheet.* 2016. Available online: http://www.farnell.com/datasheets/2625480.pdf (accessed on 18 May 2021).
33. Goodwin, S.D. Comparison of body temperatures of goats, horses, and sheep measured with a tympanic infrared thermometer, an implantable microchip transponder, and a rectal thermometer. *J. Am. Assoc. Lab. Anim. Sci.* **1998**, *37*, 51–55.

Article

Designing a Wind Energy Harvester for Connected Vehicles in Green Cities

Zuhaib Ashfaq Khan [1], **Hafiz Husnain Raza Sherazi** [2,*], **Mubashir Ali** [3], **Muhammad Ali Imran** [4,5], **Ikram Ur Rehman** [2] and **Prasun Chakrabarti** [6]

[1] Department of Electrical and Computer Engineering, Attock Campus, COMSATS University Islamabad, Punjab 43600, Pakistan; zuhaibkhan@cuiatk.edu.pk
[2] School of Computing and Engineering, University of West London, London W5 5RF, UK; ikram.rehman@uwl.ac.uk
[3] Department of Management, Information and Production Engineering, University of Bergamo, 24044 Bergamo, Italy; mubashir.ali@unibg.it
[4] School of Engineering, University of Glasgow, Glasgow G12 8QQ, UK; muhammad.imran@glasgow.ac.uk
[5] Artificial Intelligence Research Centre (AIRC), Ajman University, Ajman 20550, United Arab Emirates
[6] Techno India NJR Institute of Technology, Udaipur, Rajasthan 313003, India; drprasun.cse@gmail.com
* Correspondence: sherazi@uwl.ac.uk

Abstract: Electric vehicles (EVs) have recently gained momentum as an integral part of the Internet of Vehicles (IoV) when authorities started expanding their low emission zones (LEZs) in an effort to build green cities with low carbon footprints. Energy is one of the key requirements of EVs, not only to support the smooth and sustainable operation of EVs, but also to ensure connectivity between the vehicle and the infrastructure in the critical times such as disaster recovery operation. In this context, renewable energy sources (such as wind energy) have an important role to play in the automobile sector towards designing energy-harvesting electric vehicles (EH-EV) to mitigate energy reliance on the national grid. In this article, a novel approach is presented to harness energy from a small-scale wind turbine due to vehicle mobility to support the communication primitives in electric vehicles which enable plenty of IoV use cases. The harvested power is then processed through a regulation circuitry to consequently achieve the desired power supply for the end load (i.e., battery or super capacitor). The suitable orientation for optimum conversion efficiency is proposed through ANSYS-based aerodynamics analysis. The voltage-induced by the DC generator is 35 V under the no-load condition while it is 25 V at a rated current of 6.9 A at full-load, yielding a supply of 100 W (on constant voltage) at a speed of 90 mph for nominal battery charging.

Keywords: energy harvesting; power management; connected vehicles; wind energy harvester; smart cities; electric vehicle; IoT; Tesla; autonomous sensors

Citation: Khan, Z.A.; Sherazi, H.H.R.; Ali, M.; Imran, M.A.; Rehman, I.U.; Chakrabarti, P. Designing a Wind Energy Harvester for Connected Vehicles in Green Cities. *Energies* **2021**, *14*, 5408. https://doi.org/10.3390/en14175408

Academic Editor: Dibin Zhu

Received: 7 July 2021
Accepted: 17 August 2021
Published: 31 August 2021

1. Introduction

Vehicular ad hoc networks (VANET) [1,2] can genuinely be viewed as a game changer in repainting the future of electric vehicles. The incorporation of the concept of 5G communication modes is also being widely discussed. These communication paradigms have opened up a broad spectrum of technological advancements such as VANET [3,4], IoV [5–7], and industrial automation [8,9]–to name a few. The IoV promotes connectivity to achieve communication not only among the vehicles, but also with the road side infrastructure that promotes safety and ease of service on the road. This encourages the advancement in IoT, which focuses on the provision of communication services as well as the better control and security of household and commercial areas, paving the way for concepts such as smart cities and smart grids [10]. These practices also lead to new ideas in the field of industrial automation, promoting a new generation of robotics and automated control systems for efficient production. These innovations are going to cause a surge in power consumption in the near future, which requires to explore more ways and methods for energy generation. To this end, energy harvesting is a straightforward and elegant approach for powering on-board batteries rather than frequent battery replenishment, in order to maintain a perpetual power supply [11,12].

The constantly increasing carbon emissions caused by conventional gas-powered vehicles have led research efforts towards the inception of electric vehicles (EVs) [13–15], these being one of the greatest inventions of recent decades. This concept started to take form and become implemented during the early 21st century. The practical concept of the electric vehicle was introduced to the world by Tesla Co. The car runs on electric batteries which turn a motor in the car that rotates the axle shaft of the accelerator wheels. The technology has since modernized the automobile industry, but its range and charging are still issues to be revolutionized. The EVs are charged from static charging ports. These charging ports will receive electricity from the power grids which are already responsible for supplying power to different types of loads in residential, commercial, and industrial set ups. The relayance of EVs upon these power stations could cause a huge increase in electricity demand. There is, therefore, a need for a harvesting system for the EVs to reduce their dependence upon power grids–and if the battery is exhausted, there should be a way of feeding the vehicles with energy as it can help in disaster recovery e.g., connected police, fire-brigade and ambulance services can benefit from connectivity while coordinating their responses and services during a disaster.

Several techniques have been introduced for electricity generation in the form of moving vehicles with fans mounted at different positions, i.e., using a windmill mounted on the trunk and the top of a car or a train, however, some serious issues were highlighted as obstacles to efficient and effective generation. Previously proposed approaches were not quite compatible with all types of vehicles as they directly relied on the fan output power. However, there was limited work has been carried out covering the electrical aspects. Quartey et al. [16] first presented the concept for charging a mobile car in which a wind turbine is mounted on the top of the car to charge the car's battery using the wind energy striking on the car. The practical results had drawbacks in terms of the increase in the drag force on the car which affected the efficiency of the car [17]. Then, another work [18] highlighted the cause for the increase in the pressure on the car with the help of simulations. The simulation showed that the front and the bonnet of the car are much more reliable locations for mounting the wind turbine than the top [19].

Another attempt to make a vehicle-mounted wind turbine system was made in [20] where the authors proposed to mount a miniaturized windmill [21] on the back of a pickup truck but the design had a problem as it was not applicable to other types of vehicles [22,23]. Similarly, an energy-harvesting system was presented in [24] which exploits the motion of a moving vehicle's suspension. This system addresses the issues of low energy density, vibration dissipation, and lower energy-harvesting efficiency in current technology. The energy conversion components were introduced with inertial mass and an applied force to establish a dual-mass 2-DOF (degree of freedom) suspension dynamics model. Extending the similar lines, [25] demonstrated that the maximum power harvested by a vehicle suspension system can reach 738 W and is affected by road roughness.

Moreover, a new dual-mass piezoelectric energy harvester was designed in [26] to collect effective and practical vibration energy from vehicle tires with a maximum power of 42.08 W. Furthermore, [27] proposes an electric vehicle with an energy-harvesting system embracing vibration energy via a shock absorber. Maximum power point tracking algorithms and active output control techniques for active power converters were examined in order to determine how the energy-harvesting technique could be used to enhance the shock absorbers in vehicle suspension. An electric vehicle with a hybrid suspension system was proposed in [28] which caused the rise in EV sales. The concept, design, and idea of the new damper was provided, along with a simulation mainly focusing on the eddy current damper component. According to the authors, multi-mode energy harvesting provides an EV with a promising renewable source to generate surplus energy.

In this paper, a novel approach is presented to achieve electric generation in EVs by mounting a small-scale wind turbine (shown in Figure 1) on the top of the vehicle that can support a range of communication primitives, enabling access to plenty of applications for the Internet of Vehicles. Electrical energy is then generated by the onboard DC generator. ANSYS-based aerodynamics analysis is used to determine the optimal power generation orientation. With no load, the voltage induced by the DC generator is 35 V, while when it is fully loaded, it produces 25 V at a rated current of 6.9 A, and generates 100 W (at constant voltage) at 90 mph for nominal battery charging. This study shows interesting results in terms of newly harvested energy, making it a cost-effective solution for a number of futuristic use cases in EVs.

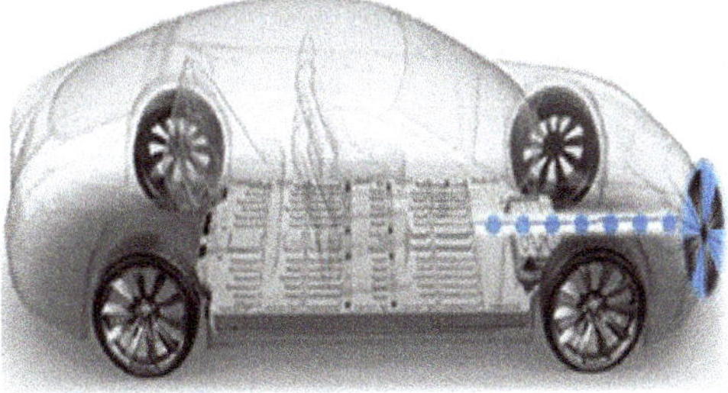

Figure 1. (**left**) Fan mounted on the front; (**right**) Side view of electric generation mechanism.

The rest of this work is structured as follows. The overall system design is proposed in Section 2. The complete methodology as well as system operation and monitoring through the Bluetooth module are presented in Section 3. The results and a detailed discussion are provided in Section 4, while Section 5 demonstrates a complete system efficiency analysis. Finally, the concluding remarks are provided in Section 6.

2. System Model

A sequential model of the system design is shown in Figure 2 given below. The basic step is the extraction of wind using fan blades, connected with the Permanent Magnet DC (PMDC) motor which generates electricity upon the rotation of its shaft primarily caused by the collision of wind. The motor is connected with a DC regulator which provides a regulated DC output to the battery. This output is measured using the voltage sensor [29] and the current sensor connected with Arduino. There is an LCD screen connected with the Arduino to display the electrical parameters of the regulated output. Then, the next phase is the protection circuitry which has two functions. One is the switching of electric power using a relay module [30] being controlled by Arduino and the decision taken upon the basis of the percentage of the charge ratio of the battery. The other function of protection circuitry is to stop the flow of current from the battery towards the motor by using a power diode [31]. After this, the electrical power is provided to the battery being charged. There is a Bluetooth module connected with Arduino to display the electrical parameters (voltage and current) on the mobile application.

The architecture of the proposed model is demonstrated in Figure 3. The first part of the system is a fan which is being used to convert the energy of the wind into the kinetic energy of the blades. This fan is connected with the shaft of the PMDC motor which, in turn, will rotate it. The DC motor will act as a generator which will generate electricity upon the rotation of its shaft. This generated electricity will be measured by using voltage and current sensors [32] in parallel and series, respectively. The output of the DC motor will be fed to the DC regulator which regulates the generated output voltage. The output of the DC regulator is again measured using voltage and current sensors so as to calculate the voltage and current, respectively, and thus power. In this way, the input and output electrical parameters of the DC regulator are used to measure the efficiency of the regulator. After this, there will be a protection circuitry installed which contains a relay module and a diode. Arduino continuously measures the voltage across the battery to calculate the percentage up to which the battery has been charged. Based upon this calculation, the Arduino will decide whether to switch on the relay. If the voltage across the battery is between the minimum and maximum set voltages, the relay will be switched on—otherwise it will be off. From a VANET perspective [33,34], a mobile app operated via a Bluetooth module was also designed, which enabled us to obtain the results of the electric generation on the app for easy access to the data. Finally, the diode before the battery will allow the current to flow only towards it and will stop the current moving in the other direction.

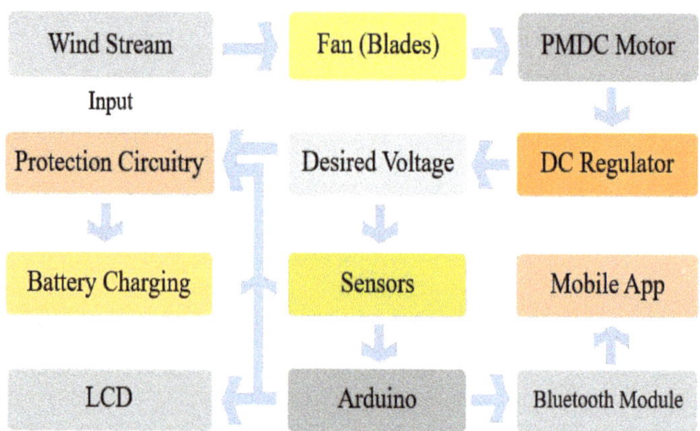

Figure 2. Wind harvesting system block diagram.

2.1. Mathematical Modeling

To make a charging system, certain parameters of the car and the electric unit for the production of the charge are required. From the mechanical aspects to the electrical parameters of the car charging system, the complete mathematical modeling of the system is required. The mathematical model of the system can be seen in Figure 4.

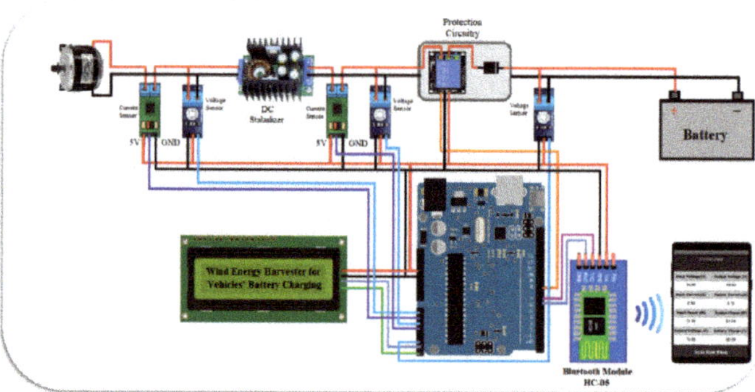

Figure 3. Proposed architecture of the entire harvesting system.

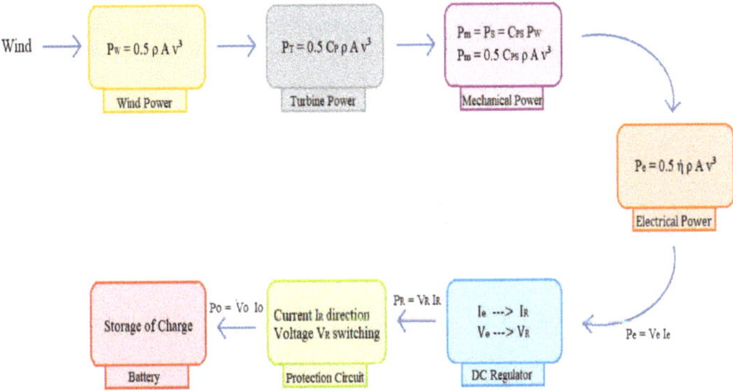

Figure 4. Mathematical model for complete energy conversion cycle.

For the calculation of various stages of power generation, the first key parameter required is the cross-sectional area of the fan, which expresses the wind that will act on this surface for turbine rotation. Under the vehicle's structural limitations, a bigger area means a larger exertion of air pressure. The second one is the air velocity that will be acting on the car due to the change in car speed: a faster car means higher air velocity.

2.1.1. Wind/Turbine Power

The air traveling with some velocity will exert kinetic energy on the turbine. Thus, by using Equation (1), the kinetic energy (KE) of the system can be calculated, and now the turbine power generated from air depends upon the mass flow rate of air which is expressed in Equation (2)—where ρ is the density of air, A is the area of the cross-section of turbine, and vs. is the air velocity [35]. Now, the turbine power is equal to the kinetic energy exerted by air given in Equation (3). The formula for the power generated by the turbine via wind power [36] is given in Equation (4), where P_r is the turbine power and C_P is the power coefficient for the turbine:

$$KE = \frac{1}{2}\,mv^2 \tag{1}$$

$$m = \rho A v \tag{2}$$

$$KE = \frac{1}{2}\,\rho A v^3 \tag{3}$$

$$P_r = \frac{1}{2}\,C_P \rho A v^3 \tag{4}$$

2.1.2. Mechanical Power

In order to find the shaft power, the first thing we need to determine is the shaft power coefficient. For this, we first need to determine the tip-to-speed ratio of the shaft which is determined by the formula:

$$\lambda = \frac{4\pi}{B} \tag{5}$$

where $B = 7$ is the number of blades:

$$\lambda = \frac{4\pi}{7} = 1.795$$

$\lambda = 1.795$ correlates to a power coefficient $C_P = 0.13$.

With an augmentation of 0.2, $C_{P_s} = 0.13 + 0.21 = 0.34$, where C_{P_s} is the shaft power coefficient. Now, the mechanical power is equal to the shaft power that is:

$$P_{mech} = P_s = C_{P_s} P_w = \frac{1}{2} C_{p_s} \rho A v^3 \tag{6}$$

For the turbine diameter of 0.36 m, the area of the turbine is determined by the formula:

$$A = \frac{\pi d^2}{4} \tag{7}$$

2.1.3. Electrical Power

The shaft power induced in the shaft of the generator is transmitted to the generator for the induction process. This induction is responsible for the generation of electric power. The formula is given in given in Equation (8), where η is the net efficiency of the generator. As power in corresponds to (speed and torque) and power out is relative to (speed and current), then the torque mechanical applied is relative to the current being drawn by the windmill. The torque on the rotor can be calculated by the formula of Equation (9), where T is the torque and ω is the angular velocity of the rotor. The ω can be calculated by the formula given in Equation (10), where R is the length of the fan blade:

$$P_{elec} = \frac{1}{2} \eta \rho A v^3 \tag{8}$$

$$P_s = T\omega \implies T = \frac{P_s}{\omega} \tag{9}$$

$$\omega = \frac{\lambda v}{R} \tag{10}$$

2.1.4. RPMs and Angle of Twist

The blade angle of the twist is the angle at which blades are set to accumulate the maximum air pressure for effective generation, where N is the rotational speed of the rotor given in revolutions per minute (rpm) [37], which can be calculated through the formula in Equation (11):

$$N = \frac{60 \lambda v}{2 \pi R} = \frac{60 \times 1.795 \times 33.33}{2 \times \pi \times 0.18} = 3174 \text{ rpm.} \tag{11}$$

2.1.5. Force Calculation

The force acting on the fan to operate is the difference between the two forces acting on it. One is the lift force (F_L), exerted by the air to rotate the fan expressed in Equation (12):

$$F_L = \frac{1}{2} C_L \rho V^2 A_t \tag{12}$$

but for every action, an opposite reaction is present. In this case, an opposite force emerges, which tries to restrict the fan from moving. This force is called drag force and is calculated by the following formula in Equation (13):

$$F_D = \frac{1}{2} C_D \rho V^2 A_t \tag{13}$$

where C_L and C_D are the lift coefficient and drag coefficient, respectively, and A_t is the fan surface area. Now, the total force on the fan is the vectorial sum of both forces, meaning that:

$$F = F_L \cos(90 - \phi) - F_D \sin(90 - \phi) \tag{14}$$

3. Methodology

As this research work brought novelty in terms of technology, there was a need to propose its design on the basis of a new analysis to be conducted. Thus, first of all, software-based simulations were used to make decisions regarding the hardware of the system. After this, the complete schematic was designed and the hardware was implemented. The components were connected in accordance with the proposed schematic. First of all, the wind turbine, whose output is given to the DC voltage regulator, has the current and voltage sensors on its input and output. The output of the DC voltage regulator is connected to the battery with a protection circuitry in its way. This protection circuitry is used to protect the battery from over-charging and low charging, by switching the circuit ON and OFF at suitable times. This decision is taken by Arduino. Moreover, this protection circuitry also stops the flow of current back from the battery towards the wind turbine. After this hardware was designed, implemented and tested for its working and it was assured that this hardware is ready to be used practically, it was then implemented on a car.

It can be noticed in Figure 5a that the fan was mounted on the front of the car. Only the fan was out of the car's body, as the air needs to strike the fan to rotate it and no other component, so it is necessary for the fan to be mounted outside the car body. Contrarily, the Permanent Magnet DC (PMDC) Motor was placed inside the car body as there should neither be any type of weight with a fan outside the car nor any drag force due to the placement of the motor. The output of the DC motor was connected to the rest of the circuit using DC wires, suitable for carrying the current through it from inside the car engine. Another view of the fan, mounted outside the car, can be seen in Figure 5b.

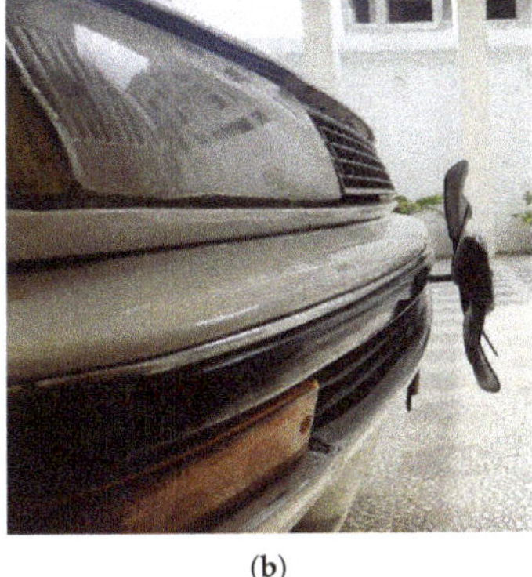

(a) (b)

Figure 5. (a) Front view of a car-mounted fan; (b) Side view of a car-mounted fan.

After the generation of electricity from the wind turbine, the output was supplied to the battery once it was stabilized, measured, and the protection was assured. The DC wires will take this electricity towards the rest of the system, which was placed inside the car. In Figure 6, it is shown that the input is provided to the system for regulation, measurements, and protection work. After this, the output is supplied to the battery so it can be charged:

3.1. Bluetooth Monitoring via Android App

In a research study where something is being measured or calculated, there is a need to display it. As in this research article, the data were displayed in LCD using Arduino. Still, there was a need to have a display that was user friendly. Hence, an Android App was developed using the MIT App Inventor, which would display the input and output electrical parameters as well as the battery percentage. These data were sent to a mobile phone from Arduino via Bluetooth connection using a Bluetooth Module, HC-05. This objective of the research was performed after achieving all the milestones of the article and therefore brings this work to two different technologies:

3.1.1. Mobile Monitoring

In this research, the car was driven and a wind turbine was used to generate electricity. This electricity was being measured and displayed on an LCD using Arduino. Furthermore, these data were being sent to the mobile phone of the user to be displayed there in an Android App, as shown in Figure 7. This mobile monitoring was achieved using Bluetooth connection between the Arduino and the mobile phone.

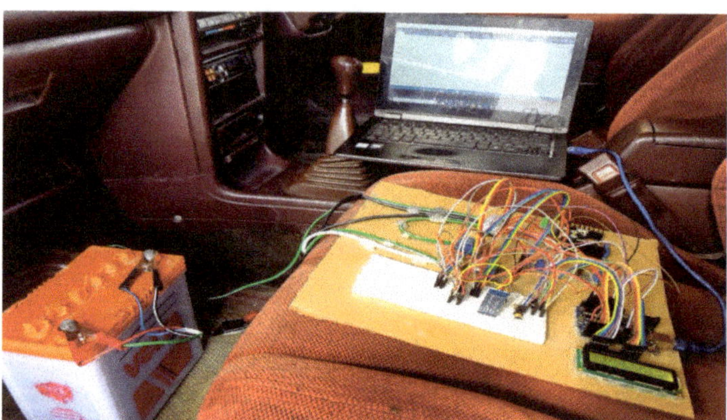

Figure 6. Hardware system implemented in the car.

3.1.2. Human–Machine Interface

As the measured data were displayed on the user's mobile phone, it had to be made user-friendly. In this Android App to be installed on the mobile phone of the user, the user will be able to open the introductory part of the app, in which they will be able to see the author's team and also watch the Demo Video of the work. Moreover, the user will be able to connect the mobile to Arduino and start receiving the information from Arduino via Bluetooth.

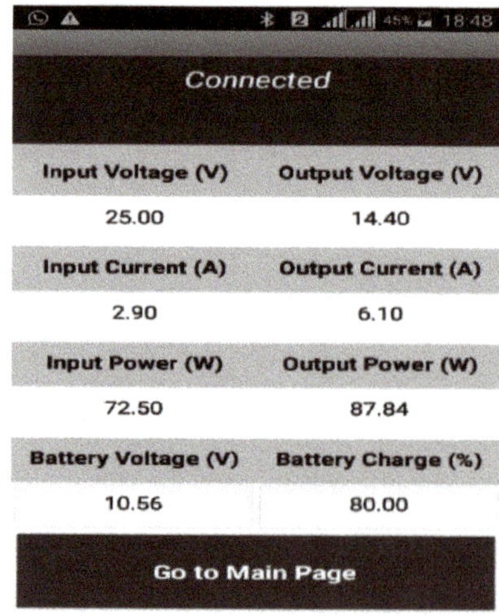

Figure 7. Measurements displayed on the App.

4. Results and Discussion

4.1. Simulation Results

For performing the software analysis, the effect of air on the car was simulated by using the ANSYS workspace 16.0 [38,39] for the evaluation of aerodynamic properties for the execution of our research. Software analysis helped us grasp an idea of the effect of air velocity on the car as well as various zones of drag coefficient present around the vehicle. For the evaluation of the air velocity and drag force, the fluent module of ANSYS was used, in which different analysis techniques were used.

4.1.1. Iteration Analysis

In this simulation analysis, the geometry of an EV was included in Fluent to perform some analysis techniques. This simulation was performed for this research work over 200 iterations at a speed of 35 m/s, which was the maximum speed at which the vehicles were driven in general. In this type of technique, an air thrust was exerted on a car in the entire axis (x, y, and z axis) of the vehicle. The x axis shows the number of iterations and along the y axis shows the scale for the residual air-stream on each coordinate of the vehicle.

4.1.2. Pathlines Analysis

This concept was further expanded via running multiple iterations of the same air model using the vectors graphical method. This gave us a comprehensive look at the effect of air on the car. The simulation in Figure 8 shows the regions in which the drag force is strongly present and where its presence is negligible. This indicates that when a car model is designed, priority is given to the aerodynamics of the car for effectively resisting the drag force. Figure 8 also shows how much velocity the air strikes the car with when the car is being driven. It can be seen that there are small lines showing the flow of air towards the car and their color indicates the air velocity in the surroundings of the car. It is clear that there is enough air velocity on the front of the car, which is necessary for the system to work properly. Too much air velocity is not required by the system as there remains a risk of its damage under too much pressure. Thus, the air velocity on the front is sufficient.

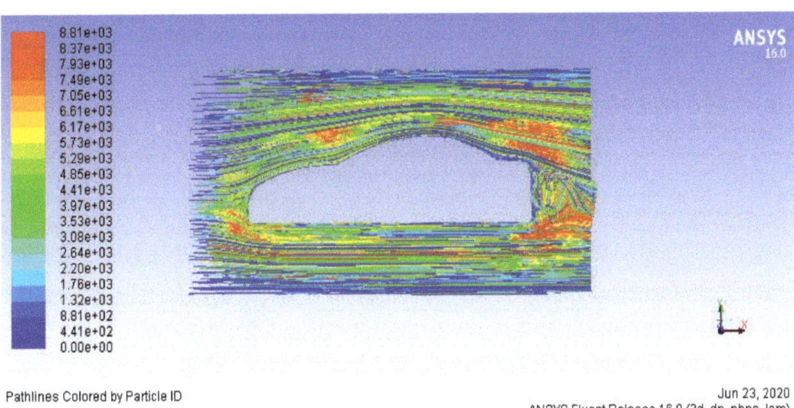

Figure 8. Air flow path-lines.

4.1.3. Contours Analysis

The following simulations used contours graphical method which measures the surface area using different iterations, with which complete structural analysis was performed to identify the effect of the air velocity on the vehicle. In Figure 9a, the front as well as sides are easily visible and indicate that the front of the EV directly confronts the maximum air as compared to the other parts of the body, whilst the sides are the second most pressure-tolerant region of the car. Simply, this analysis was performed in function of the shape of the car, and shows the parts of the car on which more pressure will be exerted by the wind when the car will be driven. It can be seen that when a car is driven, then due to its direct contact with the wind, there is maximum pressure being exerted. Moreover, the wind screen and other parts also receive a considerable amount of pressure. Another view of the same analysis can be seen in Figure 9b, where it is clear that the parts of the car which are vertical in shape receive more pressure from the wind, etc.

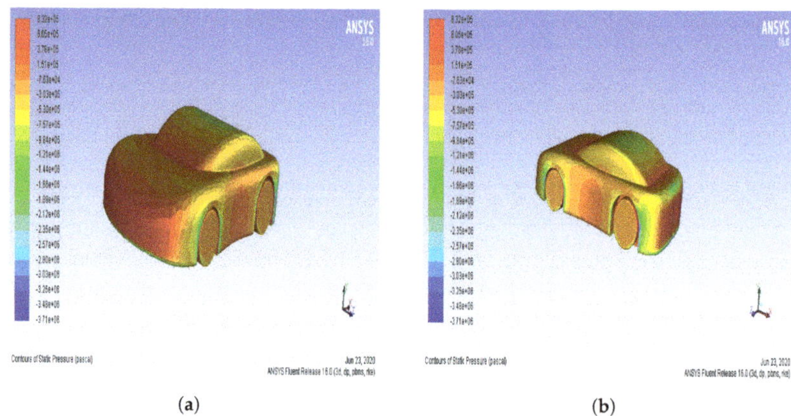

(a)   (b)

Figure 9. (**a**) Contours Analysis with front view; (**b**) Contours Analysis with side view.

4.1.4. Velocity Pathway Simulation

In Figure 10a, the air velocity acting on the car's body is evaluated. It can be seen that most of the air resistance smoothly passes around the car, while the front blocks and directly confronts the air acting on it. As it can be seen, the maximum air velocity is on the top of the car body but this location is not feasible for mounting a diffuser on the roof of the car because this will increase the weight of the vehicle. Thus, keeping this point in mind, a system has been proposed which can be adjusted on the front bumper of the car where the air velocity is according to our system needs. Basically, this analysis also tells us about the drag force on various parts of the vehicle. Now, when this analysis is carried out on a vehicle, it shows that when the air strikes the car, then obviously it has to pass across it in order to keep the vehicle moving. This analysis provides an insight about the air dynamics passing along the vehicle. In Figure 10b, it can be seen that after striking the car, most of the air is moving either from above the car or from beneath it. Hence, there is less drag on the front of the car, as the scale on the left side depicts. Thus, the front of the car is a suitable location for the wind turbine to be mounted.

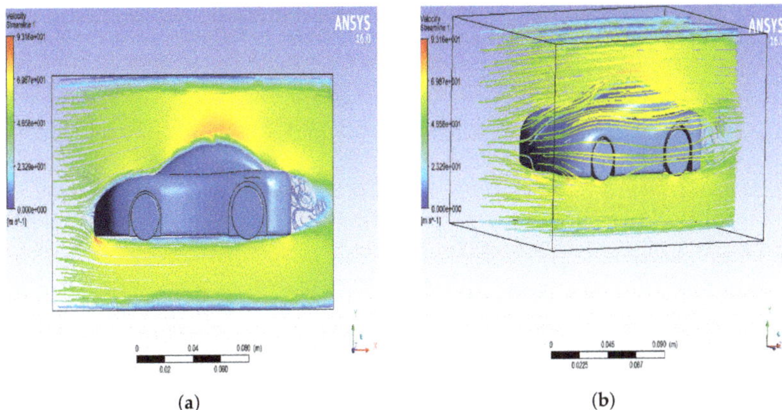

(a)   (b)

Figure 10. Air velocity flow analysis in ANSYS with (**a**) sideways view; (**b**) 3D view with least drag.

4.2. Testbed Results

4.2.1. Car Speed vs. Air Velocity

At first, calculations were made on a practical basis to help clarify the difference between the vehicle's velocity and the air velocity on it. This was so that we could clarify the fact that at the designated location, the air velocity was ample enough to run the wind generator's fan blades. The relation between them can be seen in Figure 11, which confirms that the air velocity was slightly less to allow the car to run but had enough kinetic energy to be provided to the generator to generate momentum for the fan blades.

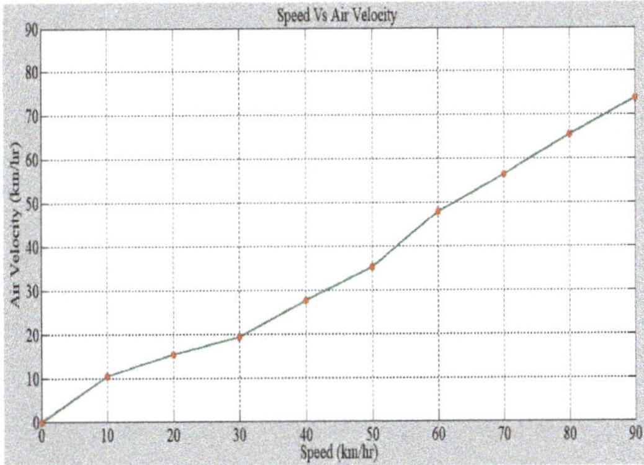

Figure 11. Air velocity against different speeds.

4.2.2. Car Speed vs. Voltage Generation

The following graph in Figure 12 was obtained as a result of a comparison between the car speed and the voltages induced in the DC generator with and without load (car battery). It can be observed that the voltage increases with the increase in the car's velocity. Without load, we obtained voltages as high as 35 V at 90 km/h, but in the presence of a load, this obtained voltage decreased to approximately 25 V at the same velocity of 90 km/h. The regulated line in the given graph indicates the voltages obtained after they passed through a DC regulator to stabilize them at a constant charging voltage, which was 14.6 V for the lead acid battery—this voltage is achieved at the speed of 45 km/h.

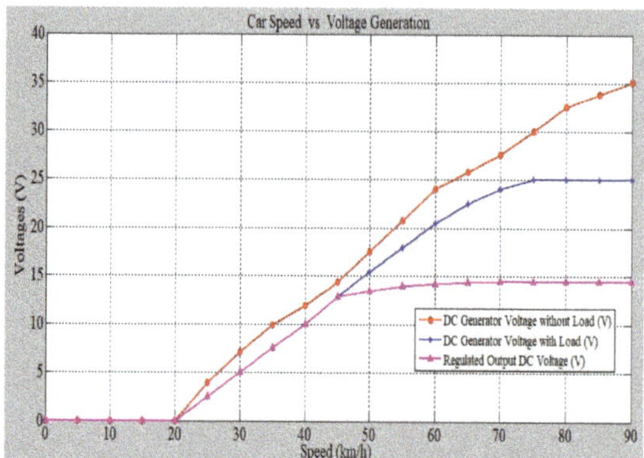

Figure 12. Voltage generation against different speeds.

4.2.3. Car Speed vs. Electrical Parameters

On the basis of the data gathered, it can be observed that at 90 km/h, the generation system has a stable voltage of 14.6 V with a charging rate of 6.9 Ah. The curves indicate that with the increase in the velocity of the car, the voltage and current are also gradually increased, except for the voltage which needs to stabilize after the required charging potential is achieved, as shown in Figure 13. At the speed of 90 km/h, the total power output obtained is approximately 100 W, as shown in Figure 14. Here, it can be noted that the charging rate can be increased with some adjustments, but it will jeopardize the life span of the battery.

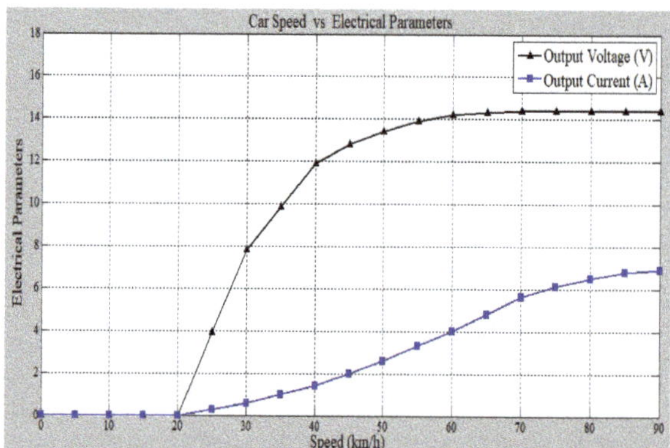

Figure 13. Electrical voltage and current output against different speeds.

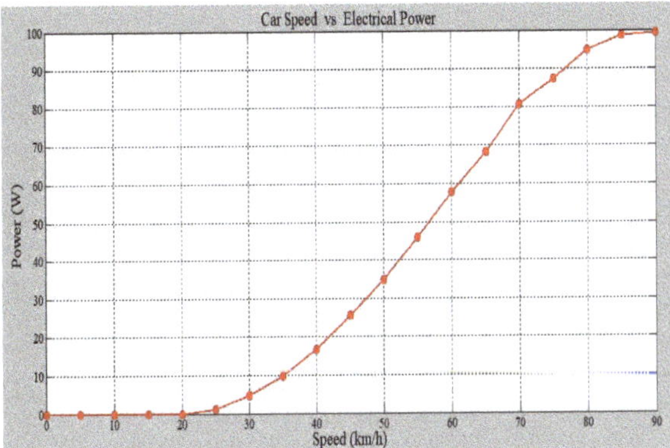

Figure 14. Electrical power output against different speeds.

4.2.4. Time vs. Battery Charging Current

The battery charging trend is presented with time in Figure 15. The chart shows the charging rate of the battery from 9 V (dead battery) up to the 13.1 V (fully charged) charging state of the battery. This is because of the battery resistance which varies. Thus, by this curve, it is to be estimated that when the battery is in a low charge condition, it has high resistance; and with the passage of time, as the battery starts charging and ionizing its charges, then the current increases with the decrease in the resistance. At one point, the maximum current drawn limit of the battery is achieved, and then the resistance of the battery increases again when it is becomes closer to the fully charged state.

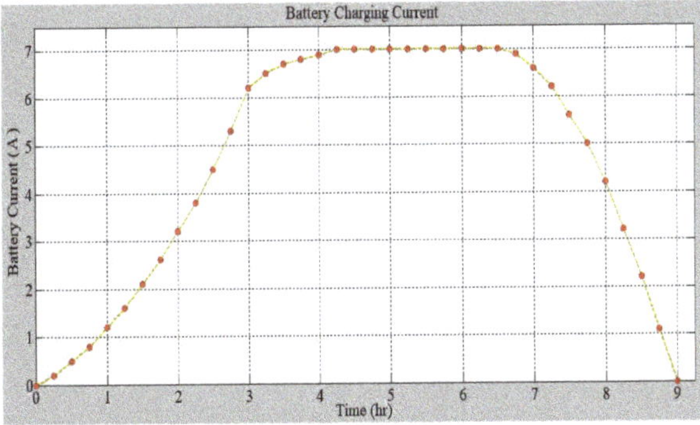

Figure 15. Battery charging trend with time.

5. System Efficiency Analysis

This system efficiency analysis was performed so as to depict the energy conversion efficiency from wind to electrical form via different transformations. First of all, the air strikes the car with mobility. As the moving car has some velocity, the air also strikes it with some velocity, but the air velocity is less than the speed of car. Thus, the car keeps on moving in the forward direction. This shows that the work being performed by the car is greater than the work performed by the wind on the car. In this research, an approach to harvest an amount of energy is proposed by utilizing the air pressure striking the car. This power can then become the turbine power after some losses. This power of the turbine is thus called because this is the energy caused by the wind turbine movement. Now, this rotating wind turbine has enough power to be used to do some useful work. Thus, this turbine is attached with a Permanent Magnet DC Motor, which has been used as a DC generator in this article, with the help of a mechanical shaft. This rotating wind turbine power will be transferred to the DC Motor and its rotor will start moving. It is important to note that the process also involves power losses during the transformation of energy from the wind turbine towards the motor.

The power available after all these transformations is the Mechanical Power which gives us an idea of how much work is being done by the generator (more specifically, by the rotor of the DC generator). As in this case, the work is being performed by the DC generator, as it will convert the mechanical power into electrical power. Electrical power is basically the measurement of the flow of current towards the load that is used for some purpose; in this case, it is a 12V lead acid battery. This electrical power, as given below, is directed towards the battery after some adjustments and protections. First of all, the DC voltages are regulated at a desired DC voltage so as to charge the battery at a constant voltage. Then, this electrical power is supplied to the battery to be stored after using protection circuitry so that it can be utilized later to feed a source. Figure 16 and Table 1 exhibit system's efficiency at different stages of the harvesting process.

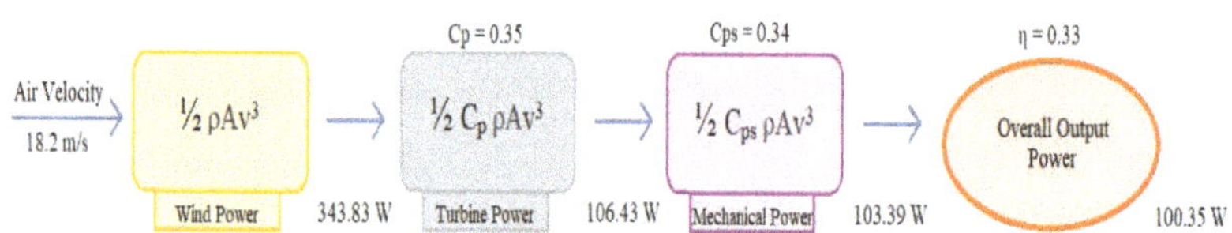

Figure 16. Efficiency of power generation at different stages.

Table 1. Power evaluation at different stages of the harvesting process.

Car Speed (km/h)	Air Velocity (m/s)	Wind Power (W)	Turbine Power (W)	Mechanical Power (W)	Electrical Power (W)
10	2.90	1.4	0.4	0.4	0.4
20	4.30	4.6	1.4	1.3	1.3
30	5.40	9.1	2.7	2.7	2.6
40	7.70	26.4	8.0	7.8	7.5
50	9.80	54.6	16.6	16.1	15.6
60	13.3	133.5	41.5	40.3	39.1
70	15.7	224.4	68.3	66.4	64.4
80	18.2	343.8	106.4	103.4	100.3
90	20.5	499.3	152.4	147.7	143.4

5.1. Betz Limit Analysis

As per Betz's law [40], no turbine can capture more than 16/27 (59.3%) of the kinetic energy in wind. The factor 16/27 is known as Betz's coefficient. The Betz limit is based on an open-circle actuator. If a diffuser is utilized to gather extra wind stream and direct it through the turbine, more kinetic energy can be separated, however, the limit still applies to the cross-sectional area of the whole structure. According to the conducted analysis, the Betz limit efficiency is 30.3% for the turbine area of 0.0791 m. For a motor of 350 W power, the overall system efficiency is 28.5%. This could be improved by increasing the area of the diffuser, but keeping in mind that the Betz limit constraints are still sufficiently critical to influence the efficiency of the system.

5.2. Drag Coefficient Analysis

In a previous research work, which was based on the effect of a drag coefficient on the car in the presence of a charging device in various locations, the need for low or negligible drag force that would add up in the original drag force due to the charging system installation was emphasized. The results conclude that a harvesting system mounted on the front of a car bumper is a more suitable and promising point for maximum production and minimum drag compared to other positions of the car, as shown in Table 2.

Table 2. Drag coefficient at the different locations on a car.

No.	Turbine Status	Drag Coefficient
1	Without wind turbine	0.39
2	At front of the car (bumper)	0.39
3	At hood of the car (bonnet)	0.45
4	At top of the car (roof)	0.51

5.3. Power Generation Analysis

According to the results gathered during analysis, the total power per hour was calculated to be approximately 100 W, which is 28.5% efficient depending on the charging requirements of the battery. As per the motor configuration, the average power that can be produced by it is in between 15% and 35% of the total power. Keeping this information, it can be concluded that the power efficiency that was acquired is among the favorable requirements.

5.4. Battery Discharge Time with respect to Car Loads

As shown in Table 3, the more frequently used electrical loads of a car along with their power and current ratings are discussed and on this basis, the rightmost column was determined by driving. This column basically shows the time required to discharge a fully charged 40 Ah battery when each load is connected to it. These loads are of different types. The first one is the lighting loads which include the headlights, back lights, and the hazard lights. Secondly, body electrical loads include the power window and the wiper/washer system. The third type is media which includes video/audio systems, the navigation and GPS systems. Another type is the HVAC loads including the air compressor of the car.

The purpose of this analysis was to verify the capability of the proposed system to overrun the loads of a car on which it is installed. Considering the electricity generation, a statistical approach was used to estimate which loads of a car can be driven on a fully charged 40 Ah battery—charged using the proposed system. Moreover, if each of them is connected to the battery, how much time will be required to use the battery completely? The answer to this is in the rightmost column. Afterwards, a very important analysis was performed. If all the loads excluding the air compressor were left to run on the battery with the specifications mentioned above, then it would keep them running all these loads for approximately half an hour, i.e., 0.52 h (31 min). However, if the air compressor was included in the loads being driven over the battery, it can still keep all these loads running for approximately 0.34 h, i.e., 20 min.

It is also pertinent to see whether a proposed system is capable of supporting the loads mentioned in Table 3. We covered all these loads one by one to understand each of them. The back lights and the hazard lights are those which could directly be supported by the proposed architecture while leaving some surplus energy. The headlights, power windows, and wiper washer system are the loads that could be supported with the full generated power being utilized, i.e., without leaving any extra energy. As far as the air compressor, the navigation and GPS systems are concerned, the harvested energy is not ample enough to support them with a direct supply. They will require an additional external electric supply as well. Moreover, the available power would be 20% and 40% of the power ratings of the air compressor and navigation/GPS system, respectively. This will help reduce their dependency on the battery or grid power. Hence, it can be concluded that the proposed system generates an ample amount of energy to directly support the frequently used loads of a car, i.e., without needing to employ a battery.

Table 3. Battery discharge time with respect to loads.

Serial No.	Loads	Power (W)	Current (A)	Discharge Time (h)
1	Headlights (2)	120	10	4.0
2	Backlights (2)	50	4.1	9.1
3	Hazard lights (4)	84	7.0	5.7
4	Power window (4)	120	10	4.0
5	Wiper and washer	140	9.8	4.0
6	Video and audio system	250	20.8	1.9
7	Navigation and GPS	150	12.5	3.2
	Sub. Total	914	76.1	0.52
8	Air compressor	500	41.6	0.9
	Total	1414	117.8	0.34

5.5. Design and Cost Analysis

As a major objective of our research, the positioning of the fan was a vital aspect of our analysis. The experimental results have shown a considerably better performance than the procedures used in previous papers. This proves our theory that the front position is the perfect spot for the adjustment of the wind turbine system for maximum utility. The fan used for wind generation worked satisfactorily with the given motor and circuit parameters. As a whole, the complete study including the hardware material for practical analysis was very cost-effective. All the materials required for the testbed work were available in the local market and were cheap to buy, while also ensuring their precision and durability. As a team, it is our strong belief that in the future, if an industrial version is created, the end product will be very user friendly in terms of cost.

6. Conclusions

With the surge to explore new dimensions in mitigating carbon emissions, today's smart cities are committed to increase their low emission zones in an effort to replace the conventional gas vehicles with electric vehicles, effectively meeting the requirements of several IoV use cases. To support the optimal operation, the design of a wind power harvester was proposed and evaluated in this article. The aerodynamic drag and air speed qualities around the vehicle with a wind turbine framework were numerically researched. The vehicle model was simulated by a Fluent module of ANSYS software to contemplate the wind stream trademark around the vehicle body and decide the drag power of the vehicle at a specific speed. Furthermore, on the basis of the obtained results, practical evaluation analysis was also performed in order to achieve the research objectives. After the simulated and practical experiments, the front bumper is recommended to be an ideal spot on the car for installing this kind of device for making an effective charging system of EVs. This article emphasized the charging of the EVs while utilizing their batteries considering a speed range from 40 to 90 km/h. This range can be improved by designing a DC voltage regulator capable of taking a larger range of input voltages to be stabilized. The authors are committed to extend this work towards automating the energy generation process inducing the integrated control and regulation of the generated voltage, along with exploring different options to transfer surplus power to the national grid for general purpose utilization.

Author Contributions: Conceptualization, Z.A.K., H.H.R.S., and M.A.; methodology, Z.A.K., H.H.R.S. and M.A.; validation, I.U.R., M.A.I. and P.C.; formal analysis, H.H.R.S.; resources, I.U.R., P.C.; writing—original draft preparation, Z.A.K., H.H.R.S. and M.A.; writing—review and editing, Z.A.K., M.A.I., H.H.R.S., P.C.; project administration, M.A.I., H.H.R.S., and P.C.; funding acquisition, M.A.I. and H.H.R.S. All authors have read and agreed to the published version of the manuscript.

Funding: This research was partly funded by EPSRC Global Challenges Research Fund–the DARE project–EP/P028764/1.

Institutional Review Board Statement: Not applicable.

Informed Consent Statement: Not applicable.

Data Availability Statement: Not applicable.

Conflicts of Interest: The authors declare no conflict of interest.

References

1. Raja, G.; Anbalagan, S.; Vijayaraghavan, G.; Dhanasekaran, P.; Al-Otaibi, Y.D.; Bashir, A.K. Energy-Efficient End-to-End Security for Software Defined Vehicular Networks. *IEEE Trans. Ind. Inform.* **2020**, *17*, 5730–5737. [CrossRef]
2. Hussain, S.A.; Iqbal, M.; Saeed, A.; Raza, I.; Raza, H.; Ali, A.; Bashir, A.K.; Baig, A. An efficient channel access scheme for vehicular ad hoc networks. *Mob. Inf. Syst.* **2017**, *2017*. [CrossRef]
3. Sherazi, H.H.R.; Raza, I.; Chaudary, M.H.; Hussain, S.A.; Raza, M.H. Multi-radio over fiber architecture for road vehicle communication in VANETs. *Procedia Comput. Sci.* **2014**, *32*, 1022–1029. [CrossRef]
4. Iwendi, C.; Uddin, M.; Ansere, J.A.; Nkurunziza, P.; Anajemba, J.H.; Bashir, A.K. On detection of Sybil attack in large-scale VANETs using spider-monkey technique. *IEEE Access* **2018**, *6*, 47258–47267. [CrossRef]
5. Azzaoui, N.; Korichi, A.; Brik, B.; el amine Fekair, M.; Kerrache, C.A. On the Communication Strategies in Heterogeneous Internet of Vehicles. In Proceedings of the Third International Conference on Smart City Applications, Casablanca, Morocco, 2–4 October 2019; Springer: Cham, Switzerland, 2019; pp. 783–795.
6. Sherazi, H.H.R.; Khan, Z.A.; Iqbal, R.; Rizwan, S.; Imran, M.A.; Awan, K. A heterogeneous IoV architecture for data forwarding in vehicle to infrastructure communication. *Mob. Inf. Syst.* **2019**, *2019*, 3101276. [CrossRef]
7. Kumar, A.; Hariharan, N. Enhanced Mobility Based Content Centric Routing In RPL for Low Power Lossy Networks in Internet of Vehicles. In Proceedings of the 2020 3rd International Conference on Intelligent Autonomous Systems (ICoIAS), Singapore, 26–29 February 2020; pp. 88–92.
8. Sherazi, H.H.R.; Imran, M.A.; Boggia, G.; Grieco, L.A. Energy harvesting in LoRaWAN: A cost analysis for the industry 4.0. *IEEE Commun. Lett.* **2018**, *22*, 2358–2361. [CrossRef]
9. Sherazi, H.H.R.; Grieco, L.A.; Imran, M.A.; Boggia, G. Energy-Efficient LoRaWAN for Industry 4.0 Applications. *IEEE Trans. Ind. Inform.* **2021**, *17*, 891–902. [CrossRef]
10. Raja, G.; Dhanasekaran, P.; Anbalagan, S.; Ganapathisubramaniyan, A.; Bashir, A.K. SDN-enabled Traffic Alert System for IoV in Smart Cities. In Proceedings of the IEEE INFOCOM 2020—IEEE Conference on Computer Communications Workshops (INFOCOM WKSHPS), Toronto, ON, Canada, 6–9 July 2020; pp. 1093–1098.
11. Citroni, R.; Di Paolo, F.; Livreri, P. Evaluation of an optical energy harvester for SHM application. *AEU Int. J. Electron. Commun.* **2019**, *111*, 152918. [CrossRef]
12. Citroni, R.; Di Paolo, F.; Livreri, P. A novel energy harvester for powering small UAVs: Performance analysis, model validation and flight results. *Sensors* **2019**, *19*, 1771. [CrossRef]
13. Sun, D.; Ou, Q.; Yao, X.; Gao, S.; Wang, Z.; Ma, W.; Li, W. Integrated human–machine intelligence for EV charging prediction in 5G smart grid. *EURASIP J. Wirel. Commun. Netw.* **2020**, *2020*, 139. [CrossRef]
14. Karfopoulos, E.L.; Hatziargyriou, N.D. A multi-agent system for controlled charging of a large population of electric vehicles. *IEEE Trans. Power Syst.* **2012**, *28*, 1196–1204. [CrossRef]
15. Xie, D.; Chu, H.; Lu, Y.; Gu, C.; Li, F.; Zhang, Y. The concept of EV's intelligent integrated station and its energy flow. *Energies* **2015**, *8*, 4188–4215. [CrossRef]
16. Quartey, G.; Adzimah, S.K. Generation of electrical power by a wind turbine for charging moving electric cars. *J. Energy Technol. Policy* **2014**, *4*, 19.
17. Ohya, Y.; Karasudani, T. A shrouded wind turbine generating high output power with wind-lens technology. *Energies* **2010**, *3*, 634–649. [CrossRef]
18. Sofian, M.; Nurhayati, R.; Rexca, A.; Syariful, S.S.; Aslam, A. An evaluation of drag coefficient of wind turbine system installed on moving car. *Appl. Mech. Mater.* **2014**, *660*, 689–693. [CrossRef]
19. Bijlani, B.; Rathod, P.P.; Sorthiya, A.S. Experimental and Computational Drag Analysis of Sedan and Square-Back Car. *Int. J. Adv. Eng. Technol.* **2013**, *4*, 63–65.
20. Awal, M.R.; Jusoh, M.; Sakib, M.N.; Hossain, F.S.; Beson, M.R.C.; Aljunid, S.A. Design and implementation of vehicle mounted wind turbine. *ARPN J. Eng. Appl. Sci.* **2015**, *10*, 8699–8860.
21. Zakaria, M.Y.; Pereira, D.A.; Hajj, M.R. Experimental investigation and performance modeling of centimeter-scale micro-wind turbine energy harvesters. *J. Wind Eng. Ind. Aerodyn.* **2015**, *147*, 58–65. [CrossRef]
22. Karthikeyan, N.; Murugavel, K.K.; Kumar, S.A.; Rajakumar, S. Review of aerodynamic developments on small horizontal axis wind turbine blade. *Renew. Sustain. Energy Rev.* **2015**, *42*, 801–822. [CrossRef]
23. Kishore, R.A.; Priya, S. Design and experimental verification of a high efficiency small wind energy portable turbine (SWEPT). *J. Wind Eng. Ind. Aerodyn.* **2013**, *118*, 12–19. [CrossRef]
24. Zhao, Z.; Wang, T.; Zhang, B.; Shi, J. Energy harvesting from vehicle suspension system by Piezoelectric harvester. *Math. Probl. Eng.* **2019**, *2019*, 1086983. [CrossRef]
25. Xie, X.; Wang, Q. Energy harvesting from a vehicle suspension system. *Energy* **2015**, *86*, 385–392. [CrossRef]

26. Xie, X.; Wang, Q. A mathematical model for piezoelectric ring energy harvesting technology from vehicle tires. *Int. J. Eng. Sci.* **2015**, *94*, 113–127. [CrossRef]

27. Lee, J.; Chun, Y.; Kim, J.; Park, B. An Energy-Harvesting System Using MPPT at Shock Absorber for Electric Vehicles. *Energies* **2021**, *14*, 2552. [CrossRef]

28. Wai, C.K.; Rong, Y.Y. Electric vehicle energy harvesting system regenerative shock absorber for electric vehicle. In Proceedings of the 2013 IEEE Conference on Sustainable Utilization and Development in Engineering and Technology (CSUDET), Selangor, Malaysia, 30 May–1 June 2013; pp. 7–10.

29. Bezanilla, F. The voltage sensor in voltage-dependent ion channels. *Physiol. Rev.* **2000**, *80*, 555–592. [CrossRef]

30. Singh, S.; Bhullar, S. Hardware implementation of auto switching and light intensity control of LED lamps. *Balk. J. Electr. Comput. Eng.* **2016**, *4*, 67–71. [CrossRef]

31. Vobecky, J.; Hazdra, P.; Homola, J. Optimization of power diode characteristics by means of ion irradiation. *IEEE Trans. Electron. Dev.* **1996**, *43*, 2283–2289. [CrossRef]

32. Gunn, C.N.; Harding, S.J.; Ricci, M.A.; Loewen, D.N.; Cowan, P.C.; Hancock, M.A. Current Sensor Assembly. U.S. Patent 7,557,563, 7 July 2009.

33. Babir, M.R.N.; Al Mahmud, S.A.; Mostary, T. Efficient M-QAM Digital Radio over Fiber System for Vehicular ad hoc Network. In Proceedings of the 2019 International Conference on Robotics, Electrical and Signal Processing Techniques (ICREST), Dhaka, Bangladesh, 10–12 January 2019; pp. 34–38.

34. Munshi, A.; Unnikrishnan, S. Vehicle to Vehicle Communication Using DS-CDMA Radar. *Procedia Comput. Sci.* **2015**, *49*, 235–243. [CrossRef]

35. Manyonge, A.W.; Ochieng, R.; Onyango, F.; Shichikha, J. Mathematical modelling of wind turbine in a wind energy conversion system: Power coefficient analysis. *Appl. Math. Sci.* **2012**, *6*, 4527–4536.

36. Ochieng, R.; Ochieng, R. Analysis of the betz criterion in wind turbine power modelling by use of "variational principle (method)" in the power equation. *Int. J. Energy, Environ. Econ.* **2015**, *23*, 283.

37. Castelli, M.R.; De Betta, S.; Benini, E. Effect of blade number on a straight-bladed vertical-axis Darreius wind turbine. *World Acad. Sci. Eng. Technol.* **2012**, *61*, 305–3011.

38. Hegde, S.S.; Thamban, A.; Bhai, S.P.M.; Ahmed, A.; Upadhyay, M.; Joishy, A.; Mahalingam, A. Highway mounted horizontal axial flow turbines for wind energy harvesting from cruising vehicles. In Proceedings of the ASME International Mechanical Engineering Congress and Exposition, Phoenix, AZ, USA, 11–17 November 2016.

39. Fotso, B.M.; Nguefack, C.F.; Talawo, R.C.; Fogue, M. Aerodynamic analysis of an electric vehicle equipped with horizontal axis savonius wind turbines. *Int. J. Recent Trends Eng. Res. IJRTER* **2019**, *5*, 17–26.

40. Gorban', A.N.; Gorlov, A.M.; Silantyev, V.M. Limits of the turbine efficiency for free fluid flow. *J. Energy Resour. Technol.* **2001**, *123*, 311–317. [CrossRef]

Article

Experimental and Numerical Characterization of a Gravitational Electromagnetic Energy Harvester

Caterina Russo *, Mirco Lo Monaco, Federico Fraccarollo and Aurelio Somà

Department of Mechanical and Aerospace Engineering, Politecnico di Torino, Corso duca degli Abruzzi 24, 10129 Torino, Italy; s270339@studenti.polito.it (M.L.M.); federico.fraccarollo@polito.it (F.F.); aurelio.soma@polito.it (A.S.)
* Correspondence: caterina.russo@polito.it

Abstract: In this paper, the dynamic experimental identification of an inductive energy harvester for the conversion of vibration energy into electric power is presented. Recent advances and requirements in structural monitoring and vehicle diagnostic allow defining Autonomous Internet of Things (AIoT) systems that combine wireless sensor nodes with energy harvester devices properly designed considering the specific duty cycle. The proposed generator was based on an asymmetrical magnetic suspension and was addressed to structural monitoring applications on vehicles. The design of the interfaces of the electric, magnetic, and structural coupled systems forming the harvester are described including dynamic modeling and simulation. Finally, the results of laboratory tests were compared with the harvester dynamic response calculated through numerical simulations, and a good correspondence was obtained.

Keywords: Autonomous Internet of Things; vibration energy harvesting; electromagnetic–mechanical modeling; autonomous sensors

Citation: Russo, C.; Lo Monaco, M.; Fraccarollo, F.; Somà, A. Experimental and Numerical Characterization of a Gravitational Electromagnetic Energy Harvester. *Energies* **2021**, *14*, 4622. https://doi.org/10.3390/en14154622

Academic Editor: Dibin Zhu

Received: 28 June 2021
Accepted: 26 July 2021
Published: 30 July 2021

Publisher's Note: MDPI stays neutral with regard to jurisdictional claims in published maps and institutional affiliations.

1. Introduction

In recent years, the widespread adoption of IoT (Internet of Things) devices and technologies has seen their application in different and various fields, including in industrial ones. These wireless devices have became more and more interesting to industries, especially for maintenance and diagnostics. By means of these devices, it is possible to create a wireless sensors network, where each node is capable of giving fundamental data for a new and deeper knowledge of the components, structure, or vehicle under monitoring. These networks provide real-time information concerning the monitored device, providing the opportunity to record a continuous data stream, which can be crucial to opening new frontiers in predictive maintenance and condition monitoring. Among all the features a node should have, the most crucial ones are a long lifecycle, a reduced size, and a solid wireless connection. A long lifecycle is necessary due to the maintenance cost associated with the battery replacement and disposal. Furthermore, a small and compact size is required to realize nodes suitable for easy integration in the system, even in places inaccessible to or unsafe for human intervention. Moreover, each node needs to be solidly connected to the network to send data when required. As mentioned earlier, in these applications, the use of conventional batteries has become a disadvantage as their replacement requires human intervention and their disposal is an environmental and safety issue [1]. Considering this, the need for an alternative power supply for these sensor networks emerges. For this reason, the research topic of energy harvester devices continues to grow, especially in the Autonomous Internet of Things (AIoT) field. The process of the conversion of the unused energy in the form of electricity, called harvesting, could help provide unlimited energy for the lifespan of the electronic device, generating new categories of AIoT systems. The electric energy can be provided by many sources such as light, wind, temperature gradients, radio frequency waves, the kinetic energy of sea waves, and mechanical vibrations available in the ambient environment [2,3]. It is possible to classify, as schematically described in Figure 1, the different energy-harvesting devices by their energy sources such as solar, thermal, and vibration. Concentrating the study of this paper on the mechanical vibration energy source, the most common conversion principles are piezoelectric, magnetic inductive, and capacitive.

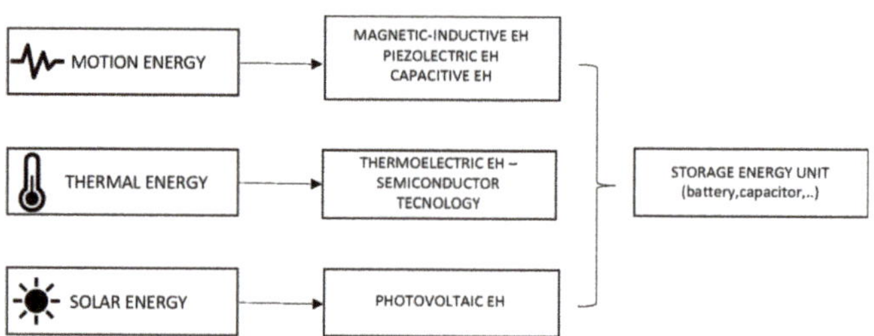

Figure 1. Schematic list of energy harvesting from environmental energy sources.

In previous works [4–6], the authors have analyzed the performances of different energy harvesters: piezoelectric, electrostatic-capacitive, and magnetically levitated; they also patented some dedicated devices for energy harvesting. Among them, the magnetic inductive energy-harvesting strategy was preferred in this work for the conversion of vehicle mechanical vibrations into electricity. The reasons were mainly related to the compatible dynamic response of these generators (low resonance frequency, long travels) and the better ratio between generated power and device volume (higher power density). The considered generators were designed to tune their dynamic response to the vehicle excitation, which means lowering the resonance frequency using very low stiff suspensions. For this purpose, traditional springs or mechanical suspensions were replaced with magnetic suspensions, which can considerably reduce the force/displacement ratio by preserving the long travels and velocities of the oscillating permanent magnet. The most crucial design issues of the inductive harvester are the dynamic response of the magnetic suspension and the electromagnetic coupled system dimensioning. The experimental characterization of different harvesters with different numbers of the coil is needed to identify the behavior of the generator.

Energy Harvesting Technology for a Power Supply Monitoring Sensor Network

As described in the Introduction, the devices powered by energy harvesters can be used to provide vital information on the operational and structural circumstances by placing them in inaccessible locations. They represent a valid alternative to batteries and cables and are promising for opening new opportunities for the development of mobile and wireless devices. Dedicated devices for harvesting the energy dissipated by the mechanical vibrations of machines have shown their applicability in supplying autonomous distributed sensing systems [7,8]. Energy harvesters will allow using, for example, wireless sensors in many applications such as industrial and structural monitoring, transportation and logistics, and energetic efficiency control [9]. Recent applications are also in the fields of wearable device and human walking monitoring [10,11]. They can provide help in the on-site charging of rechargeable batteries and supercapacitors. The sensors can be coupled with energy harvesting systems to obtain self-powered wireless sensing units; the sensing unit can measure environmental parameters and send the processed information by radio-frequency modules without physical connections, as shown in Figure 2. The conversion of unused energy into electricity has motivated many academic and industrial researchers during the last few years.

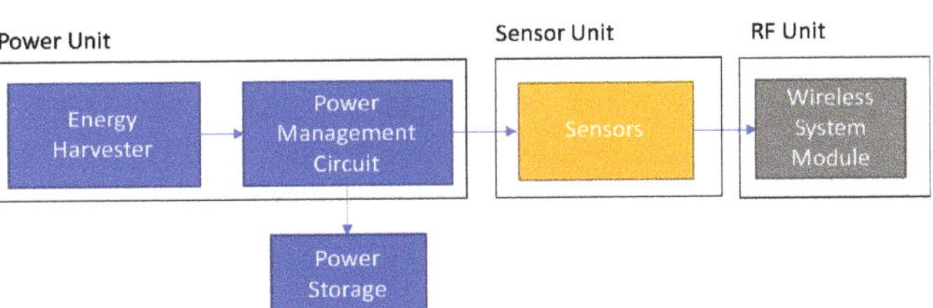

Figure 2. Architecture of a self-powered sensing wireless node system.

The focus of this paper is to present the characterization of a vibrational linear energy harvester and the realization of its numerical model in MATLAB/Simulink to estimate the power obtained by varying the resistive load or the excitation frequency. In the literature, these harvesters are widely studied and presented in different configurations of dimensions and springs. Obviously, the more the harvester size is reduced, the more it is suitable to be integrated into a sensor network, but using small masses and volumes makes harvesting the right amount of power more complex. On the other hand, a spring is chosen considering the proper stiffness to guarantee the tuning of the resonance frequency with the characteristic frequency of the application. Here we report some examples extracted from the literature that show different small-sized harvesters with symmetrical springs. First, in Yuen et al. [12], an example of an AA battery-sized energy harvester was proposed. This represents an interesting miniaturized system capable of 120 µW with 30 kΩ. Furthermore, in a more recent research work [13,14], a C battery-sized harvester was proposed so that the system was suitable for an integrated electronic device. The harvester was a two-degree-of-freedom nonlinear electromechanical harvester with a magnetic spring that, with the impact of the two masses, could provide a broadening of the power frequency response thanks to the impact of the two masses. The obtained output power was 0.178 mW with a high figure of merit of 2.6 at 11.5 Hz. The figure of merit used in these papers was the volumetric one; it compares the power output of the harvester represented with the power output of an harmonic oscillator of the same volume and mass, with the mass made of gold, the half volume occupied by the mass, and the other being free of displacement. These examples perfectly demonstrate the effort of the researchers to obtain the maximum power from harvesters of a very reduced size in order to insert them into a wireless network.

One of the things that distinguishes our model from others is the presence of a non-symmetrical magnetic spring with only one fixed magnet at the lower end of the harvester; in this way, the moving magnet returns to the equilibrium position by gravitational force. In fact, many harvesters proposed in the literature present a symmetrical magnetic spring with two fixed magnets collocated at the ends and an oscillating magnet. An example was shown in Vishwas Bedekar [15], where they generated 3 mW at 5 Hz and 1 mW at 3.5 Hz with a device the size of a pen. The integrated pen harvester prototype was found to generate continuous power of 0.46–0.66 mW under the excitation of usual human actions such as jogging and jumping, which is enough for a small-scale pulse rate sensor. Furthermore, Dallago [16] proposed an analytic model of a vibrating electromagnetic harvester, considering nonlinear effects. The model can predict the induced voltage power output with different loads applied. The structure of the harvester consisted of two fixed magnets on the ends and two moving magnets, arranged in a configuration that created a repulsive force. The output power produced from this device was 6 mW. Moreover, Christopher Lee [17] proposed a nonlinear magnetic suspension, modeled using the Duffing equation and using an experimental test to compare it to the numerical model, obtaining 16 mW. Furthermore, in the work of Marco P. Soares dos Santos [18], a magnetic suspension was used, and a combined experimental and theoretical approach was presented. The model provides unique insight into the fundamental mechanisms of energy transduction and enables both the geometric optimization of harvesters before manufacturing and the rational design of intelligent energy harvesters. Finally, in D.F. Berdy [19], a different configuration of the magnetic suspension was proposed with a parallelogram-shaped oscillating magnet.

The friction issue was highlighted and described also in the Simulink model, and 410 μW at 6.7 Hz of power was obtained. In this paper, a gravitational magnetic suspension energy harvester with cylindrical magnets is presented. In Section 2, the structure and the geometry of the device are described. In Section 3, the experimental apparatus is presented, both for the static and the dynamic tests. The fundamental equation of the system and the numerical model are presented in Section 4, and in Section 5, the experimental results and their comparison to the numerical one are shown. At the end of Section 6, the results and the conclusion are discussed.

2. Harvester Configuration

The harvester scheme, reported in Figure 3, included a fixed magnet, a suspended magnet, and four coils. One characteristic of this harvester is the presence of an asymmetrical spring. The fixed magnet was present only at the lower end of the harvester, and the moving magnet returns to the equilibrium position by gravitational force. Due to this characteristic, a bumper was placed to avoid impacts within the moving magnet and the upper end of the harvester.

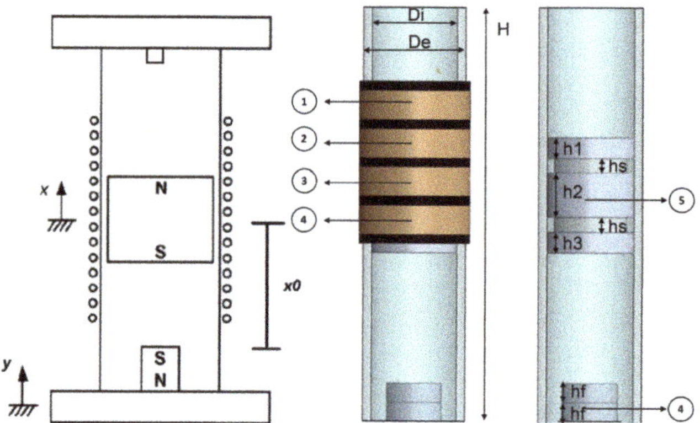

Figure 3. Energy harvester scheme.

In Table 1 are reported the main components of the energy harvester. In particular, it is important to underline the structure of the moving magnet. It was composed by three magnets assembled using their attraction forces and spacers with proper dimensions to obtain the requested global height and mass of the magnets. The maximum travel of the moving magnet was measured from the top surface of the fixed magnet (x_0 on the left of Figure 3) and, in the same way, the other measurements were made as well, such as the height of each coil and the equilibrium position. The last one is the distance between the top face of the fixed magnet and the center of mass of the moving one.

The main characteristics of the two harvesters are reported in the tables below: Tables 1–3.

Table 1. Description of the components of the energy harvester.

N	Element
1	Coil 1
2	Coil 2
3	Coil 3
4	Coil 4
5	Moving Magnet
6	Fixed Magnet

Table 2. Characteristics of the fixed and moving magnet.

	Moving Magnet	**Fixed Magnet**
Magnetization	N42	N42
Magnetic flux density (T)	1.29–1.32	1.32–1.37
Total mass (g)	186	10
Diameter (mm)	30	15
Height h1,h3 (mm)	7	
Height h2 (mm)	15	
Spacer height,hs (mm)	5	
Total height (mm)	39	
Height hf		9.5

For this study, two types of energy harvester were compared, Harvester 1 (EH 1) and Harvester 2 (EH 2), keeping the same magnets, but changing the number of turns of each coil. In particular, Harvester 1 had four coils, each of two-hundred eighty turns, while Harvester 2 had four coils, each of four-hundred thirty-five turns. Moreover, for Harvester Coils 2 and 4 were tested to find out the effect of the axial coil location on the output power. The main characteristics of the two harvesters are reported in the table below.

Table 3. Dimensions and common electric characteristics of Harvesters 1 and 2.

	General Characteristics of Harvester
Cylinder height,H (mm)	188.5
Cylinder external diameter,De (mm)	36
Cylinder internal diameter,Di (mm)	30.2
Max moving magnet travel (mm)	176.6
Axial coil length (mm)	11
Equilibrium position of the magnet,x_0 (mm)	59.6
External coil diameter (mm)	39
Internal coil diameter (mm)	36
Wire diameter (mm)	0.1
Wire cross section area (m^2)	$7.9 \cdot 10^{-9}$
Copper resistivity (Ωm)	$1.7 \cdot 10^{-8}$

In Table 4, the height of each coil is reported using as the zero reference the top end of the fixed magnet.

Table 4. Dimensions and electric characteristics specific to Harvesters 1 and 2.

	Harvester 1	**Harvester 2**
Coil 1 first turn height (mm)	53.8	51.1
Coil 2 first turn height (mm)	68.3	68.1
Coil 3 first turn height (mm)	84.6	84.1
Coil 4 first turn height (mm)	103.9	102.1
Coil resistance (Ω)	70	110
Number of turns	280	435

3. Experimental Setup

For the static characterization of the magnetic suspension, a test bench including a DC voltage generator and two LK-G82 KEYENCE (Mexico City, Mexico) laser sensors (50 kHz sampling frequency, 0.2 mm, 70.05% accuracy) with an LK-G signal controller was used. The harvester was integrated with a support beam having a plate on the top for the application of the calibrated masses. The equilibrium position of the moving magnet was evaluated leaving the supporting beam free from the calibrated masses, then the other positions of the moving magnet were computed with respect to the equilibrium position by applying fourteen different masses with increasing values.The measurements of the moving magnet position were conducted on a seismic table to reduce the effects of external noise on the experimental results. The laser sensors were supplied by the DC voltage generator and used to measure the different positions of the moving magnet under the weight of the calibrated masses. The measured data were sent to a signal controller and then to a PC for postprocessing. The schematic of the static tests setup is represented in Figure 4a, and the picture of the static experimental bench is shown in Figure 4b.

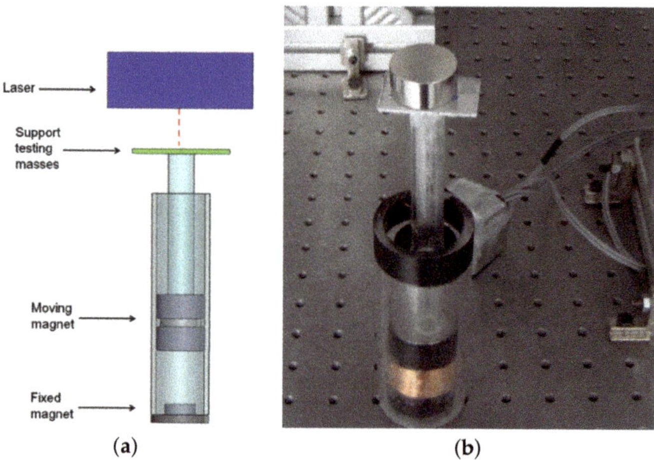

Figure 4. Picture and scheme of the experimental static measures. (**a**) Scheme of the experimental static measures. (**b**) Picture of the experimental static measures.

The dynamic characterization of the suspension and the measurement of the harvester performances were conducted with a shaker (TIRA TV51120) with its amplifier (BAA 500) with variable gain control (TIRA GmbH, Schalkau, Germany). LabView (National Instruments, Austin, TX, USA) was used to supply a sinusoidal voltage signal to the moving base of the shaker; the closed-loop feedback signal was provided by an IMI 608A11 piezoelectric accelerometer (PCB Piezotronics, Depew, NY, USA). The output voltage from the harvester was measured across the load resistor, whose value was chosen in order to obtain the maximum power. The scheme and a representation of the test bench are reported in Figures 5 and 6.

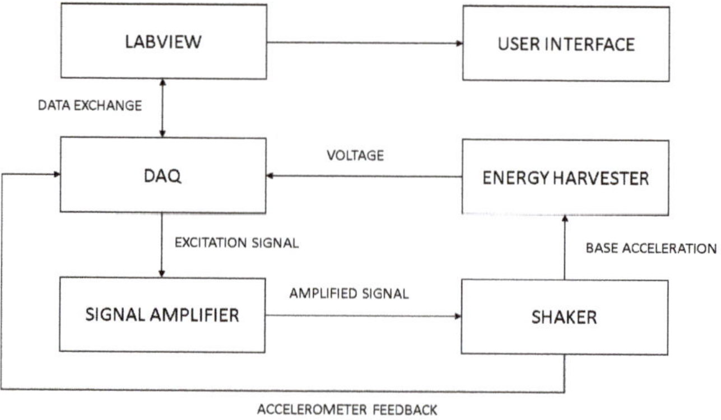

Figure 5. Input/output scheme of the experimental apparatus.

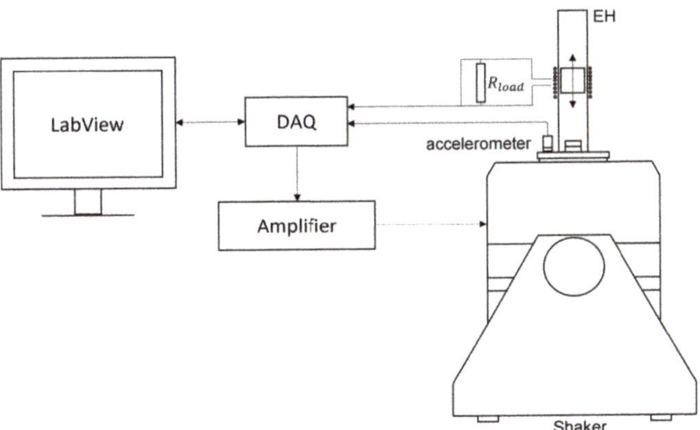

Figure 6. Input/output scheme of the experimental apparatus.

4. Mathematical Model

Roundy et al. [8] proposed the basic principles of the functionality of a vibration energy harvester. These generators consisted of a seismic mass m attached by means of a spring and a viscous damper to a moving base (Figure 7), and its analytical model was a 1 d.o.f mass–spring–damper system. The moving magnet had a relative motion z with respect to the base, which received an imposed sinusoidal input base displacement y. The vibrational external source made the seismic mass move out of phase with respect to the box. This relative motion between the mass and the base could drive a suitable transducer to generate electrical energy.

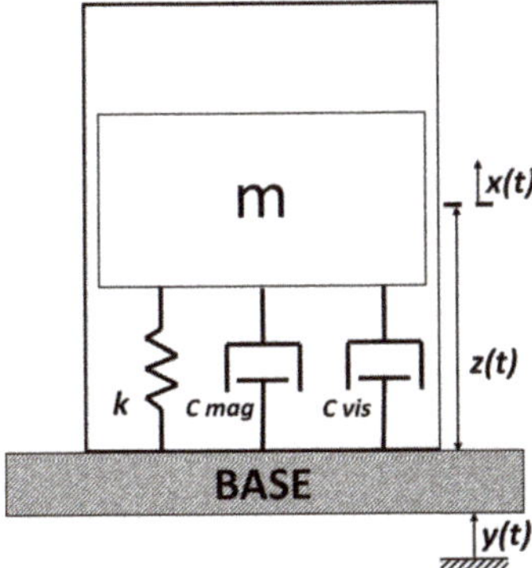

Figure 7. The 1 d.o.f mass–spring–damper scheme model.

The magnetic interaction within the two magnets can be studied as a nonlinear magnetic spring. Moreover, the damping coefficient had two main components, the viscous one caused by the air trapped in the tube and the friction between the magnets and the internal surface of the tube; then, there was also the magnetic damping due to induction interaction between the moving magnet and the coils. The equation of motion can be written in terms of the relative motion, considering the starting position as the equilibrium of the moving magnet, when the repulsive magnetic force is equal to its weight. Thus, the absolute motion coordinate of the moving magnet x is related to the static equilibrium condition of the moving magnet x_0.

$$m\ddot{x} + c_{tot}(\dot{x} - \dot{y}) + k(x - y) = 0 \tag{1}$$

$$z = x - y \tag{2}$$

$$m\ddot{z} + c_{tot}\dot{z} + kz = -m\ddot{y} \tag{3}$$

$$\ddot{y} = -\omega^2 y_0 sin(\omega t) = -Y_0 sin(\omega t) \tag{4}$$

Then, the form to introduce this equation into MATLAB Simulink is:

$$\ddot{z} = -\frac{c_{tot}}{m}\dot{z} - \frac{k}{m}z + Y_0 sin(\omega t) \tag{5}$$

where:

- m is the mass of the moving magnet;
- c_{tot} is the sum of viscous damping c_{vis} and magnetic damping c_{mag};
- k is the magnetic nonlinear stiffness;
- Y_0 is the acceleration amplitude (in this case, it is sinusoidal);
- ω is the excitation frequency.

In Table 5, the main Simulink blocks, identified by the blue numbered labels in Figure 8, are described to have a better understanding of the Simulink model implementation.

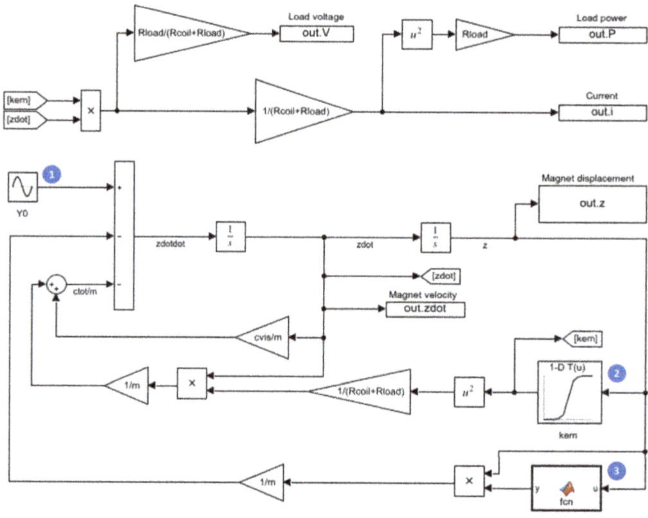

Figure 8. Simulink model of the energy harvester.

Table 5. Main Simulink blocks.

N	Type	Description
1	Sine Wave	Excitation with Y_0 amplitude and ω frequency
2	1D Lookup Table	Electromagnetic coefficient–moving magnet displacement relation
3	MATLAB Function	Stiffness–moving magnet displacement relation

The model built in MATLAB/Simulink allowed studying the dynamic behavior of the system and the interaction between its mechanical and electrical parts in a single simulation. Moreover, it was possible to introduce the nonlinear characteristic of the system into the model in a straightforward way. In detail, the stiffness characteristic was modeled as a MATLAB function block, which received as the input the instantaneous moving magnet position (z) and computed the output stiffness, as explained in the next Section 4.1. To introduce the electromagnetic coupling coefficient, a 1D lookup table was used, which received as the input the instantaneous moving magnet position (z) and computed the output as is explained in the next Section 4.2. To give a complete description of the simulation model, the settings for the main model parameters are reported in Table 6.

Table 6. MATLAB/Simulink model settings.

Name	Value
Simulation Time	20 (s)
Solver Type	ode45 variable step
Max Step Size	10^{-3}
Min Step Size	auto
Relative Tolerance	10^{-3}

4.1. Mechanical Model

The stiffness of the magnetic suspension is directly influenced by the properties of the two magnets. By varying their dimensions, masses, and magnetization, it is possible to obtain different nonlinear characteristics of the spring. The experimental data regard the stiffness values of the suspension when the moving magnet is pressed versus the fixed one by the application of the calibrated masses. Since the equilibrium position was assumed as the origin of the z-axis, these stiffness values refer to a negative position of the moving magnet. The stiffness numerical model was obtained by finding the best fit of these experimental results. The chosen relation between the magnetic force and the displacement of the moving magnet is an exponential equation, as is shown in the the Figure 9.

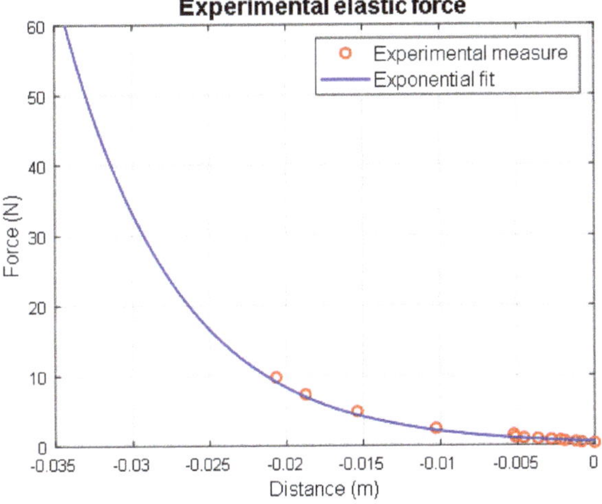

Figure 9. Overlap of the experimental and exponential fit of the magnetic stiffness.

The zero position coincides with the equilibrium position of the magnet x_0. The magnetic force equation is expressed in the form:

$$F(z) = ae^{-bz} \tag{6}$$

Subsequently, the stiffness can be obtained from the derivative of this equation:

$$k(z) = -bae^{bz} \tag{7}$$

This allows computing the stiffness in the equilibrium position and obtaining the natural frequency of the linear system.

$$k(z = 0) \approx 72 N/m \tag{8}$$

$$\omega = \sqrt{\frac{k_{lim}}{m}} \approx 3.4 Hz \tag{9}$$

However, the introduction of the previously described exponential stiffness also for the positive values of the z-axis into the MATLAB/Simulink model resulted in an inaccurate description of the real behavior. This inaccuracy was brought to light by a too low stiffness, causing resonance peaks at lower values with respect to the experimental results. To overcome this problem, the stiffness for the positive z-axis was interpolated as a polynomial equation between the value of stiffness in z = 0 and a supposed value at the extreme position of the cylinder, found by trial and error to have the best superimposition of the numerical and experimental resonance peaks, the result is shown in Figure 10. The final expression of the stiffness of the magnetic spring results:

$$k(z) = -bae^{bz} z \leq 0 \tag{10}$$

$$k(z) = p_1^3 + p_2^2 - p_3 + p_4 z > 0 \tag{11}$$

where:

- $a = 0.53$
- $b = -137.9$
- $p_1 = 6.387 \cdot 10^4$
- $p_2 = 1.858 \cdot 10^4$
- $p_3 = 1906$
- $p_4 = 71.71$

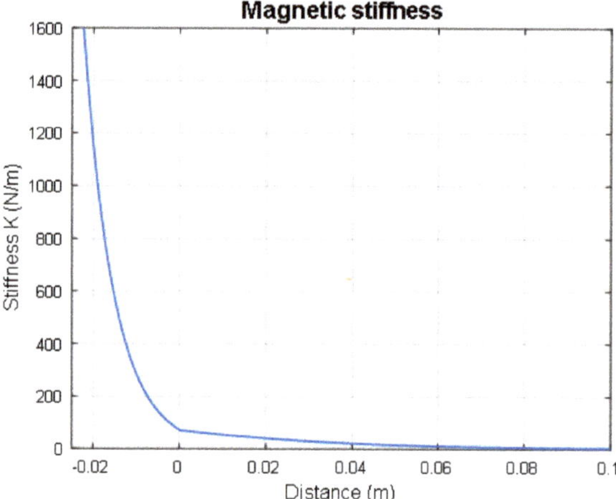

Figure 10. Exponential magnetic stiffness and cubic polynomial fit when the magnet moves away from the fixed one.

The mechanical action of the air and the friction between the cylinder's inner surface and the moving magnet were modeled as viscous damping. The damping ratio coefficient was evaluated experimentally by exciting the harvester with a step input on the dynamic test bench. The output voltage generated in time by the device was then processed, and the coefficient was obtained through its logarithmic decrement. Its value increased when the moving magnet velocity decreased, leading to nonlinear viscous damping depending on the oscillation amplitude of the moving magnet. This phenomenon was caused by tolerance errors in the inner surface of the cylinder, leading to variable friction with the moving magnet during its motion. The nonlinear viscous damping had a minimum when the moving magnet was in resonance and a maximum when it oscillated around the equilibrium position. For the sake of simplicity in the Simulink model, constant viscous damping having the value obtained by the mean logarithmic decrement was introduced, obtaining a good approximation. Since Harvesters 1 and 2 had the same cylinder and magnets, they had the same viscous damping ratio, equal to $\zeta = 0.04$.

4.2. Electromagnetic Model

The purpose of this study is to evaluate the output power generated on the load both by numerical and experimental analysis. The tests on the experimental bench returned the load voltage value, and the corresponding power could be computed. By applying the Kirchoff laws to the circuital part of the system (Figure 11), it was possible to relate the voltage produced by the harvester to the output power on the load.

$$i = \frac{V_{eh}}{(R_{load} + R_{coil})} \tag{12}$$

$$P_{load} = V_{load} i \tag{13}$$

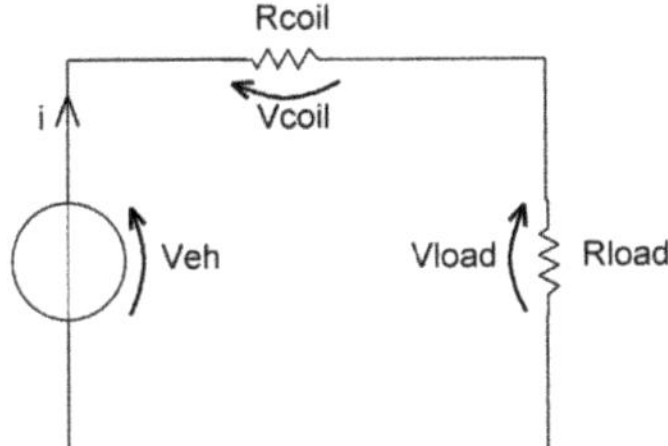

Figure 11. Electrical circuit of the energy harvester.

The electromagnetic damping was introduced to the nonlinear model of the energy harvester by analyzing its circuital part. Another damping force component can be introduced in the equation of motion considering its dependence on the velocity state.

$$k_{em}(z) = -\frac{d\Phi}{dz} \tag{14}$$

$$V_{eh} = -\frac{d\Phi}{dt} = -\frac{d\Phi}{dz}\frac{dz}{dt} = k_{em}\dot{z}(t) \tag{15}$$

$$c_e(z) = \frac{k_{em}^2(z)}{R_{load} + R_{coil}} \tag{16}$$

$$F_e = k_{em}(z) \cdot i(t) = k_{em}^2(z)\frac{\dot{z}(t)}{R_{load} + R_{coil}} \tag{17}$$

where:
- k_{em} is the electromagnetic coupling coefficient;
- $\frac{d\Phi}{dz}$ is the variation of the magnetic flux on the z-axis;
- V_{eh} is the voltage generated by the harvester;
- c_e is the electromagnetic damping;
- F_e is the electromotive force.

The electromagnetic coupling coefficient k_{em} links the electric with the mechanical variables of the system. Considering it as a constant leads to an approximation, because, as the moving magnet travels in the tube during vibration, its flux linkage with the coils varies. In addition, the coils have a non-negligible extension along the axial direction, so the coefficient has another degree-of-freedom related to the position of the coil compared to that of the moving magnet. For this reason, the axial location of the coil is an important variable that needs to be studied to find the optimum value of the output power. To derive the coefficient values, the magnetic flux produced by the moving magnet through the winding should be evaluated. Considering the radial magnetic field produced by the cylindrical magnet, assumed to be a dipole, the relation between the electromagnetic coefficient and the position of the magnet is obtained [20,21]:

$$k_{em}(x) = \frac{1}{2} \frac{m \mu_0 l_w}{V_c} \sum_{i,j=1}^{2} (-1)^{i+j} \left[ln\left(r_i + \sqrt{r_i^2 + (a_j - x)^2} \right) - \frac{r_i}{\sqrt{r_i^2 + (a_j - x)^2}} \right] \tag{18}$$

$$m = \frac{1}{\mu_0} B_r V_m \tag{19}$$

where:

- l_w is the total length of the coil;
- V_c is the volume of the coil;
- $r_{1,2}$ are respectively the inner and outer radius of the coil;
- $a_{1,2}$ are respectively the height of the first and last turn of the coil;
- $\mu_0 = 4\pi \cdot 10^{-7}$ is the magnetic constant;
- V_m is the moving magnet volume;
- m is the magnetic moment of the dipole.

In Figures 12 and 13, the plots of the electromagnetic coefficient and damping with respect to the magnet position for Harvester 2 are reported. The height and axial location of the coils have a direct effect on the electromagnetic circuit in terms of shifting these curves along the x-axis, changing thevalue of the coefficients at a certain magnet position during vibration.

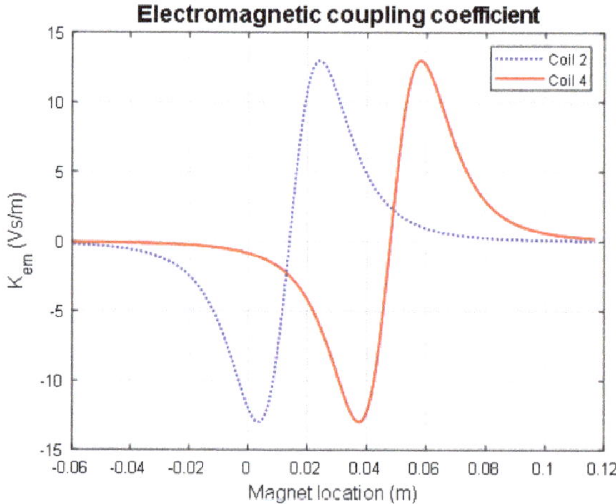

Figure 12. Electromagnetic coupling coefficient for Coils 2 and 4 over the extension of the harvester.

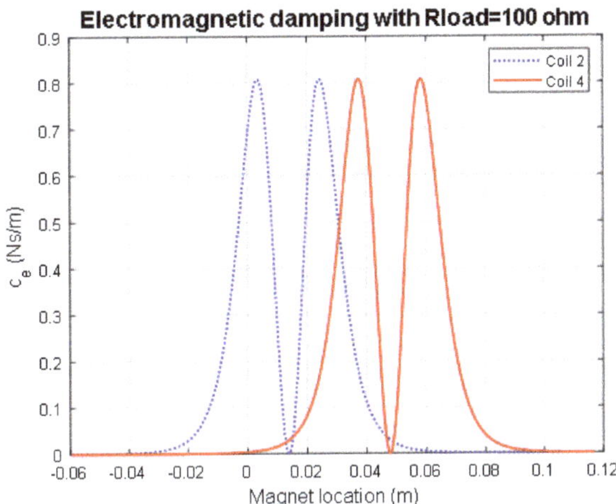

Figure 13. Electromagnetic damping for Coils 2 and 4 over the extension of the harvester.

5. Experimental Results

In this section, the experimental results are given and their comparison with the numerical model is proposed. The most relevant experimental results are the frequency response function (FRF) and the load response. The experimental tests returned as the output the voltage on the load, whose value could be used to derive the current, and finally, the power. The FRF is expressed in terms of RMS power, and it was computed using an optimum resistive load chosen during the load tests. The FRF curves were obtained by varying the frequency of the sinusoidal excitation and keeping the amplitude constant. Figure 14 below shows the FRFs obtained for different excitation amplitudes, from 0.1 g to 0.4 g, on Coil 2 of Harvesters 1 and 2, with an optimum resistive load respectively of 100 Ω and 150 Ω. Through this comparison, it is possible to point out the nonlinear behavior of the harvester. In the literature, there are several symmetrical spring energy harvesters, which have two fixed magnets on the two ends of the tube and present a symmetrical stiffness characteristic. In this case, as described in the previous paragraph, the stiffness is nonsymmetrical due to the presence of only one fixed magnet, and this produces a slightly softening spring behavior. In this case, a harvester with the same magnetic characteristic, but with two magnets on the top and lower end is expected to produce a reduced amount of power because keeping the same dimension of the case, the axial displacement of the moving magnet is reduced due to the presence of the fixed magnet on the top end. In line with the spring stiffness characteristic, as seen Section 4.1, the magnetic force decreases rapidly as the moving magnet moves away from the fixed one. This behavior causes the decrease of the resonance frequency with increasing excitation amplitude because, for larger resonant oscillation, the spring force decreases. This characteristic leads to a nonlinear response that can be observed by the sudden drop in the power amplitude for frequencies slightly below the resonance peak. Moreover, from the comparison between the FRFs of Harvester 1 and Harvester 2, it is possible to observe that Harvester 2, with a coil of 435 turns, generated more power at every excitation. In particular, for 0.4 g, the maximum power for Harvesters 1 and 2 was 31 mW and 44 mW. For different excitations the results obtained are reported in Table 7.

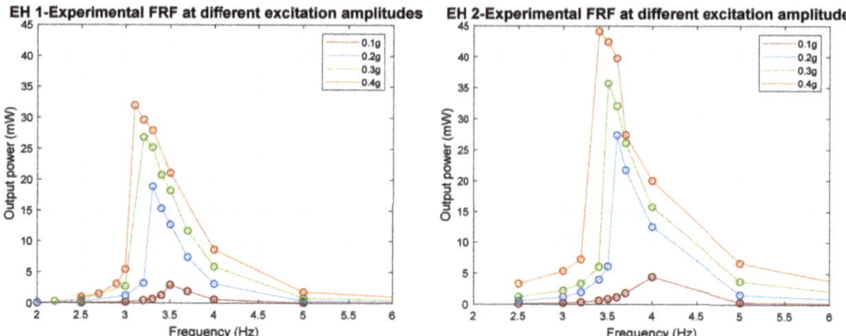

Figure 14. Experimental RMS power FRF for different excitation amplitudes, on the left, for EH 1 and, on the right, for EH 2.

In Figures 15 and 16, the first comparison between the model and the experimental device for the coil of Harvesters 1 and 2 is presented. In particular, the plots below show the FRF obtained experimentally and from the numerical model for different excitation amplitudes, from 0.1 g to 0.4 g, with an optimum resistive load of 100 Ω and 150 Ω, respectively. A good correspondence both in the peak amplitude and resonance frequency was obtained. Considering the different model implementations of the viscous damping with respect to the real behavior of the harvester, as seen in Section 4.1, errors on the resonance peak amplitudes between the numerical and experimental results could be found. In particular, this occurred when the numerical peaks had lower amplitudes than the experimental ones because the constant viscous damping was greater than the real nonlinear one in resonance and, vice versa, when the experimental peak was greater than the numerical one. In this study, we focused mainly on the resonance frequency tuning of the model with the real device, obtaining a maximum difference of 0.3 Hz between the numerical and experimental peaks. In future works, the nonlinear behavior of the viscous damping will be correctly implemented in the numerical model, even though this simplification would not lead to significant inconsistencies. For 0.4 g and 0.1 g excitation, a slight difference in resonance frequency between the numerical model and experimental results occurred. This could be the result of the modeling procedure of the stiffness, as explained in Section 4.1, but mainly because the experimental data were gathered around a frequency of 3 Hz, which was near the lower end of the shaker's working range.

Table 7. RMS output power for EH 1 and EH 2.

Excitation	EH 1 (mW)	EH 2 (mW)
0.1 g	2.6	4.4
0.2 g	18.8	27.4
0.3 g	26.7	35.7
0.4 g	31.6	44.1

The load resistance is a variable of the circuit that needs to be optimized because it directly affects the output power of the system. It was introduced into the expression of the electromagnetic damping (see Equation (16)), and therefore in the generated current Equation (12). The tests on the experimental bench confirmed the dependence of the electrical quantities on the load resistance and returned the value that generated the optimized power. In order to optimize the output power of the harvester, the reduction of the electrical damping is one of the possible solutions. In the literature [22,23], it was found that this can be obtained by matching the mechanical and electrical load; the result presented it for a linear model, but after the test on the four different harvesters with an increasing number of turns, it was possible to obtain a result in line with this statement. In fact, it can be pointed out that the load resistance optimum almost coincided with the coil resistance of each solution. The numerical model results properly fit the experimental values of the voltage, current, and power, as can be seen in the plots in Figure 17 for Coil 2 of Harvesters 1 and 2, with an excitation amplitude of 0.2 g and at their resonance frequency.

EH 1 – FRF at different excitation

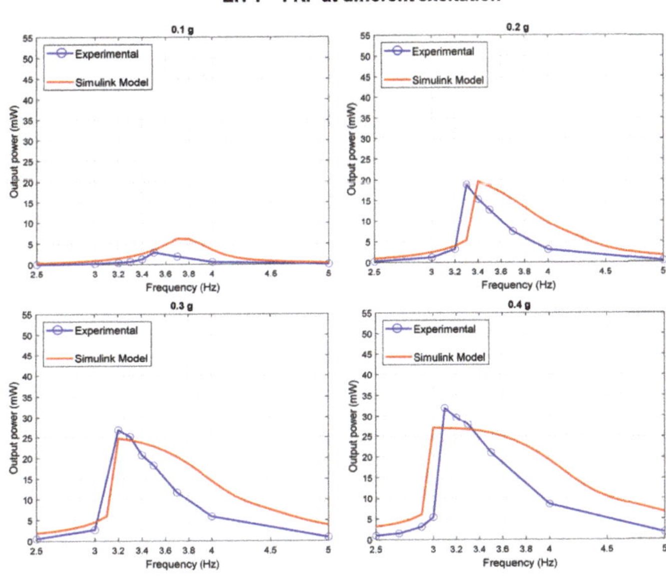

Figure 15. EH 1: Experimental and model comparison of the RMS output power FRF for different excitation amplitudes.

EH 2 – FRF at different excitation

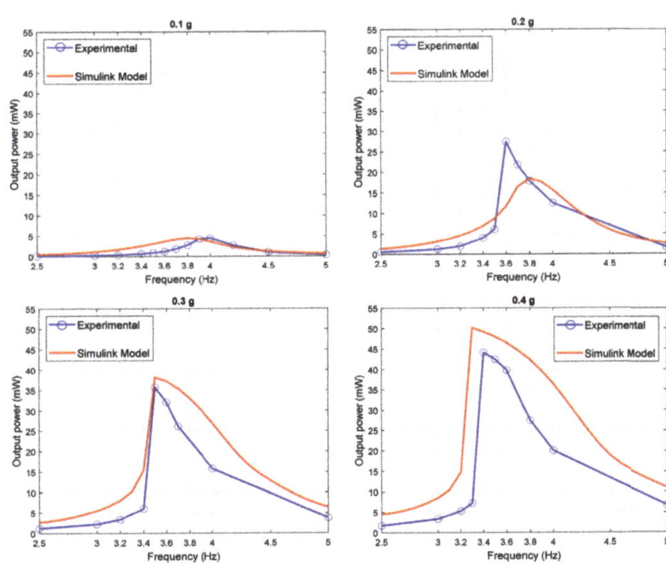

Figure 16. EH 2: Experimental and model comparison of the RMS output power FRF for different excitation amplitudes.

The optimization of the output power required the design of the coil position and length: they both affect the flux linkage during the magnet oscillation. For this purpose, various experimental tests changing the coil axial location and the number of turns (see Table 8) should be carried out to compare the results of the output power at resonance and find a trend. The tests on the axial location were implemented on the same harvester, with a fixed excitation amplitude of 0.3 g, a constant number of turns, and consequently, coil resistance and optimized resistive load. The axial location tests (see Figure 18) showed that the highest power was generated by Coil 2, whose location almost coincided with the equilibrium position of the moving magnet.

Table 8. Effect of the number of turns on the output power at a 0.2 g excitation amplitude.

N_{turns}	$R_{coil}(\Omega)$	Optimum $R_{load}(\Omega)$	Optimum Power 0.2 g (mW)
280	70	100	18.8
435	110	150	27.4
550	140	180	28.6
870	220	270	15.8

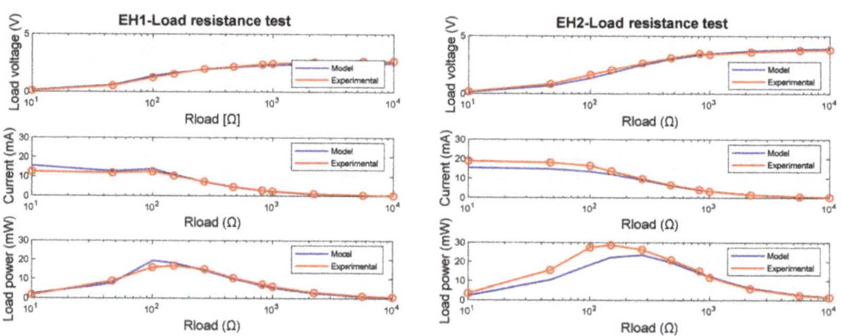

Figure 17. Experimental and numerical comparison of the load tests.

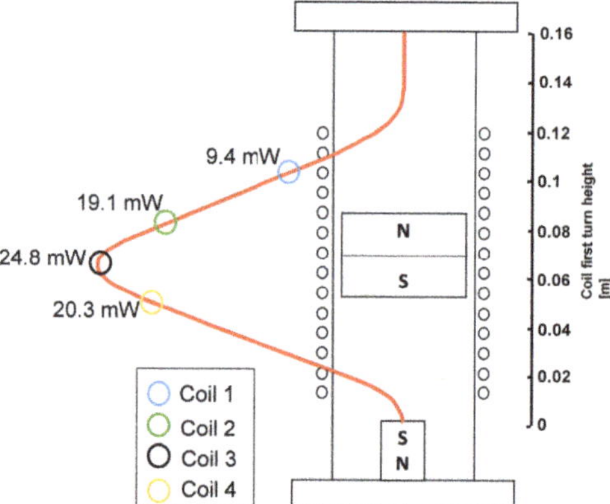

Figure 18. The optimum power output with respect to the height of the harvester.

For each test on the number of turns, the RMS power was evaluated with the optimized resistive load, considering that the coil resistance changes, keeping the excitation amplitude at 0.4 g and connecting Coil 2. For both analyses, the numerical model results fit the experimental ones well. The number of turns tests (see Figure 19) showed that the optimum value of the number of turns was 550, with a corresponding R_{coil} = 140 Ω. The output power of the harvester with Coil 2, a number of turns of 550, optimized R_{load} = 180 Ω, at a resonance frequency of 3.3 Hz, and an excitation amplitude of 0.4 g was the maximum obtained on the experimental tests and was equal to 46 mW.

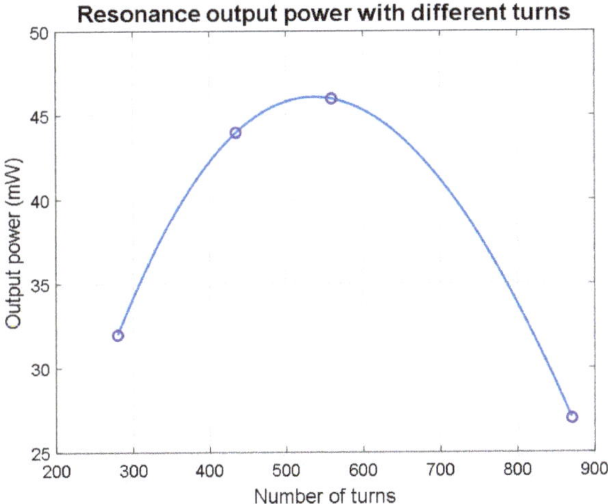

Figure 19. The optimum power with respect to the number of turns.

6. Discussion and Conclusions

Many electromagnetic energy harvester solutions, as the one presented in this paper, can be found in the literature. The applications of these devices are many, and consequently, so are their configurations and generated power. A comparison of the obtained results from energy harvesters similar to the one described in this work is shown in Table 6. The table compares the harvesters with respect to their power density (PD), which is the ratio between the output power and the volume of the harvester and their normalized power density, which is the power density divided by the excitation squared [24]. This last index offers a fairer comparison among the harvesters, taking into account also the amplitude of the excitation at which they produce the considered power. The harvester of this paper generated one of the highest powers at the lowest frequencies among the considered solutions. As seen in Table 9, this harvester presented a normalized power density (NPD) of 1.44 mW/cm^{3}g^2, which is a good value because it offers a high power considering the low resonance frequency (3.4 Hz). This characteristic needs to be pointed out because, for this research, it is important to maintain a low frequency for future industrial application, for example in the railway field.

Table 9. Literature comparison.

Year of Publication	Reference	R.M.S (mW)	Frequency (Hz)	Excitation (g)	Volume (cm^3)	Power Density (mW/cm^3)	NPD (mW/cm^{3}g^2)
2007	[25]	90	37.0	2	150	0.600	0.5
2011	[26]	1.18	9.0	1.6	7.4	0.159	0.06
2013	[27]	4.48	6.0	0.5	9.05	0.495	2.14
2014	[19]	0.41	6,.7	0.1	7.7	0.053	5.3
2015	[28]	2.15	5.17	2.06	6.47	0.33	0.078
2016	[29]	2.06	11.50	0.4	8.12	0.25	1.58
2018	[30]	20.6	5.0	6	6.77	3.04	0.08
2021	This work	44.1	3.4	0.4	191	0.23	1.44

So far, the main drawback of our harvester was its dimensions, which have to be optimized in order to be more suitable for IoT applications. Further developments may address bandwidth amplification strategies and active resonance tuning of the harvester to improve the global efficiency of the device.

In conclusion, this paper carried out a study on the dynamic behavior of a gravitational electromagnetic generator based on asymmetrical magnetically levitated suspensions addressed to structural monitoring applications. The design of the interfaces of the electric, magnetic, and structural coupled systems forming the harvester were described including the dynamic modeling and simulation. The characterization activity on the energy harvester prototypes showed very promising results in terms of output power and conversion efficiency. The frequency and acceleration identified as the typical operative working condition were addressed to applications in the vehicle field. The experimental tests had the purpose of evaluating the magnetic stiffness curve and the dependency of the electrical power on the number of turns and their height. The F-z curve was then used to model the device dynamics with a single-degree-of-freedom model. The frequency response function (FRF) was obtained by measuring the oscillation amplitude at variable frequencies and the voltage output over an optimum resistance R_{load}. The output power was evaluated for different harvester prototypes tuned to the frequency of the dominant component of the vibration frequency range of the vehicle dynamics. The results of the laboratory test and the harvester dynamic response were compared with those obtained using numerical simulations built in MATLAB/Simulink, and a good correspondence was obtained, in terms of both the FRF and the load curve. The results of the present paper can be used in the system engineering approach to design novel AIoT systems, taking into account the specifications and requirements of the duty cycle and power spectral density available in the environment.

Author Contributions: The authors contributed equally to this manuscript. All authors read and agreed to the published version of the manuscript.

Funding: This research received no external funding.

Institutional Review Board Statement: Not applicable.

Informed Consent Statement: Not applicable.

Data Availability Statement: Not applicable.

Conflicts of Interest: The authors declare no conflict of interest.

References

1. Dewan, A.; Karim, M.N.; Beyenal, H. Alternative power sources for remote sensors: A review. *J. Power Sources* **2014**, *10*, 142–149. [CrossRef]
2. Beeby, S.; O'Donnell, T.P. *Electromagnetic Energy Harvesting*; Priya, S., Inman, D.J., Eds.; Springer: Boston, MA, USA, 2009.
3. Chalasani, S.; Conrad, J.M. A survey of energy harvesting sources for embedded systems. In Proceedings of the IEEE Southeastcon, Huntsville, AL, USA, 3–6 April 2008; pp. 442–447. [CrossRef]
4. De Pasquale, G.; Soma, A.; Zampieri, N. Design, Simulation, and Testing of Energy Harvesters With Magnetic Suspensions for the Generation of Electricity From Freight Train Vibrations. *J. Comput. Nonlinear Dyn.* **2012**, *7*, 41011. [CrossRef]
5. De Pasquale, G.; Somà, A.; Fraccarollo, F. Comparison between piezoelectric and magnetic strategies for wearable energy harvesting. *J. Phys. Conf. Ser.* **2013**, *476*, 12097. [CrossRef]
6. De Pasquale, G.; Somà, A.; Fraccarollo, F. Piezoelectric energy harvesting for autonomous sensors network on safety-improved railway vehicles. *J. Mech. Eng. Sci.* **2012**, *226*, 1107–1117. [CrossRef]
7. Approaching the Horizon of Energy Harvesting, Texas Instruments. Available online: https://www.ti.com/lit/wp/sszy004/sszy004.pdf?ts=1627464210392&ref_url (accessed on 28 June 2021).
8. Roundy, S.; Wright, P.; Rabaey, J.K. A study of low level vibrations as a power source for wireless sensor nodes. *Comput. Commun.* **2003**, *26*, 1131–1144. [CrossRef]
9. Bosso, N.; Magelli, M.; Zampieri, N. Application of low-power energy harvesting solutions in the railway field: A review. *Veh. Syst. Dyn.* **2021**, *59*, 841–871. [CrossRef]
10. Shi, H.; Liu, Z.; Mei, X. Overview of human walking induced energy harvesting technologies and its possibility for walking robotics. *Energies* **2019**, *13*, 86. [CrossRef]
11. Mocera, F.; Aquilino, G.; Somà, A. Nordic Walking Performance Analysis with an Integrated Monitoring System. *Sensors* **2018**, *18*, 1505. [CrossRef]
12. Yuen, S.; Lee, J.; Li, W.; Leong, P. An AA-Sized Vibration-Based Microgenerator for Wireless Sensors. *Pervasive Comput. IEEE* **2007**, *6*, 64–72. [CrossRef]
13. Nico, V.; Boco, E.; Frizzell, R.; Punch, J. A C-battery scale energy harvester, part a: System dynamics. In *Smart Materials, Adaptive Structures and Intelligent Systems*; American Society of Mechanical Engineers: New York, NY, USA, 2015; Volume 57304.
14. Boco, E.; Nico, V.; Frizzell, R.; Punch, J. A C-battery scale energy harvester, part b: Transducer optimization and modeling. In Proceedings of the ASME 2015 Conference on Smart Materials, Adaptive Structures and Intelligent Systems, 21–23 September 2015; Springs: Colorado, CO, USA, 2015; Volume 2, pp. 1–10.

15. Bedekar, V.; Oliver, J.; Priya, S. Pen harvester for powering a pulse rate sensor. *J. Phys. Appl. Phys.* **2009**, *42*, 64–72. [CrossRef]
16. Dallago, E.; Marchesi, M.; Venchi, G. Analytical Model of a Vibrating Electromagnetic Harvester Considering Nonlinear Effects. *IEEE Trans. Power Electron.* **2010**, *25*, 1989–1997. [CrossRef]
17. Lee, C.; Stamp, D.; Kapania, N.R.; Mur-Miranda, J.O. Harvesting vibration energy using nonlinear oscillations of an electromagnetic inductor. *Energy Harvest. Storage* **2010**, *7683*, 76830.
18. Dos, Santos, M.P.S.; Ferreira, J.A.; Simões, J.A.; Pascoal, R.; Torrão, J.; Xue, X.; Furlani, E.P. Magnetic levitation-based electromagnetic energy harvesting: a semi-analytical nonlinear model for energy transduction. *Sci. Rep.* **2016**, *6*, 18579. [CrossRef] [PubMed]
19. Berdy, D.F.; Valentino, D.J.; Peroulis, D. Design and optimization of a magnetically sprung block magnet vibration energy harvester. *Sens. Actuators Phys.* **2014**, *218*, 69–79. [CrossRef]
20. Sneller, A.; Mann, B. On the nonlinear electromagnetic coupling between a coil and an oscillating magnet. *J. Phys. Appl. Phys.* **2010**, *43*, 295005. [CrossRef]
21. Donoso, G.; Ladera, C.; Martín, P. Magnet fall inside a conductive pipe: Motion and the role of the pipe wall thickness. *Eur. J. Phys.* **2009**, *310*, 855–869 [CrossRef]
22. Williams, C.B.; Yates, R.B. Analysis of a micro-electric generator for microsystems. *Sens. Actuators A* **1996**, *52*, 8–11. [CrossRef]
23. Stephen, N.G. On energy harvesting from ambient vibration. *J. Sound Vib.* **2006**, *293*, 409–425. [CrossRef]
24. Beeby, S.; Torah, R.; Tudor, M.; Glynne-Jones, P.; O'Donnell, T.; Saha, C.; Roy, S. Micro electromagnetic generator for vibration energy harvesting. *J. Micromech. Microeng.* **2007**, *17*, 1257–1265. [CrossRef]
25. Constantinou, P.; Mellor, P.H.; Wilcox, P. A model of a magnetically sprung vibration generator for power harvesting applications. In Proceedings of the 2007 IEEE International Electric Machines & Drives Conference, Antalya, Turkey, 3–5 May 2007; Volume 1, pp. 725–730.
26. Foisal, A.R.M.; Lee, B.C.; Chung, G.S. Fabrication and performance optimization of an AA size electromagnetic energy harvester using magnetic spring. In Proceedings of the SENSORS, 2011 IEEE, Limerick, Ireland, 28–31 October 2011; pp. 1125–1128.
27. Munaz, A.; Lee, B.C.; Chung, G.S. A study of an electromagnetic energy harvester using multi-pole magnet. *Sens. Actuators Phys.* **2013**, *201*, 134–140. [CrossRef]
28. Halim, M.A.; Park, J.Y. A non-resonant, frequency up-converted electromagnetic energy harvester from human-body-induced vibration for hand-held smart system applications. *J. Appl. Phys.* **2014**, *115*, 9. [CrossRef]
29. Nico, V.; Boco, E.; Frizzell, R.; Punch, J. A high figure of merit vibrational energy harvester for low frequency applications. *Appl. Phys. Lett.* **2016**, *108*, 013902. [CrossRef]
30. Zhao, X.; Cai, J.; Guo, Y.; Li, C.; Wang, J.; Zheng, H. Modeling and experimental investigation of an AA-sized electromagnetic generator for harvesting energy from human motion. *Smart Mater. Struct.* **2018**, *27*, 085008. [CrossRef]

Article

Quad-Trapezoidal-Leg Orthoplanar Spring with Piezoelectric Plate for Enhancing the Performances of Vibration Energy Harvester

Yan Liu [1,*], Shuting Mo [1], Siyao Shang [1], Hai Wang [1], Peng Wang [2] and Keyuan Yang [3]

[1] Key Laboratory of Electronic Equipment Structure Design, Ministry of Education, Xidian University, Xi'an 710071, China; stmo@stu.xidian.edu.cn (S.M.); syshang@stu.xidian.edu.cn (S.S.); wanghai@mail.xidian.edu.cn (H.W.)

[2] Shaanxi Key Laboratory of Industrial Automation, School of Mechanical Engineering, Shaanxi University of Technology, Hanzhong 723001, China; wangpeng@snut.edu.cn

[3] Academy of Space Information Systems, Xi'an 710100, China; ykymail@126.com

* Correspondence: liuy@xidian.edu.cn

Received: 14 October 2020; Accepted: 11 November 2020; Published: 13 November 2020

Abstract: To validate the potentials of unequal-length section-varied geometry in developing a orthoplanar spring-based piezoelectric vibration energy harvester (PVEH), a modified spring with quad-trapezoidal-leg configuration is designed, analyzed, and fabricated. A basic quad-trapezoidal-leg orthoplanar spring (QTOPS) is theoretically analyzed, and the structural effective stress and eigenfrequency are formulated to determine the main dimension parameters. Then, an improved QTOPS with additional intermediations is constructed and simulated. Porotypes with different leg geometries and mass configurations are fabricated and tested. The results of QTOPS and a conventional rectangular-shaped spring are compared. It is verified that the proposed approach provides the structure with an enlarged effective stress and lower resonant frequency, which makes it more suitable to construct a high-performance PVEH than the orthoplanar spring with equal-length or rectangular legs.

Keywords: piezoelectric harvester; orthoplanar spring; trapezoidal leg; vibration energy

1. Introduction

Due to the extensive existence of mechanical vibration in ambient environment, the vibration energy harvester (VEH) has been regarded as an important alternative for powering the long-service electronic systems [1,2]. However, the ambient vibration usually features low frequency, varied direction, and weak amplitude, which is a great handicap for harvesting the energy efficiently [3]. Although several energy-harvesting technologies have been utilized, developing a high-performance VEH still faces challenges and attracts great research attention [4].

In particular, research studies regarding the PVEH are actively being conducted, which are targeted at lowing the resonant frequency, broadening the efficient bandwidth, and sensing the weak vibrations. Many excellent solutions have been executed, including multi-mode structure, nonlinearity, frequency up-conversion, etc. [3,5]. The multi-mode approach renders close resonant peaks to PVEH and makes the device advantageous for harvesting the energy in a random environmental vibration [6,7]. The multiple-element array can adapt different target vibrations by combining several basic harvesting units, but the device size is also greatly increased. Then, a multi-mode scheme can also be achieved by using additional masses, by which the structural DOF is increased and peaks close to each other can be realized [8–10]. The nonlinear approach, usually realized by structural nonlinearity and magnetic coupling, works well in broadening the device bandwidth [11,12]. The proposed devices can be monostable, bistable, and tristable, and they are often modeled by the Duffing equation [13–16].

Frequency up-conversion, either in through contact or in a contactless manner, can transmit the low-frequency vibration to the harvesting beam and stimulate it to vibrate at its resonant point, which benefits the device in the operating bandwidth and harvesting efficiency [17–19].

Recently, orthoplanar spring (OPS) is utilized to construct high-performance PVEHs, due to its monolithic planar configuration and capacity of generating normal out-of-plane motion [20]. A series of OPS-based energy harvesters have been reported with multi-mode and nonlinear features [21–23]. The analysis and experimental results showed that favorable multiple nonlinear vibration modes were achieved by the designed compliant orthoplanar springs. Moreover, a quad-leg spring with atypical legs was also studied to show the effect of structural asymmetry [24]. Nevertheless, some supplementary work could be done to further optimize the configuration of OPS. In former literatures, it has been proven that the promotion in effective stress can be fulfilled by introducing unequal-length beams into multi-beam structures [25]. Meanwhile, varying the cross-section of the cantilever has also been utilized to improve the performance of PEVHs [26]. However, most previous works only focus on the symmetrical geometries with equal-length rectangular legs, and the effect of leg shape and length is still not systematically studied.

To validate the potential of unequal-length section-varied geometry in developing OPS-based PVEH and make a contribution to the previous works, this paper focuses on the design and analysis of quad-trapezoidal-leg orthoplanar spring (QTOPS) and explorers its application in PVEHs. First, a basic quad-leg structure with trapezoidal legs is formulized to analyze the structural stiffness, stress, and resonant frequency, validating the promotion effect of QTOPS. Then, a modified design for PVEH is constructed and its properties are evaluated via finite element analysis (FEA). After that, a brief description of the device prototype and experimental setup is presented. Finally, characterization experiments are conducted, and conclusions are summarized.

2. Design and Analysis

2.1. Modeling of the Basic Structure

The basic QTOPS consists of four trapezoidal legs and a central plate. As shown in Figure 1, the legs are arranged around the central plate in a decussate form. A single plan-view for the trapezoidal leg is also defined with proper dimension parameters. The legs can be divided into two groups: two vertical ones with a length of L_1 and two longitudinal ones with a length of L_2. The QTOPS structure can be regarded as a classical cross-beam-mass pattern, in which the central plate is treated as a rigid mass and supported by the four legs. Based on the displacement of a conventional cross-beam-mass structure and setting $L_1 = L_2 = L$ for simplicity, the stiffness of QTOPS K can be correspondingly written as [27]

$$K = \frac{4Et^3a}{AL^3} \tag{1}$$

where E is the Young's modulus of the material, and t, a, and L are the thickness, base width, and length of the trapezoidal leg, respectively. A is a correction factor, which is mainly determined by the ratio of top width b and base width a. Combining the formula derived by Sader et al. and the structural response, the factor A can be approximately expressed as

$$A = \frac{3b/a}{(b/a-1)^3}\left[(1-b/a) - \frac{(1-b/a)^2}{2b/a} + (b/a)\ln(b/a)\right]\left[0.54 + 0.78b/a - 0.28(b/a)^2\right] \tag{2}$$

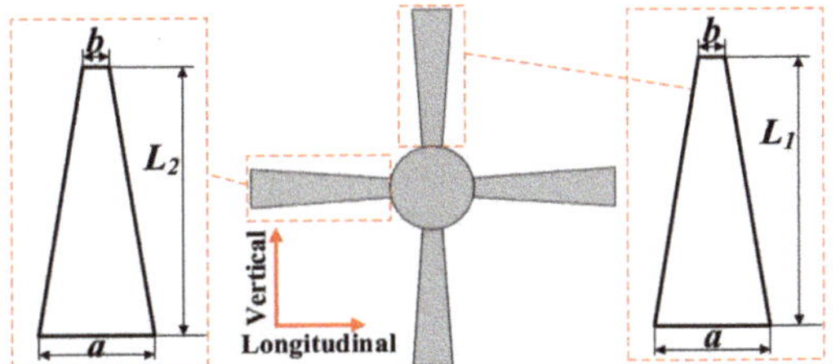

Figure 1. Sketch for the basic quad-trapezoidal-leg orthoplanar spring (QTOPS).

Equation (2) can be divided into two terms:

$$A_1 = \frac{3b/a}{(b/a-1)^3}\left[(1-b/a) - \frac{(1-b/a)^2}{2b/a} + (b/a)\ln(b/a)\right] \tag{3}$$

$$A_2 = 0.54 + 0.78b/a - 0.28(b/a)^2 \tag{4}$$

A_1 is derived from the derived by Sader et al., considering the effect from the shape of the cantilever beam [28,29]. A_2 is a supplementary part to accurately represent the cross-beam-mass model of QTOPS, which is obtained by fitting the relationship between the simulated displacement and loaded force of structures with different b/a. With the obtained function, the structural stiffness can be plot vs. b/a with E = 198 GPa, t = 0.2 mm, a = 15 mm, and L = 50 mm. Figure 2 shows the calculated stiffness of the basic structure and the results from the force–displacement relationship simulated by finite element analysis (FEA). It can be seen that correction factor A realizes a favorable precision in bridging the calculated and simulated stiffness. The obtained curve indicates the direct proportion between K and b/a.

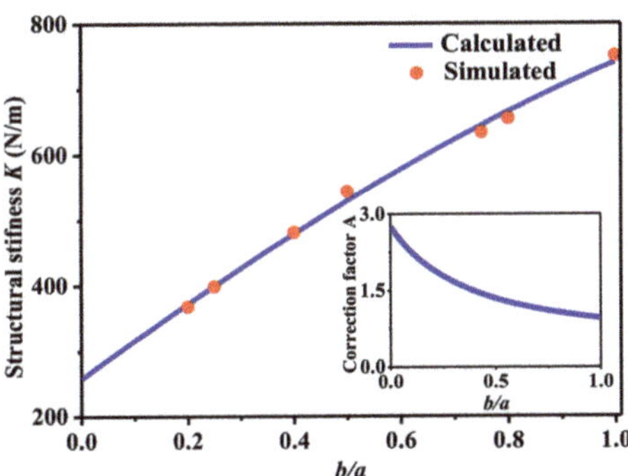

Figure 2. Stiffness and correction factor for the basic structure.

For a certain structure, there are two operating parameters closely related to the performance of PVEH: structural resonant frequency impacts the efficient frequency for PVEH and structural stress under external load determines the energy conversion capability. Sometimes, the frequency and stress can be utilized to evaluate the potential of the proposed structure in constructing a high-performance PVEH [1,30]. With the stiffness function, the eigenfrequency and stress of the basic QTOPS can be easily calculated. A central mass weighing 25×10^{-3} kg is added to the model, and an 10 m/s^2 acceleration is applied. Figure 3 plots the results for structural eigenfrequency and the maximum stress of a leg under different combinations of leg length ratio and b/a. In the analysis, the length of two vertical legs (L_1) is set as 50 mm and the two longitudinal legs are shortened to achieve different length ratios (L_2/L_1). A lower frequency can be observed under a larger L_2/L_1 ratio and smaller b/a, and a larger stress can be obtained under a smaller L_2/L_1 and b/a. The results verify the enhancing capacity of unequal-length configuration and a trapezoidal-shaped leg. Based on the results, the basic structure with the combination of $b/a = 0.5$ and $L_2/L_1 = 0.8$ is chosen as the fundamental frame to construct the desired PVEH.

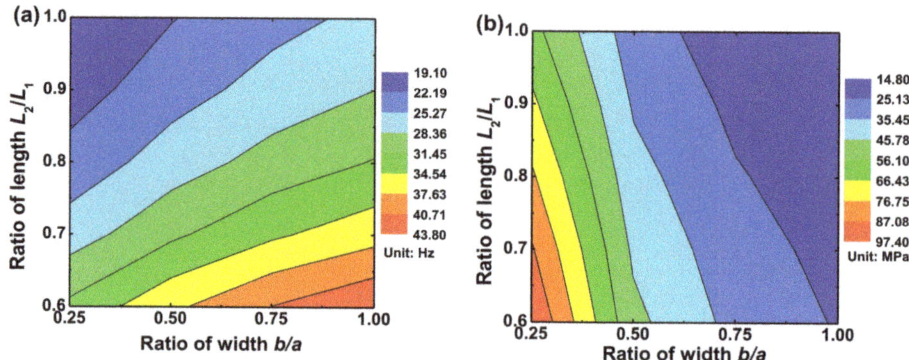

Figure 3. Calculated eigenfrequency (**a**) and maximum stress (**b**) of the basic QTOPS.

2.2. Construction of the PVEH

To fully exploit the features of the basic structure, an improved geometry with four additional intermediations is proposed to construct the desired PVEH. Figure 4 depicts the proposed QTOPS-based schemes for PVEH and their nomenclature. Benefiting from the introduction of intermediations, more than one mass can be easily placed in the device, including the Mc (center), M1/M2 (intermediation at short leg), and M3/M4 (intermediation at long leg). In the proposed PVEH schemes, the piezoelectric plates could be arranged at the ends of legs for performance comparison, as shown in Figure 4. When put into practical applications, the PVEH will be mounted onto the targets by fixing the ends indicated by the red triangles, which is also utilized as a constraint condition in the following simulation and prototype characterization.

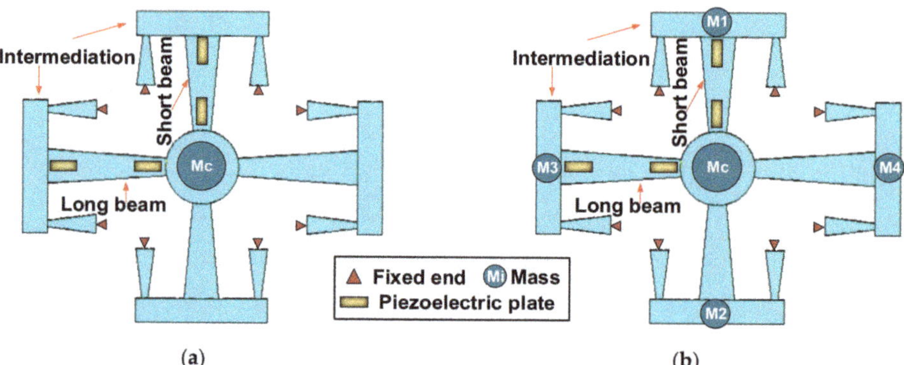

Figure 4. The energy harvester based on the QTOPS with Case 1 (**a**) and Case 2 (**b**) configurations.

After introducing the intermediations and necessary ancillaries, the structural complexity is significantly increased, making it difficult to theoretically formulate the resonant frequency and stress of the proposed scheme. Herein, the finite element analysis (FEA), an effective and widely used method in structural engineering, is employed to analyze the target features [31,32]. In the analysis, two different combinations of mass (Case 1: only Mc and Case 2: Mc + M1–M4) are used. The finite element models of the proposed QTOPS and conventional quad-rectangular-leg OPS (QROPS) were developed using commercial package ANSYS 14.5 (ANSYS Inc., Canonsburg, PA, USA). The whole structures were meshed with 3D SOLID95 elements, in which every element was defined by 20 nodes with three degrees of freedom per node. The whole structures of Case 1-based and Case 2-based QTOPS were discretized with 55,018 and 47,214 elements respectively; the values were 45,361 and 38,541 for Case 1-based and Case 2-based QROPS. In all simulations, the constraint condition was loaded by constraining all the freedoms of fixed ends indicated by the red triangles in Figure 4. Firstly, static analysis is conducted to study the stress distribution along the legs, and the results can help to further refine the positions of piezoelectric plates. Then, modal analysis is conducted to evaluate the resonant frequency. The parameters for simulation are shown in Table 1. A 10 m/s^2 acceleration is loaded, and the weight of masses is settled as Mc = 25×10^{-3} kg, M1 = M2 = M3 = M4 = 10×10^{-3} kg.

Table 1. Parameters for finite element analysis (FEA) simulation.

Material Parameters			Dimension				
Young Modulus (GPa)	Poisson's Ratio	Density (kg/m^3)	Length of Long Leg (mm)	Base Width (mm)	Thickness (mm)	L_2/L_1	b/a
198	0.3	7810	50	15	0.2	0.8	0.5

Figure 5a presents the stress distribution in the proposed QTOPS under Case 1. For easy description, the leg end connecting with the central plate is denominated as the top end, and the other one is named as the base end. It can be seen that more considerable stresses develop at the top end of legs. More clarity for the stress distribution is plotted in Figure 5b by mapping the normal stress onto the preset paths, which are coincident with the centerline of their corresponding legs (also shown in Figure 5a). In Figure 5b, the sharp variation of every curve presents the stress decrease in the thick plate region, which has the same length in every curve but is modulated by the different scales for long and short legs. Compared with conventional QROPS, the stress in the top end of the short and long leg is lifted to 152.8% and 150.4%, respectively. Meanwhile, the stress distribution curves also verify the enhancing effect of the proposed unequal-length configuration. The maximum stress of the short leg in QTOPS is 41.8% larger than that of the long leg. Based on the results, the piezoelectric plates should cover the top end of the short leg to make better use of the energy seized by QTOPS.

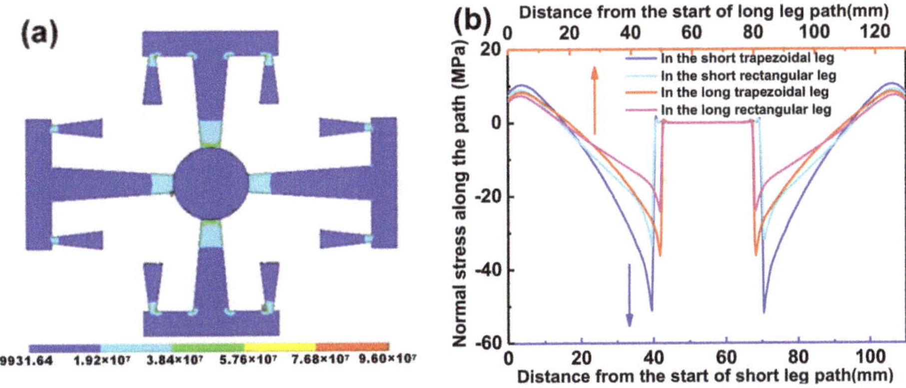

Figure 5. Stress distribution for Case 1: (**a**) von Mises stress and (**b**) normal stress along the leg paths.

The stress results of Case 2-based QTOPS are shown in Figure 6. Similarly, the stress concentrates more at the top end of the legs, and the trapezoidal unequal scheme also promotes the stress values. Compared with QROPS, the maximum stress in the short and long trapezoidal legs is lifted to 167.5% and 165.3%, respectively. Meanwhile, the short leg in Case 2 also features a stress that is 1.36 times the value of that in the long leg. These results under Case 2 further validate the superiority of QTOPS in promoting the effective stress for harvesting the vibration efficiently.

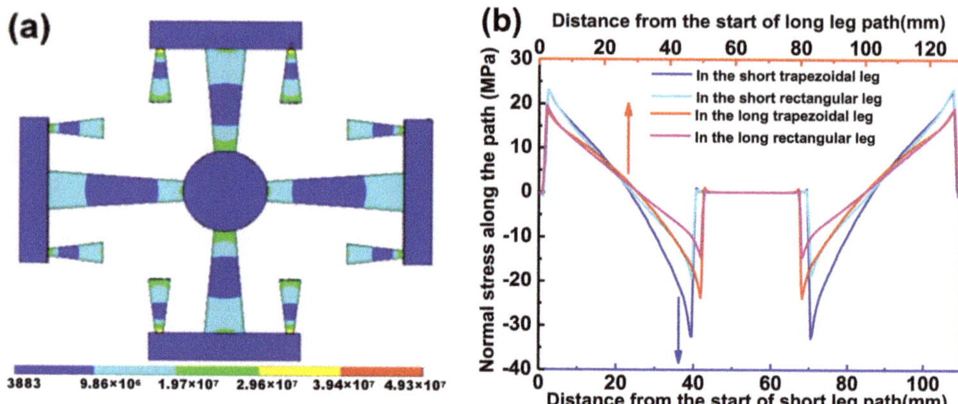

Figure 6. Stress distribution for Case 2: (**a**) von Mises stress and (**b**) normal stress along the leg paths.

Modal analysis is also conducted to evaluate the dynamic response of the proposed QTOPS-based PVEH. When only utilizing the central mass (namely Case 1), the OPS presents just one mode below 50 Hz (17.6 Hz), as shown in Figure 7. When adding additional four intermediate masses (Case 2), five modes below 50 Hz (20.7, 33.9, 34.2, 34.6, and 36.4 Hz) can be achieved. The introduction of intermediate mass produces a potential in realizing a multi-mode vibration energy harvester. Meanwhile, the trapezoidal scheme exhibits lower resonant frequencies than the rectangular scheme in both Case 1 and Case 2, further distinguishing the superiority of QTOPS-based PVEH.

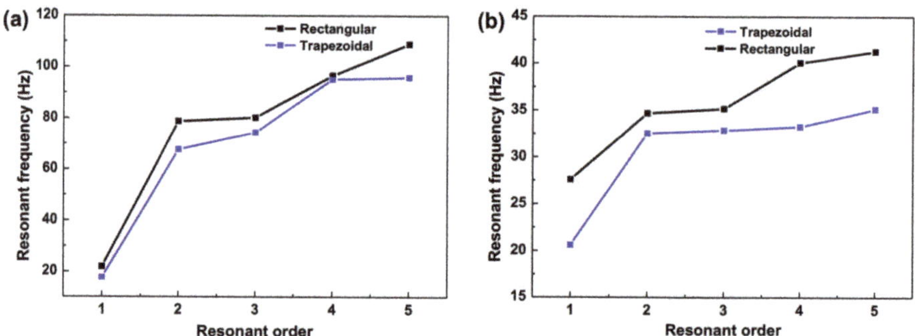

Figure 7. Resonant frequency of Case 1-based (**a**) and Case 2-based (**b**) devices.

3. Prototypes and Experiments

To verify the designed potentials of proposed QTOPS in PVEH, several porotypes are fabricated with a 0.2 mm-thick spring steel (Metal grade 65 Mn) film, whose parameters are the same as the values in Table 1. The long leg is set at 50 mm for length, 15 mm for base width, and 7.5 mm for top width. Correspondingly, the short leg is set at 40 mm for length, 15 mm for base width, and 7.5 mm for top width. The length ratio between the short and long leg is 0.8, and the width ratio between the base and top is 0.5, which is consistent with the design parameters. The masses are added by attaching proper iron blocks in the central plate and intermediations, and piezoelectric plates made of lead zirconate titanate ceramics (Type: PZT-5A) [4] with the dimensions of 7 mm × 15 mm × 0.23 mm (width × length × thickness) are adhered onto the top end of the legs. An additional prototype with QROPS is also built for comparison, whose leg lengths are the same as with the trapezoidal approach, but the width is kept as 15 mm.

An experimental setup, as shown in Figure 8, is built to test the prototypes. Stimulating vibration is produced by a shaker system, in which the function generator (SDG1020, SIGLENT, Shenzhen, China) provides sinusoidal voltage with varied frequency and the signal is amplified by the power amplifier (SD1492A, BDHSD, Qinhuangdao, China) to excite the shaker (SD1482A, BDHSD, Qinhuangdao, China). An oscilloscope (GDS-1072B, GWINSTEK, Suzhou, China) is utilized to measure the generated voltage by PVEH prototypes and monitor the output of the accelerometer (CA-YD-180, BDHSD, Qinhuangdao, China). The current source (SD14T03, BDHSD, Qinhuangdao, China) is used to transform the output of the accelerometer into voltage form.

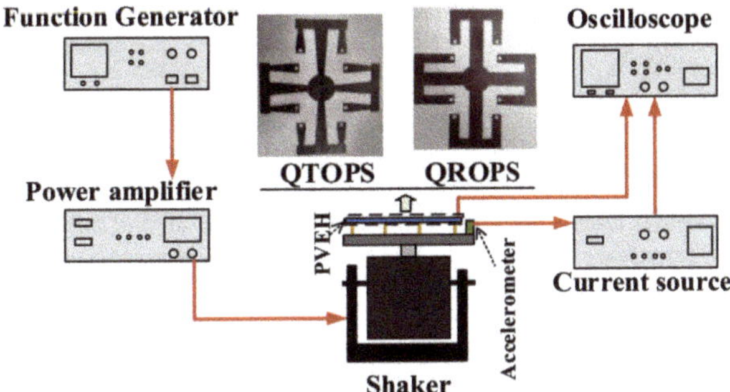

Figure 8. Experimental setup.

4. Results

In this section, we present the voltage and power outputs of the QTOPS-based PVEH. An additional QROPS-based prototype is also measured as a contrast. The characteristics of devices under different mass locations (Case 1 and Case 2) are tested, and the target vibration is imitated by a sinusoidal acceleration with an amplitude of 10 m/s^2 and different frequencies. In the experiments, the prototypes are firstly stimulated by the loaded vibrations to obtain their working frequencies (resonance frequencies) by testing their generated voltages at different frequencies. Then, the power tests are conducted by recording the voltage over the loaded resistors when the devices are stimulated at their working frequencies. All the experiments are conducted at room temperature of 25 °C and a humidity of 55% RH.

4.1. Results for Case 1-Based Prototypes

First, the open-circuit output voltage of Case 1 configurations is tested, and the voltage–frequency (V-f) curve is shown in Figure 9a. The curves with same color indicate the results of the same prototype, but the piezoelectric plates are respectively located at the short leg (solid markers) and long leg (hollow makers). The QTOPS resonates at 16.4 Hz with corresponding peak voltages of 13.45 V and 9.25 V for the short and long leg, respectively. Meanwhile, the QROPS has a resonant frequency of 19.2 Hz, and its peak voltages are 8.70 V and 6.42 V for the short and long leg, respectively. Compared with QROPS, the utilization of the trapezoidal leg provides a 1.55× increase to the generated voltage and a decline of 14.6% to the resonant frequency. Meanwhile, the short leg features a higher voltage in both QTOPS and QROPS, and the promotion can be up to 145.4%. However, there is only one resonant peak in the range of 0–50 Hz for both QTOPS and QROPS devices.

Figure 9b presents the power–load resistance curve of these devices at their resonance frequencies of 16.4 and 19.2 Hz, respectively. The best matching resistance for QTOPS-based PVEH is about 296 kΩ, and the corresponding output power is 192.9 and 84.9 μW for short and long legs; the values of QROPS-based PVEH are about 256 kΩ for optimized resistance, and 89.1 and 42.7 μW for output

3. Yildirim, T.; Ghayesh, M.H.; Li, W.; Alici, G. A review on performance enhancement techniques for ambient vibration energy harvesters. *Renew. Sustain. Energy Rev.* **2017**, *71*, 435–449. [CrossRef]

4. Yang, Z.; Zhou, S.; Zu, J.; Inman, D. High-performance piezoelectric energy harvesters and their applications. *Joule* **2018**, *2*, 642–697. [CrossRef]

5. Tran, N.; Ghayesh, M.H.; Arjomandi, M. Ambient vibration energy harvesters: A review on nonlinear techniques for performance enhancement. *Int. J. Eng. Sci.* **2018**, *127*, 162–185. [CrossRef]

6. Luo, C.; Xu, H.; Wang, Y.; Li, P.; Hu, J.; Zhang, W. Unsymmetrical, interdigital vibration energy harvester: Bandwidth adjustment based on mode separation technique. *Sens. Actuators A Phys.* **2017**, *257*, 30–37. [CrossRef]

7. Deng, H.; Du, Y.; Wang, Z.; Zhang, J.; Ma, M.; Zhong, X. A multimodal and multidirectional vibrational energy harvester using a double-branched beam. *Appl. Phys. Lett.* **2018**, *112*, 213901. [CrossRef]

8. Aldraihem, O.; Baz, A. Energy harvester with a dynamic magnifier. *J. Intell. Mater. Syst. Struct.* **2011**, *22*, 521–530. [CrossRef]

9. Li, X.; Upadrashta, D.; Yu, K.; Yang, Y. Analytical modeling and validation of multi-mode piezoelectric energy harvester. *Mech. Syst. Signal Process.* **2019**, *124*, 613–631. [CrossRef]

10. Tang, L.; Yang, Y. A multiple-degree-of-freedom piezoelectric energy harvesting model. *J. Intell. Mater. Syst. Struct.* **2012**, *23*, 1631–1647. [CrossRef]

11. Daqaq, M.F.; Masana, R.; Erturk, A.; Dane Quinn, D. On the role of nonlinearities in vibratory energy harvesting: A critical review and discussion. *Appl. Mech. Rev.* **2014**, *66*, 040801. [CrossRef]

12. Pellegrini, S.P.; Tolou, N.; Schenk, M.; Herder, J.L. Bistable vibration energy harvesters: A review. *J. Intell. Mater. Syst. Struct.* **2012**, *24*, 1303–1312. [CrossRef]

13. Andò, B.; Baglio, S.; Bulsara, A.R.; Marletta, V. A bistable buckled beam based approach for vibrational energy harvesting. *Sens. Actuators A Phys.* **2014**, *211*, 153–161. [CrossRef]

14. Fan, K.; Tan, Q.; Liu, H.; Zhang, Y.; Cai, M. Improved energy harvesting from low-frequency small vibrations through a monostable piezoelectric energy harvester. *Mech. Syst. Signal Process.* **2019**, *117*, 594–608. [CrossRef]

15. Shan, G.; Wang, D.F.; Song, J.; Fu, Y.; Yang, X. A spring-assisted adaptive bistable energy harvester for high output in low-excitation. *Microsyst. Technol.* **2018**, *24*, 3579–3588. [CrossRef]

16. Zhou, S.; Cao, J.; Litak, G.; Lin, J. Numerical analysis and experimental verification of broadband tristable energy harvesters. *tm-Tech. Messen* **2018**, *85*, 521–532. [CrossRef]

17. Deng, H.; Wang, Z.; Du, Y.; Zhang, J.; Ma, M.; Zhong, X. A seesaw-type approach for enhancing nonlinear energy harvesting. *Appl. Phys. Lett.* **2018**, *112*, 213902. [CrossRef]

18. Kwon, D.-S.; Ko, H.-J.; Kim, M.-O.; Oh, Y.; Sim, J.; Lee, K.; Cho, K.-H.; Kim, J. Piezoelectric energy harvester converting strain energy into kinetic energy for extremely low frequency operation. *Appl. Phys. Lett.* **2014**, *104*, 113904. [CrossRef]

19. Kathpalia, B.; Tan, D.; Stern, I.; Erturk, A. An experimentally validated model for geometrically nonlinear plucking-based frequency up-conversion in energy harvesting. *Smart Mater. Struct.* **2018**, *27*, 015024. [CrossRef]

20. Parise, J.J.; Howell, L.L.; Magleby, S.P. Ortho-planar linear-motion springs. *Mech. Mach. Theory* **2001**, *36*, 1281–1299. [CrossRef]

21. Nabavi, S.; Zhang, L. Nonlinear multi-mode wideband piezoelectric mems vibration energy harvester. *IEEE Sens. J.* **2019**, *19*, 4837–4848. [CrossRef]

22. Dhote, S.; Yang, Z.; Zu, J. Modeling and experimental parametric study of a tri-leg compliant orthoplanar spring based multi-mode piezoelectric energy harvester. *Mech. Syst. Signal Process.* **2018**, *98*, 268–280. [CrossRef]

23. Dhote, S.; Zu, J.; Zhu, Y. A nonlinear multi-mode wideband piezoelectric vibration-based energy harvester using compliant orthoplanar spring. *Appl. Phys. Lett.* **2015**, *106*, 163903. [CrossRef]

24. Dhote, S.; Li, H.; Yang, Z. Multi-frequency responses of compliant orthoplanar spring designs for widening the bandwidth of piezoelectric energy harvesters. *Int. J. Mech. Sci.* **2019**, *157–158*, 684–691. [CrossRef]

25. Wang, P.; Zhao, Y.; Tian, B.; Liu, Y.; Wang, Z.; Li, C.; Zhao, Y. A piezoresistive micro-accelerometer with high frequency response and low transverse effect. *Meas. Sci. Technol.* **2017**, *28*, 015103. [CrossRef]

26. Savarimuthu, K.; Sankararajan, R. Design and analysis of cantilever based piezoelectric vibration energy harvester. *Circuit World* **2018**, *44*, 78–86. [CrossRef]

27. Bao, M.-H. *Micro Mechanical Transducers: Pressure Sensors, Accelerometers and Gyroscopes*; Elsevier: Amsterdam, The Netherlands, 2000.

28. Sader, J.E.; White, L. Theoretical analysis of the static deflection of plates for atomic force microscope applications. *J. Appl. Phys.* **1993**, *74*, 1–9. [CrossRef]

29. Slattery, A.D.; Blanch, A.J.; Shearer, C.J.; Stapleton, A.J.; Goreham, R.V.; Harmer, S.L.; Quinton, J.S.; Gibson, C.T. Characterisation of the material and mechanical properties of atomic force microscope cantilevers with a plan-view trapezoidal geometry. *Appl. Sci.* **2019**, *9*, 2604. [CrossRef]

30. van Kempen, R.B.; Herder, J.L.; Tolou, N. A Strain Distribution Based Classification of Ortho-Planar Springs: An Outlook to Piezoelectric Applications. In *ASME 2016 International Design Engineering Technical Conferences and Computers and Information in Engineering Conference*; ASME: Charlotte, NC, USA, 2016; Volume 4, p. V004T008A013.

31. Wang, L.; Ding, J.; Jiang, Z.; Luo, G.; Zhao, L.; Lu, D.; Yang, X.; Ryutaro, M. A packaged piezoelectric vibration energy harvester with high power and broadband characteristics. *Sens. Actuators A Phys.* **2019**, *295*, 629–636. [CrossRef]

32. Wang, L.; Zhao, L.; Jiang, Z.; Luo, G.; Yang, P.; Han, X.; Li, X.; Maeda, R. High accuracy comsol simulation method of bimorph cantilever for piezoelectric vibration energy harvesting. *AIP Adv.* **2019**, *9*, 095067. [CrossRef]

Publisher's Note: MDPI stays neutral with regard to jurisdictional claims in published maps and institutional affiliations.

Article

Self-Oscillating Boost Converter of Wiegand Pulse Voltage for Self-Powered Modules

Xiaoya Sun [1], Haruchika Iijima [1], Stefano Saggini [2] and Yasushi Takemura [1,*]

[1] Electrical and Computer Engineering, Yokohama National University, Yokohama 240-8501, Japan; sun-xiaoya-tk@ynu.jp (X.S.); iijima-haruchika-xy@ynu.jp (H.I.)

[2] Department of Electrical, Management and Mechanical Engineering, University of Udine, 33100 Udine, Italy; stefano.saggini@uniud.it

* Correspondence: takemura-yasushi-nx@ynu.ac.jp

Abstract: This paper introduces a new method of electricity generation using a Wiegand sensor. The Wiegand sensor consists of a magnetic wire and a pickup coil wound around it. This sensor generates a pulse voltage of approximately 5 V and 20 μs width as an induced voltage in the pickup coil. The aim of this study is to generate a DC voltage of 5 V from the sensor, which is expected to be used as a power source in self-powered devices and battery-less modules. We report on the design and verification of a self-oscillating boost converter circuit in this paper. A DC voltage obtained by rectifying and smoothing the pulse voltage generated from the Wiegand sensor was boosted by the circuit. A stable DC output voltage in the order of 5 V for use as a power supply in electronics modules was successfully obtained. A quantitative analysis of the power generated by the Wiegand sensor revealed a suitable voltage-current range for application in self-powered devices and battery-less modules.

Keywords: self-powered device; battery-less modules; energy harvesting; Wiegand sensor; self-oscillating boost converter

Citation: Sun, X.; Iijima, H.; Saggini, S.; Takemura, Y. Self-Oscillating Boost Converter of Wiegand Pulse Voltage for Self-Powered Modules. *Energies* **2021**, *14*, 5373. https://doi.org/10.3390/en14175373

Academic Editor: Dibin Zhu

Received: 2 July 2021
Accepted: 25 August 2021
Published: 29 August 2021

Publisher's Note: MDPI stays neutral with regard to jurisdictional claims in published maps and institutional affiliations.

1. Introduction

In the Internet of Things (IoT) society, electronic devices and modules can be connected to the internet and exchange information through various sensors [1]. These are generally called IoT devices. The number of IoT devices is increasing rapidly and is expected to reach 80–120 billion by 2025 [2]. A significant number of batteries that need expensive and time-consuming maintenance are required for these devices, which also cause environmental pollution. Energy harvesting, such as collecting small amounts of energy from the surroundings and converting them into electrical energy, is expected to solve this problem [3,4]. Energy from the surrounding environment, if available, can continuously supply electrical power, and thus, be used as an independent power source for long periods of time, without replacing the power harness and battery. With developments in miniaturization and energy-saving approaches, low-power power supplies can gradually satisfy the many requirements of IoT devices [5]. This research introduces the use of a self-oscillating boost converter circuit for electricity generation using a Wiegand sensor [6,7] as an energy-harvesting element.

The Wiegand sensor generates pulse voltages that do not depend on the frequency of the external magnetic field [8]. These pulse voltages are generated with a constant intensity, even under ultra-slow changes in the magnetic field. Therefore, the Wiegand sensor has attracted significant attention as a power supply for the battery-less operation of electronic devices and for energy harvesting [9]. The contribution of this research involves the development of a DC power supply for electronic devices and modules using the Wiegand sensor. It is essential to build a DC power supply of 5 V because it can be used for multiple IoT devices. The power generated by the Wiegand sensor is in the order of 1 mW, even when the frequency of the applied magnetic field is as low as 1 kHz [8]. In this study, we designed and verified a self-oscillating boost converter circuit [10–12] connected to the Wiegand sensor as a power generator. DC power generation of 5 V was realized using the Wiegand sensor; this may not be achievable using other methods under an excitation frequency of 1 kHz.

The remainder of this paper is organized as follows: Following the introduction of the Wiegand sensor and its pulse generation in Section 2, the circuits for DC conversion of the pulse voltage and the self-oscillating boost converter are presented in Section 3. In Section 4, we present the circuit properties of the self-oscillating boost converter connected to the Wiegand sensor, derived both experimentally and through simulations. Finally, the conclusions of this study are presented in Section 5.

2. Wiegand Sensor as a Voltage Source

2.1. Measurement of Pulse Voltage from the Wiegand Sensor

We used a magnet wire composed of iron–cobalt–vanadium (FeCoV) with a diameter of 0.25 mm and a length of 11 mm. The Wiegand sensor used in this study consisted of a wire and a pickup coil with 3000 turns wound around it. The magnetic properties of this wire are essentially the same as those we have previously reported in detail [13]. Its magnetic structure is shown in Figure 1. The outer layer and inner core exhibit soft and hard magnetic properties with lower (1.6 kA/m) and higher (6.4 kA/m) coercive forces, respectively. The direction of magnetization of these layers can be either in parallel or antiparallel configurations, as shown in Figure 1—a specific feature of the Wiegand wire.

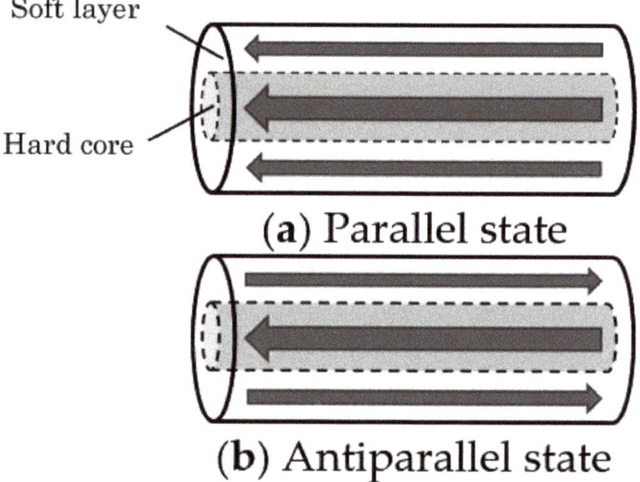

Figure 1. Two states of magnetization direction of the Wiegand wire: (**a**) parallel and (**b**) antiparallel states of the soft layer and hard core.

When a magnetic field exceeds the coercive force of the soft layer, the latter exhibits a fast magnetization reversal, which is called the Wiegand effect [6]. A pulse voltage is induced in the pickup coil wound around the wire [8,14]. The Wiegand sensor consists of a Wiegand wire and a pickup coil. As fast magnetization reversal is initiated independently from the changing ratio of the applied magnetic field, the intensity and width of the pulse are constant [15]. Figure 2 shows the measured waveform of the pulse voltage generated from the Wiegand sensor. We measured the waveform of the open-circuit voltage across both ends of the pickup coil using an oscilloscope [8,9]. The intensity and frequency of the alternating applied magnetic field were 3.2 kA/m and 1 kHz, respectively. An excitation coil with 25 mm length, 22 mm diameter, and 90 turns was used. An alternating magnetic field was applied to the Wiegand sensor by using the excitation coil, a signal generator, and a bipolar amplifier. Positive and negative pulses with widths of 20 µs were alternatingly induced in the pickup within 1 ms. These were attributed to electromagnetic induction caused by the change in magnetic flux corresponding to the alternating magnetization reversal of the soft layer.

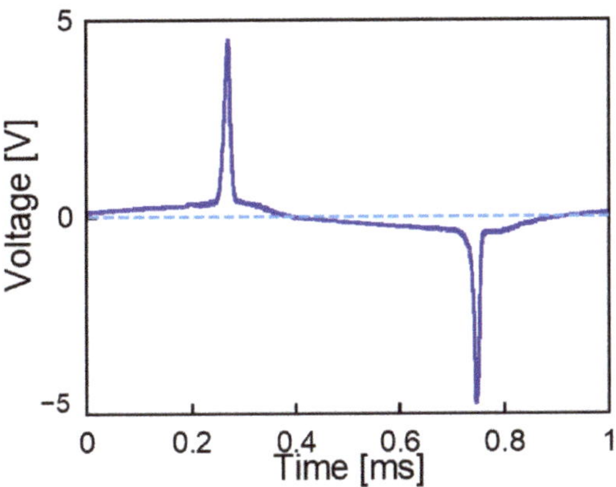

Figure 2. Measured waveform of the pulse voltage generated by the Wiegand sensor. The frequency of the applied magnetic field was 1 kHz.

2.2. Pulse Voltage from the Wiegand Sensor Used as a Voltage Source in Simulation

As previously reported, we can determine the equivalent circuit model of the Wiegand sensor [16]. The intrinsic pulse voltage, V_{in}, of the Wiegand sensor was defined to evaluate the application circuits of the Wiegand sensor through MATLAB®/Simulink® simulations. Figure 3 shows the waveform of the intrinsic pulse voltage, V_{in}, generated from the Wiegand sensor. The performances of the simulated and experimental circuits were in agreement when employing the equivalent circuit model of the Wiegand sensor, which consisted of V_{in} as a voltage source, an internal resistance of 180 Ω, and an inductance of 17 mH [16].

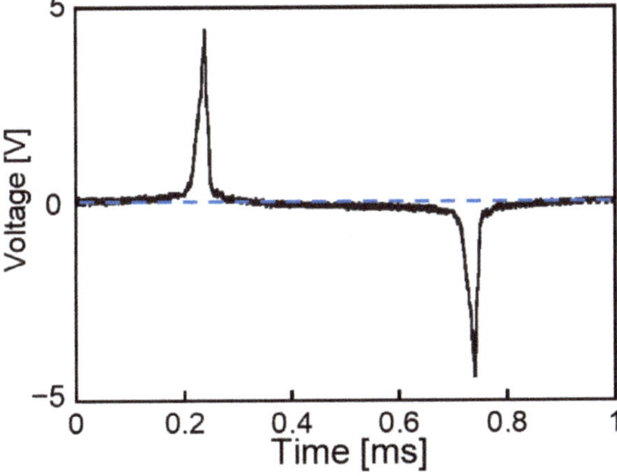

Figure 3. Waveform of the intrinsic pulse voltage, V_{in}, generated from the Wiegand sensor, used to simulate the circuit through MATLAB®/Simulink®. The excitation frequency is 1 kHz.

2.3. DC Conversion of Wiegand Pulse Voltage

AC–DC conversion is used to obtain a DC voltage from the Wiegand pulse voltage, as shown in Figure 4. The alternatingly positive and negative pulse voltages are rectified by the rectifier circuit using diodes. A smoothing filter circuit using a capacitor converts the pulse voltages to DC. Figure 5 shows the DC conversion circuit, diode parameters, capacitor, and resistor used in our experiments and simulations. D_1–D_4, indicated in Figure 5, represent the diodes (RBR3MM30A) for rectification. R_{Load} is a 5.5 MΩ load resister. C_1 was used as a smoothing capacitor in the range of 1–220 nF to analyze a processed and constant DC voltage. C_1 = 1, 10, 20, 50, 100, or 220 nF was connected to the full-wave bridge rectifier, and waveforms of the output voltage, V_{out}, were measured. Figure 6a shows that V_{out} saturates at 2.77 V, regardless of the capacitance of C_1. The relaxation time of the saturation is longer for a smaller C_1. Figure 6b shows the simulated waveforms of the output voltage, which agree with the experimental results. LTspice® was used for the circuit simulation [16].

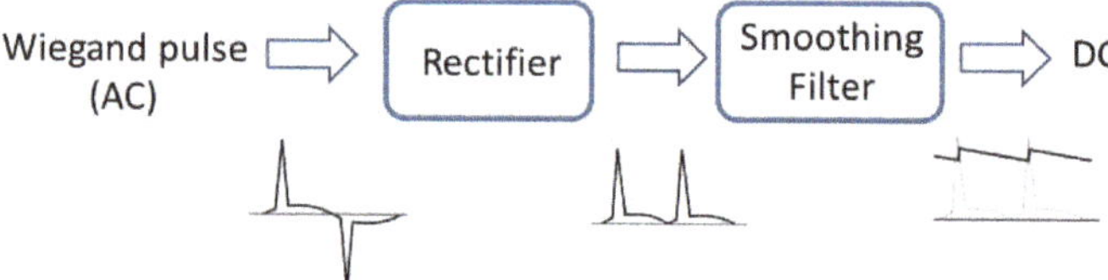

Figure 4. Block diagram for DC conversion of the Wiegand pulse voltage.

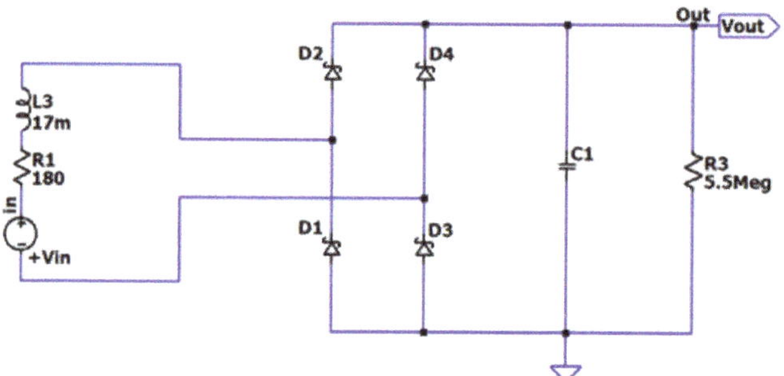

Figure 5. DC conversion circuit connected to the equivalent circuit model of the Wiegand sensor.

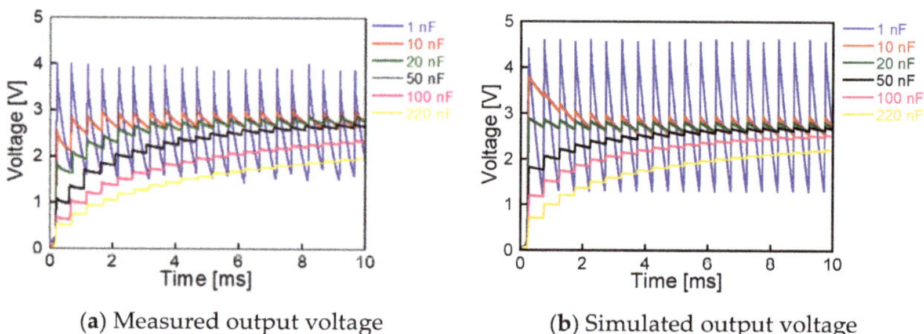

(**a**) Measured output voltage   (**b**) Simulated output voltage

Figure 6. (**a**) Measured and (**b**) simulated output voltages after DC conversion of the Wiegand pulse voltage.

The frequency of the ripple was 2 kHz, i.e., twice the excitation frequency, because of the full-wave rectifier. The output voltage ripple is high for $C_1 \leq 20$ nF, and it is very low for $C_1 > 20$ nF. The ripple rate, Ripple, was calculated using the following equation:

$$Ripple = \frac{V_{\text{max}} - V_{\text{min}}}{V_{\text{average}}} \times 100 \ [\%] \tag{1}$$

where V_{max}, V_{min}, and V_{average} are the maximum, minimum, and average voltages, respectively, applied for 1 ms during one cycle of excitation. The ripple rates calculated from the experimental and simulated output voltages are shown in Figure 7.

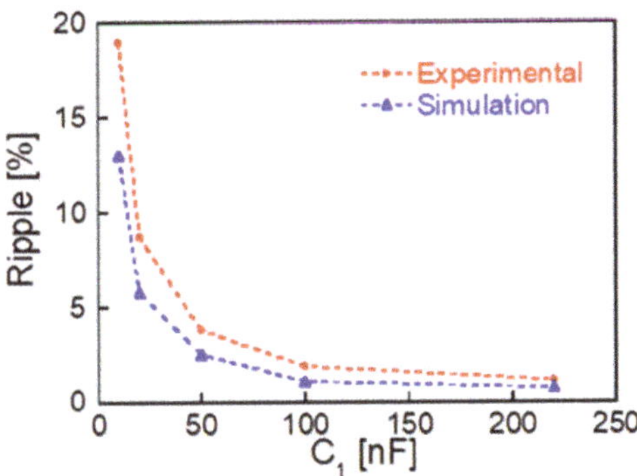

Figure 7. Ripple of the output voltage.

3. Self-Oscillating Boost Converter Circuit

As mentioned in Section 2.2, the maximum voltage obtained at the smoothing capacitor after DC conversion of the Wiegand pulse voltage is 2.77 V. It is fundamental to obtain a DC voltage of 5 V for operating several electronics modules. In this study, we apply a self-oscillating boost converter circuit for a Wiegand pulse voltage. The feature of the booster converter is that the energy stored in an inductor increases the output voltage, which then exceeds the input voltage. Figure 8 shows the typical circuit of a boost converter. The alternating sequence of storing energy in the inductor and transmitting it back to the circuit boosts the voltage. The energy is stored in inductor L when M, a field-effect transistor (FET), is in the ON state, whereas the stored energy is transferred from L to capacitor C when M is in the OFF state. As a result, V_{out} higher than V_{in} is obtained. Generally, the switching ON/OFF of M, controlled by an external signal, is used to apply this alternating sequence [10]. As this study aims to develop self-powered electronic modules, the external signal for an alternating sequence cannot be used. Therefore, a self-oscillating boost converter is employed. Figure 9 shows the self-oscillating boost converter circuit used in this study for a Wiegand pulse voltage. The input voltage, V_{in}, of a 20-μs-wide pulse used as the power source generates an oscillating voltage at V_{C1}. The frequency of these oscillations corresponds to a resonant frequency determined by the inductor L and capacitors C_1 and C_{gs} [11,12]. This oscillation voltage at V_{C1} switches the consecutive ON and OFF states of M, as shown in Figure 9.

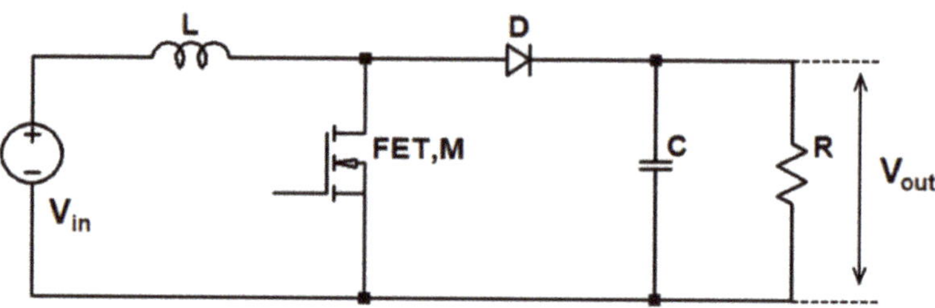

Figure 8. Boost converter circuit.

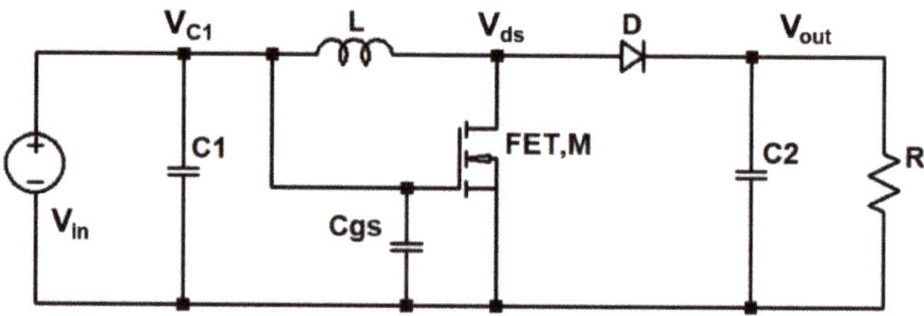

Figure 9. Self-oscillating boost converter circuit.

- ON state of M
  When the oscillating pulse voltage at V_{C1} exceeds the gate threshold voltage, V_{th}, of the FET, the FET is turned ON, and V_{ds} drops to the ground level. This allows current I_L to flow in L, where the energy is stored. The current flows through the diode, D, and supplies the output voltage, V_{out}.
- OFF state of M
  When the oscillating pulse voltage at V_{C1} is below V_{th}, the FET is turned OFF. The energy stored in L during the FET is transmitted to the capacitor of the output, C_2, through diode D.

D prevents a backflow current to L. By repeating the ON/OFF switching sequence of the FET, a DC output voltage exceeding V_{in} is obtained.

4. Experimental Results and Discussion

4.1. Design of a Self-Oscillating Boost Converter for Wiegand Pulse

In this study, we designed and fabricated a rectifying and boosting circuit for the Wiegand pulse voltage in Figure 10. The circuit consists of a bridge rectifier with diodes and self-oscillating boost converter components, such as inductors, capacitors, n-channel FET, and diodes, as described in the previous section and in Figures 8 and 9. Details of the parameters of the circuit elements are indicated in Figure 10 and Table 1. The input voltage is supplied from a Wiegand sensor. An alternating magnetic field of 3.2 kA/m was applied to the wire. The frequency of this field was 1 kHz. The Wiegand sensor is advantageous in terms of its efficient power generation at low frequency ranges below 1 kHz [8].

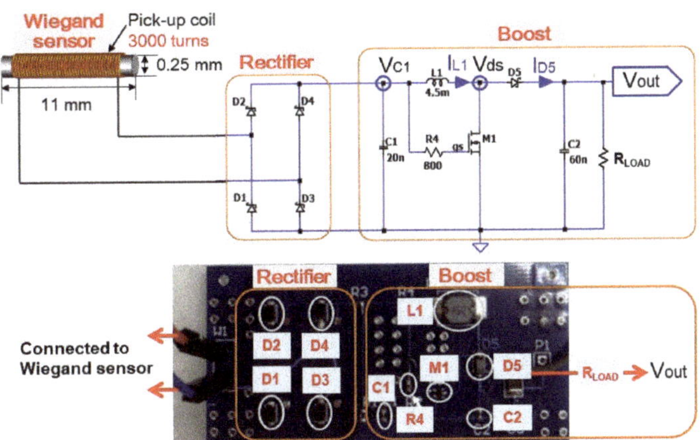

Figure 10. Experimental design of the circuit for the Wiegand sensor with a self-oscillating boost converter.

Table 1. Parameters of the circuit elements for the Wiegand pulse voltage with a self-oscillating boost converter.

Component	Value/Type (Model Name)
Capacitance: C_1	20 nF
Capacitance: C_2	60 nF
Inductance: L_1	4.5 mH
Resistance: R_4	800 Ω
Diode: D_1–D_5	low V_F, Schottky (RBR3MM30A)
MOSFET: M_1	n-channel (RE1C002UN)
Resistance: R_{LOAD}	1 kΩ–5 MΩ

Figure 11 shows the simulated waveforms of the voltages and currents in the self-oscillating boost converter. The rectified voltage of the Wiegand pulse is oscillated by a resonance of C_1 and L_1. This oscillated voltage, V_{C1}, switches the FET ON/OFF. V_{ds} confirms the ON/OFF status of the FET. As a result of the oscillated I_{L1} and I_{D5} and the smoothing capacitor C_2, a constant DC voltage is obtained as the output. V_{out} is 5.1 V, which is higher than the input voltage of V_{C1}, and a DC voltage of 2.77 V is obtained without the booster circuit, as shown in Figure 6.

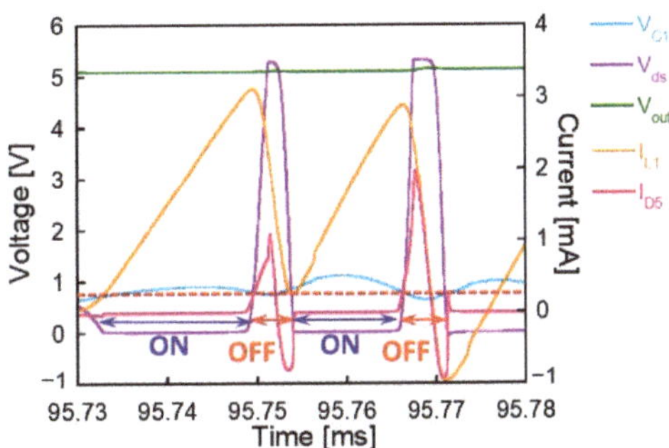

Figure 11. Simulated waveforms of voltages and currents in the self-oscillating boost converter.

V_{out} depends on the circuit parameters of L_1 and C_1, as shown in Figure 12. V_{out} increases as C_1 decreases. The combination of L_1 = 4.5 mH and C_1 = 20 nF is optimum for producing a DC voltage of approximately 5.1 V, thus meeting the aim of this study to generate a DC voltage of 5 V that can be used as a power source for various electronic modules. The dependency of the output voltage ripple on capacitor C_2 was also studied. As shown in Figure 13, V_{out} is not dependent on C_2, and is stable with fewer ripples when $C_2 > 60$ nF.

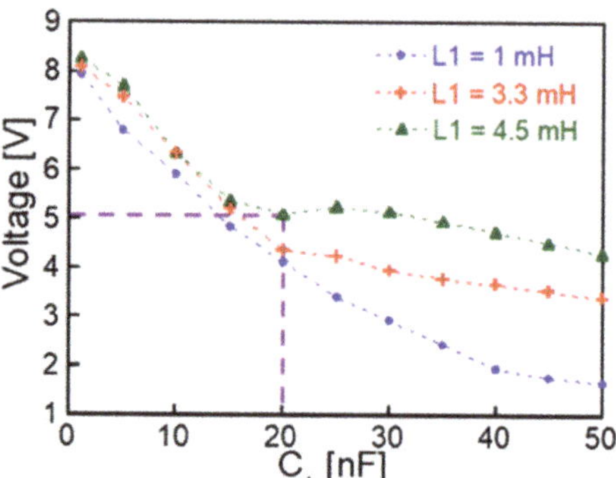

Figure 12. Dependence of the output voltage, V_{out}, on parameters L_1 and C_1.

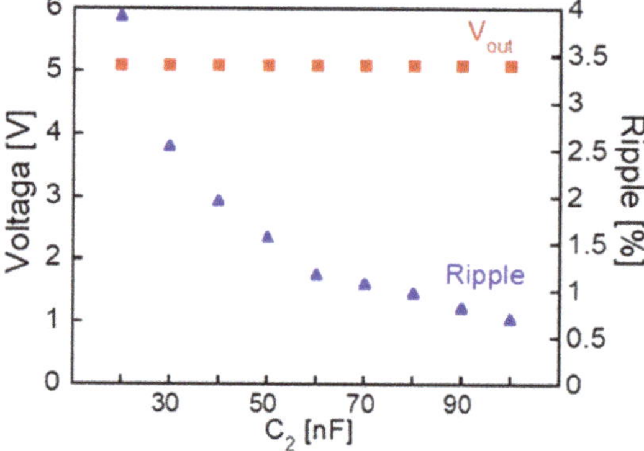

Figure 13. Dependence of the output voltage, V_{out}, on parameter C_2 and its ripple rate.

4.2. Experimental and Simulated Results of the Self-Oscillating Boost Converter

Figure 14 shows the experimental and simulated waveforms for the output and other voltages of the self-oscillating boost converter circuit presented in Figure 10 and Table 1. We measured the waveforms of voltages at V_{out}, V_{ds}, and V_{C1}, as indicated in the circuit diagram in Figure 10, by using an oscilloscope. Since the applied field frequency was 1 kHz, the full-time scale of 0.5 ms in Figure 14 corresponds to one cycle of the generated Wiegand pulse. The observed oscillations of V_{ds}, V_{C1}, and V_{out} agreed with the corresponding simulated values.

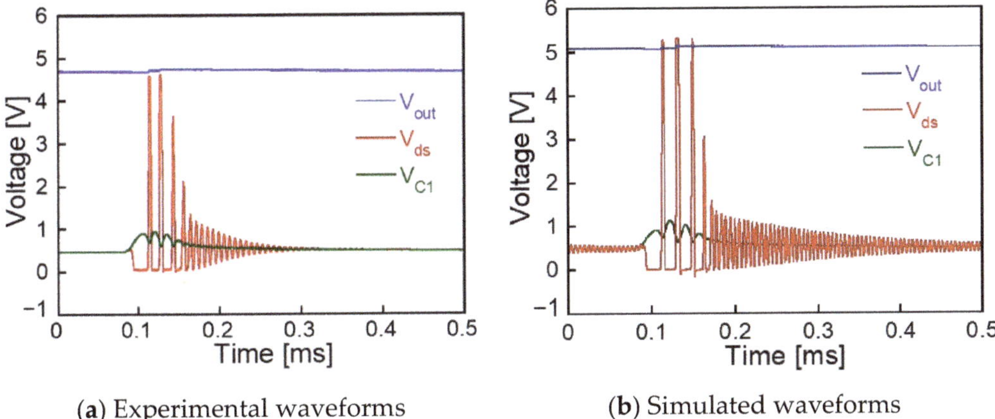

Figure 14. Waveforms of V_{out}, V_{ds}, and V_{C1} in the self-oscillating boost converter.

Figure 15 shows the time dependency of V_{out}. The simulated and observed saturated voltages are almost equivalent. We have discussed the experimental and simulated results of the circuit shown in Figure 10. The load resistance $R_{Load} = 5\,\text{M}\Omega$ was used, corresponding to an almost "open circuit condition" for the output. Figure 16 shows the dependence of I_{out} and V_{out} on the load resistance R_{Load}. V_{out} decreases as R_{Load} increases. Figure 12 shows that V_{out} changes with C_1, reaching 5 V on adjusting C_1. However, the ripple of V_{out} degrades at $C_1 < 10$ nF.

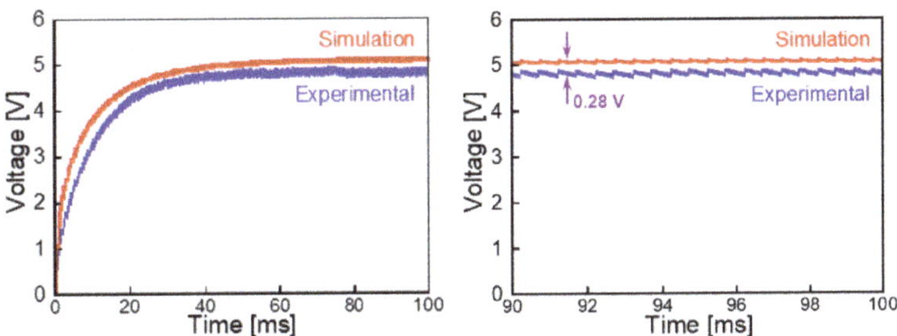

Figure 15. Simulated and experimental DC output voltage, V_{out}.

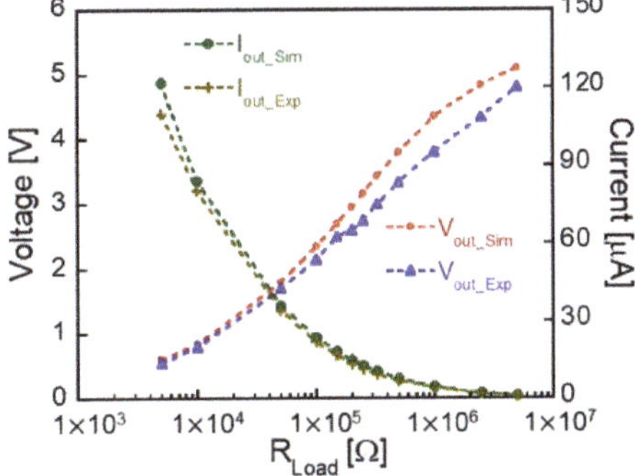

Figure 16. Dependence of simulated and measured V_{out} and I_{out} on load resistance, R_{Load}.

Figure 17 shows the electric power P_{out} utilized at R_{Load}. A maximum power of 63 µW was experimentally obtained at R_{Load} = 10 kΩ, which does not match with the resistance of the pickup coil, such as 180 Ω for the Wiegand sensor [9]. This mismatch is attributed to the elements and operation of the self-oscillating boost circuit. In fact, we have reported that the maximum power was obtained at a load resistance of 2 kΩ, higher than the DC coil resistance for the Wiegand sensor connected with rectifying and smoothing circuits [8].

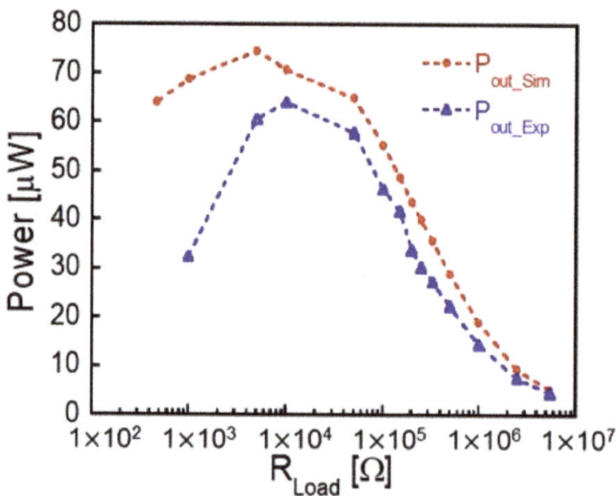

Figure 17. Dependence of the output power, P_{out}, on load resistance, R_{Load}.

In this study, an alternating magnetic field is externally applied to the Wiegand sensor as excitation energy, leading to the generation of the Wiegand pulse voltage. An attractive feature of the Wiegand sensor is that the generated pulse voltage is independent of the frequency of the applied alternating magnetic field. Figure 18 shows the measured V_{out} and its ripple rate function under an excitation frequency of 1 kHz and lower; V_{out} decreases with the frequency. However, V_{out} of approximately 5 V and a low ripple rate are obtained at a frequency range of up to 0.6 kHz. When the frequency is 0.4 kHz, the output voltage still reaches 3.3 V with a ripple rate lower than 5%. This result indicates that the self-oscillating boost converter can be used in practical applications as a power source for electronic modules.

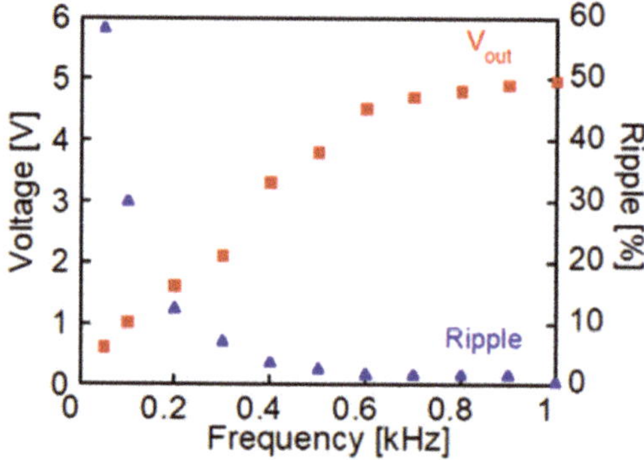

Figure 18. Dependence of measured V_{out} on the frequency and its ripple rate.

4.3. Application of the Wiegand Sensor as a Power Source

Figure 19 summarizes the relationship between V_{out} and I_{out}, obtained using the Wiegand sensor with a self-oscillating boost converter. It shows the voltage and current range functions for the load resistance used for practical application as a power source. A stable output of 5 V is maintained for currents up to 1 μA. This voltage/current range is used in low-energy IoT devices [17]. Furthermore, it is compatible with the existing energy-harvesting IC, such as power-storing buck DC–DC converters used for photovoltaic and vibration power generation elements [18]. Typically, a DC–DC converter is used in combination with storage batteries to ensure high efficiency and a maximized current supply in the order of 1 mA [19]. Therefore, the developed circuit system with the Wiegand sensor can be used with a storage battery; it allows for a higher capability of current consumption and can be used as a power supply for IoT devices.

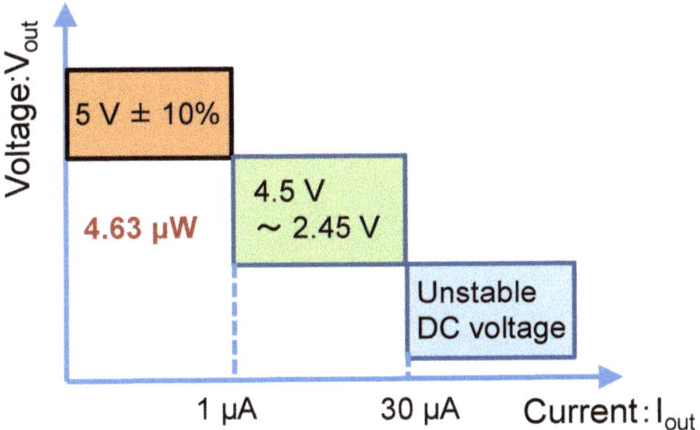

Figure 19. Mapping of V_{out}–I_{out} for practical applications of the Wiegand sensor with a self-oscillating boost converter.

5. Conclusions

We designed a self-oscillating boost converter circuit connected to the Wiegand sensor. The Wiegand sensor consists of an FeCoV magnetic wire with a diameter of 0.25 mm, length of 11 mm, and pickup coil with 3000 turns wound around the wire. This magnetic wire, i.e., the Wiegand wire, generates a peak pulse voltage of 4.62 V and 20 μs width during the magnetization reversal of its outer layer under a lower coercive field. An alternating magnetic field of 3.2 kA/m at 1 kHz was applied to the Wiegand sensor and alternating positive and negative pulse voltages were induced in the pickup coil. A DC voltage of 2.77 V was obtained by a bridge rectifier and a smoothing capacitor connected to the Wiegand sensor. This DC voltage could be intensified to approximately 5 V through a self-oscillating boost converter circuit. The experimental results of the voltage/current and ripple characteristics agreed with the simulation results. This study represents a significant development pertaining to the use of the Wiegand sensor as a power source for battery-less devices and modules.

Author Contributions: Conceptualization, Y.T.; Methodology, Y.T. and S.S.; Investigation, X.S. and H.I.; Writing—Original Draft Preparation, X.S.; Writing—Review and Editing, Y.T. All authors have read and agreed to the published version of the manuscript.

Funding: This research was partially funded by the JSPS KAKENHI, grant number 19K21965.

Conflicts of Interest: The authors declare no conflict of interest.

References

1. Dhananjay, S.; Gaurav, T.; Antonio, J. A survey of Internet-of-Things: Future vision, architecture, challenges and services. In Proceedings of the IEEE World Forum on Internet of Things, Seoul, Korea, 6–8 March 2014; pp. 287–292.
2. Balakrishna, S.; Thirumaran, M.; Solanki, V.K. A framework for IoT sensor data acquisition and analysis. *EAI Endorsed Trans. Internet Things* **2018**, *4*, 1–13. [CrossRef]
3. Citroni, R.; Di Paolo, F.; Livreri, P. Evaluation of an optical energy harvester for SHM application. *Int. J. Electr. Comm.* **2019**, *111*, 152918. [CrossRef]
4. Sharpes, N.; Vučković, D.; Priya, S. Floor tile energy harvester for self-powered wireless occupancy sensing. *Energy Harvest. Syst.* **2016**, *3*, 43–60. [CrossRef]
5. Morimura, H.; Oshima, S.; Matsunaga, K.; Shimamura, T.; Harada, M. Ultra-low-power circuit techniques for mm-size wireless sensor nodes with energy harvesting. *IEICE Electron Express* **2014**, *11*, 1–12. [CrossRef]
6. Wiegand, J.R.; Velinsky, M. Bistable Magnetic Device. U.S. Patent 3820090, 25 April 1972.
7. Dlugos, D.J.; Small, D.; Siefer, D.A. Wiegand Effect Energy Generator. U.S. Patent 6191687, 24 September 1998.
8. Takahashi, K.; Takebuchi, A.; Yamada, T.; Takemura, Y. Power supply for medical implants by Wiegand pulse generated from a magnetic wire. *J. Mag. Soc. Jpn.* **2018**, *42*, 49–54. [CrossRef]
9. Takemura, Y.; Fujinaga, N.; Takebuchi, A.; Yamada, T. Battery-less Hall sensor operated by energy harvesting from a single Wiegand pulse. *IEEE Trans. Magn.* **2017**, *53*, 4002706. [CrossRef]
10. Erickson, R.W.; Maksimovic, D. *Fundamentals of Power Electronics*; Springer: Boston, MA, USA, 1997.
11. Liu, K.H.; Lee, F. Zero-voltage switching technique in DC/DC converters. *IEEE Trans. Power Electron.* **1990**, *5*, 293–304. [CrossRef]
12. Saggini, S.; Ongaro, F.; Corradini, L.; Affanni, A. Low-power energy harvesting solutions for Wiegand transducers. *IEEE J. Emerg. Sel. Top. Power Electron.* **2015**, *3*, 766–779. [CrossRef]
13. Yang, C.; Sakai, T.; Yamada, T.; Song, Z.; Takemura, Y. Improvement of pulse voltage generated by Wiegand sensor through magnetic-flux guidance. *Sensors* **2020**, *20*, 1408. [CrossRef]
14. Chang, C.-C.; Chang, J.-Y. Novel Wiegand-effect based energy harvesting device for linear magnetic positioning system. *Microsyst. Technol.* **2020**, *26*, 3421. [CrossRef]
15. Tanaka, H.; Takemura, Y.; Abe, S.; Kohno, S.; Nakamura, H. Constant velocity of domain wall propagation independent of applied field strength in Vicalloy wire. *IEEE Trans. Magn.* **2007**, *43*, 2397–2399. [CrossRef]
16. Sun, X.; Yamada, T.; Takemura, Y. Output characteristics and circuit modeling of Wiegand sensor. *Sensors* **2019**, *19*, 2991. [CrossRef]
17. Newell, D.; Duffy, M. Review of power conversion and energy management for low-power, low-voltage energy harvesting powered wireless sensors. *IEEE Trans. Power Electron.* **2019**, *34*, 9794–9805. [CrossRef]
18. R1801K. Available online: https://www.mouser.co.uk/datasheet/2/792/r1801k001a_eev-1917814.pdf (accessed on 6 May 2021).
19. Drew, J. Powering a Dust Mote from a Piezoelectric Transducer. Available online: https://www.analog.com/jp/technical-articles/powering-a-dust-mote-from-a-piezoelectric-transducer.html (accessed on 6 May 2021).

Article

Numerical Simulation of Acoustic Resonance Enhancement for Mean Flow Wind Energy Harvester as Well as Suppression for Pipeline

Liuyi Jiang [1], Hong Zhang [2,*], Qingquan Duan [1] and Xiaoben Liu [2]

1   College of Safety and Ocean Engineering, China University of Petroleum, Beijing 102249, China; jiangliuyi409@163.com (L.J.); dqq@cup.edu.cn (Q.D.)
2   National Engineering Laboratory for Pipeline Safety, MOE Key Laboratory of Petroleum Engineering, Beijing Key Laboratory of Urban Oil and Gas Distribution Technology, China University of Petroleum, Beijing 102249, China; xiaobenliu@cup.edu.cn
*   Correspondence: hzhang@cup.edu.cn

Abstract: Acoustic resonance in closed side branches should be enhanced to improve the efficiency of wind energy harvesting equipment or thermo-acoustic engine. However, in gas pipeline transportation systems, this kind of acoustic resonance should be suppressed to avoid fatigue damage to the pipeline. Realizable k-ε delayed detached eddy simulations (DDES) were conducted to study the effect of different branch pipe shapes on acoustic resonance. At some flow velocities, the pressure amplitude of the simulation results is twice as large as that of the experimental results, but the simulation can accurately capture the flow velocity range where acoustic resonance occurs. The results prove the feasibility of the method of the equivalent diameter of the circular cross-section pipe and the square cross-section pipe to predict acoustic resonance. The pressure pulsation amplitude of acoustic resonance in a square cross-section pipe is significantly increased than that in a circular square cross-section pipe, indicating that the square cross-section branch configuration can be more condupive to improving the efficiency of wind energy harvesting. The influence of the angle between the branch and the main pipe on the acoustic resonance was studied for the first time, which has an obvious influence on the acoustic resonance. It is found that the design of a square wind energy harvester is better than that of a circular one; meanwhile, changing the branch angle can increase or suppress the acoustic resonance, which can improve the utilization efficiency of the acoustic resonance and provide a new method for suppressing the acoustic resonance.

Keywords: acoustic resonance; closed side branch; DDES; wind energy harvester

Citation: Jiang, L.; Zhang, H.; Duan, Q.; Liu, X. Numerical Simulation of Acoustic Resonance Enhancement for Mean Flow Wind Energy Harvester as Well as Suppression for Pipeline. *Energies* **2021**, *14*, 1725. https://doi.org/10.3390/en14061725

Academic Editor: Dibin Zhu

Received: 23 December 2020
Accepted: 15 March 2021
Published: 19 March 2021

1. Introduction

Acoustic resonance occurs as a gas flows through a closed side branch at a certain velocity. In industry, this kind of acoustic resonance may exhibit the two sides like a coin, a valuable side and a harmful side.

1.   For wind-energy harvesting, acoustic resonance can output standing waves in the closed side branch without moving parts, providing continuous pressure fluctuation for wind harvesting equipment [1–4] or temperature difference for thermo-acoustic engines [5,6], the acoustic resonance, in this case, needs to be strengthened to improve utilization efficiency [7].

2.   Closed side branches exist widely in industry, such as natural gas transmission pipelines [8,9], safety valves of the steam pipeline in nuclear power plant [10–15], a vertical-lift system of fighters [16–18]. Acoustic resonance in these cases can induce severe vibration and noise, cause fatigue damage to the pipeline, threaten the safe operation of the pipeline, and even cause safety accidents, which need to be

avoided or suppressed. Therefore, it is of practical significance to enhance or suppress acoustic resonance.

The mechanism of acoustic resonance has been clarified [19,20]. When the gas flows through the branch mouth, the wall friction of the main pipe disappears, and the shear layer falls off from the upstream wall, forming periodic vortices, which spread to the downstream of the main pipe into the branch. As the vortex reattaches to the downstream wall, the acoustic energy accumulates into the branch pipe, and the vortex interacts with the gas in the branch, which further promotes to form of a large vortex. The sound wave generated by the vortex propagates to the branch and reflects at the bottom of the branch. The incident wave and the reflected waveform a standing wave in the closed side branch, resulting in great pressure fluctuations [21–27] (Figure 1). From the perspective of the acoustic resonance mechanism, the branch shape affects the shedding and reattachment of the acoustic vortex, so the branch shape can affect the acoustic resonance, which is worthy of in-depth study.

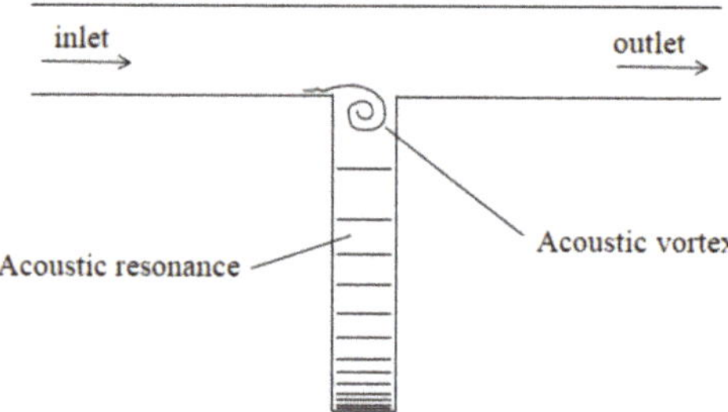

Figure 1. The gas flows through a closed side branch.

The cross-section shapes of the pipes studied before are square and circular. The square pipe is convenient for the visualization of research. After the tracer particles are injected into the flow, the flow mode can be directly captured, avoiding the influence of the intersection line at the T-shaped intersection of the circular pipe. In general, square pipes or circular pipes were studied separately. Both square and circular pipes were studied by Bruggeman et al. [28]. Bruggeman et al. pointed out that the mechanisms of acoustic resonance in square and circular pipes are the same and proposed that the results studied in a circular pipe can roughly evaluate whether acoustic resonance occurs in the range of Strouhal number in an equivalent square pipe, in which the equivalent diameter is calculated as $d_{eff} = \pi d/4$, where d is the circular branch diameter. The validity of this equivalence method means that the visualization results in a square pipe can be used to predict the acoustic resonance in an equivalent circular pipe. Moreover, there is no curve intersecting lines in the flow area of square pipes, so it is easier to obtain high-quality orthogonal grids. This not only saves the workload of grid generation in Computational Fluid Dynamics (CFD) preprocessing but also reduces the hardware resources and time cost in the computer solving process. However, the acoustic resonance characteristics of the equivalent diameter configurations have not been studied.

The previous studies focus on the mechanism, characteristics, and suppression methods of the acoustic resonance in closed side branches, in which all the branches are perpendicular to the main pipeline [8–11,13–20,23,24,26,27]. Some visualization studies of acoustic resonance in square cross-section pipe are to capture the acoustic vortex mode [9,13]. The mechanism and characteristics of acoustic resonance in the high-pressure natural gas transmission pipeline [29], the mean flow wind energy harvester (atmospheric air pipeline) [1,2], and the steam pipeline of nuclear power plant [10,12] are studied. The main purpose of changing the branch shape is to suppress acoustic resonance. Baldwin et al. [11] proposed that the internal structure of the safety valve should be designed out of the *St*-range at which acoustic resonance occurs. Jungbauer et al. [30] proposed to increase the diameter of the branch at the connection with the main pipe and use a reducer to connect the original branch pipe to change the *St*-range of acoustic resonance, thereby suppressing acoustic resonance. These two methods are effective, but the velocity in the main velocity is limited to a certain range once the pipeline structure is designed and constructed.

Compared with the branch cross-section and the branch diameter, the branch angle has a greater influence on acoustic resonance and has more potential in suppressing or enhancing acoustic resonance. However, all the pipeline structures in the previous study are branch pipelines perpendicular to the main pipeline [19,20]. In the field of mean flow wind energy harvesting, all the research only focused on the structure of the circular cross-section and the branch perpendicular to the main pipe [2–6,31]. There is currently a lack of research on the influence of branch angle on acoustic resonance.

Therefore, in this paper, the acoustic resonance of a circular pipe and an equivalent diameter square pipe are simulated with the realizable k-ε DDES model. In addition, the influences of the intersecting line of the pipe on the acoustic resonance are discussed. Furthermore, the influence of the branch angle on acoustic resonance is studied. Current simulations show that these changes can improve mean flow wind harvesting and suppress the acoustic resonance in industrial pipelines.

2. Mathematical Methods

The method workflow of the simulation is shown in Figure 2. In this section, the CFD method is introduced in the order shown in Figure 2. The higher the pressure potential energy, the more beneficial it is to improve the efficiency of the wind energy harvester. Therefore, the pressure distribution determines the installation position of piezoelectric devices in the wind energy harvester. The wider the velocity range and the more consistent the frequency are, the more convenient it is to harvest wind energy. Moreover, the velocity field can show the mechanism of pressure amplitude enhancement. Therefore, velocity, pressure and vorticity are analyzed.

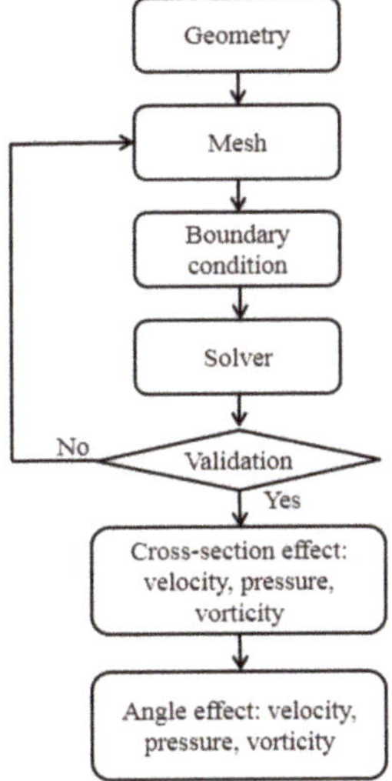

Figure 2. The method workflow of the simulation.

2.1. Geometry Model

The geometric model is shown in Figure 3, which is the flow area in the pipe. D is the main pipe internal diameter, L is the main pipe length, the left is the inlet, and the right is the outlet. The closed side branch is located at the symmetry center of the main pipe. l is the length of the branch (from the bottom of the branch to the inner wall of the main pipe), d is the diameter of the branch (or the width of the equivalent square pipe branch), and the center point $p1$ at the bottom of the branch is the pressure monitoring point. α is the angle between the branch and the pipe.

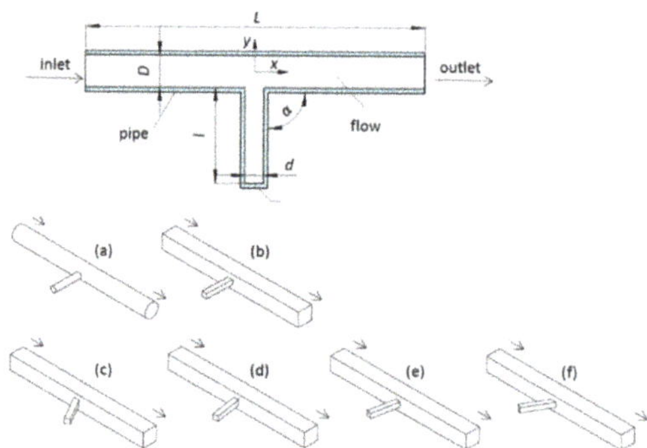

Figure 3. Geometry model schematic of pipe shapes, (**a**) case 1: circular pipe, $\alpha = 90°$, (**b**) case 2: square pipe, $\alpha = 90°$, (**c**) case 3: square pipe, $\alpha = 60°$, (**d**) case 4: square pipe, $\alpha = 80°$, (**e**) case 5: square pipe, $\alpha = 100°$, and (**f**) case 6: square pipe, $\alpha = 120°$.

Six cases of different structures of pipes were configured (Table 1). In all cases, the shapes of main pipes and branch pipes were, respectively, configured as circular or square. As the unstable shedding of the boundary layer at the branch mouth was the cause of acoustic resonance, the shape configurations of the main pipe were based on the same Reynolds number to ensure that the flow in the boundary layer of the main pipeline was similar. The shape configurations of branch pipe were based on the equivalent method proposed in reference [28]. The geometric size of case 1 (Figure 3a) was consistent with the experimental pipe size in reference [27], which was convenient for comparison and validation of the results. In case 1, $D = 0.089$ m, $D/d = 0.135$, $L = 0.9$ m, $l = 0.17$ m. Different shapes of the main pipe and branch pipe were arranged in cases 1–2 (Figure 3a,b) to study the influence of cross-section shape on acoustic resonance, and different angles were set in cases 3–6 (Figure 3c–f) to study the influence of branch angle on acoustic resonance. In order to ensure that the acoustic resonance frequency was basically the same, the branch lengths of all geometric models were kept the same to avoid the influence of the branch length on the acoustic resonance. Among them, case 3 and case 6 were mirror bodies about the XOZ plane, and case 4 and case 5 were mirror bodies about the XOZ plane. Using this strategy, case 3 and case 6 could share a set of grids, which reduces the meshing work. In the simulation solution, only the inlet and outlet boundary conditions need to be exchanged; case 4 and case 5 had the same effect.

Table 1. Configuration of the cases.

Case	Main Pipe Cross-Section Shape	Branch Cross-Section Shape	Branch Angle A
1	Circular	Circular	90°
2	Square	Square	90°
3	Square	Square	60°
4	Square	Square	80°
5	Square	Square	100°
6	Square	Square	120°

2.2. Grid Partition

Only the grid details of case 1 are shown here because the mesh densities of cases 2–6 are similar. The grid-type was a hexahedral structured grid, and all grid blocks of the flow area were O-shaped to improve the grid quality (Figure 4). The thickness of the first boundary layer was 0.3 mm, and the internal grid size increased with a rate of 1.2. In the branch entrance and its adjacent area, the flow turbulence was more intense, so the grid in this area was refined, and the grid size was kept at about 0.9 mm.

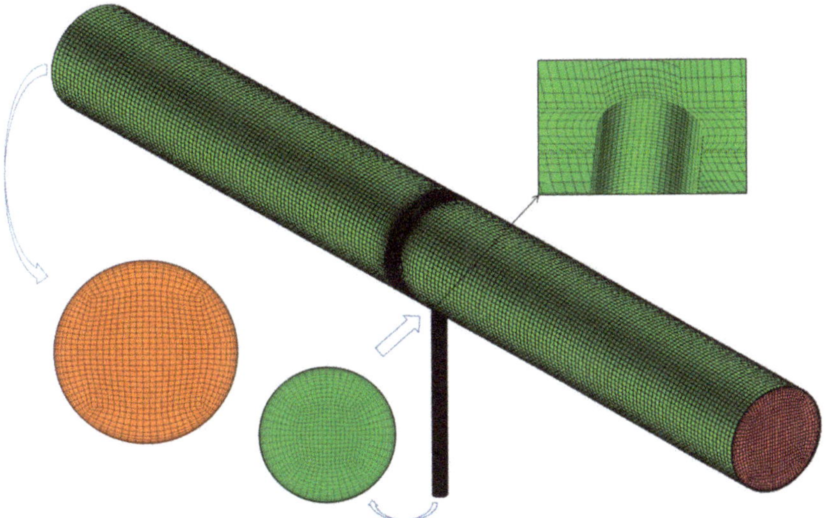

Figure 4. Grid partition of case 1.

The branch angle and the intersection line of the main pipe and the branch pipe of the circular cross-section had an influence on the mesh quality. The grid quality of all cases was as follows: the mesh quality of the main pipe with a circular cross-section was the worst, but it was guaranteed to be greater than 0.5 (case 1); the mesh quality of the main pipe with square cross-section was better, as the branch angle $\alpha = 90°$, all grids were orthogonal, and the mesh quality was equal to 1 (case 2); as the branch angle $\alpha = 80°$ and $\alpha = 100°$, the mesh quantities were the same, greater than 0.9 (case 4 and case 5), as the branch angle $\alpha = 60°$ and $\alpha = 120°$, the mesh quantities were the same, greater than 0.85 (case 3 and case 6).

2.3. Boundary Conditions

The inlet boundary condition was the mass-flow inlet, the outlet boundary condition was the pressure outlet, the outlet pressure value was set to 0, and the reference pressure was set to 0.35 MPa.

2.4. CFD Turbulence Model and Solver

In CFD numerical simulation, the choice of turbulence model affects the accuracy of numerical calculation and the cost of computing hardware and time. The LES model can provide more accurate calculation results, but its calculation cost is high. For example, Morita et al. [32] used the LES turbulence model to simulate the acoustic resonance in the closed side branch successfully, but the number of grids was greater than 5 million, which lead to the numerical calculation cost high. Radavich et al. [33] used the RANS turbulence model to simulate the acoustic resonance phenomenon and obtained the mode of acoustic resonance, but the pressure fluctuation simulated by Radavich is smaller than the experimental results. Because the wall is rigid in the simulation, it may not be reasonable that the simulation results are smaller than the experimental results. In addition, our simulation experience shows that the pressure fluctuation results of the RANS model are easy to decay, and it is not easy to simulate the acoustic resonance phenomenon that should exist. In this paper, we used delayed detached-eddy simulation (DDES), a hybrid model of LES-RANS extended by Spalart et al. [34], which is widely and effectively used [35–39]. Due to the advantages of accuracy and economy, the k–ε turbulence model is the most commonly used CFD model, among which the realizable k–ε model shows the best performance in simulating separated flows [40]. The excitation source of acoustic resonance in the closed side branch is the separated shear layer of the main pipe at the branch mouth. Therefore, in the present simulation, the RANS region is solved by the realizable k–ε model.

The commercial CFD software ANSYS Fluent v19.2 was used to simulate the acoustic resonance in the closed side branch. The following equations (the mass, momentum, and energy conservation equations as well as ideal gas law) were solved in the simulation process.

$$\frac{\partial \rho}{\partial t} + \frac{\partial}{\partial x_j}(\rho u_i) = 0 \tag{1}$$

$$\frac{\partial}{\partial t}(\rho u_i) + \frac{\partial}{\partial x_j}(\rho u_i u_j) = -\frac{\partial p}{\partial x_i} + \frac{\partial \tau_{ij}}{\partial x_j} \tag{2}$$

$$\frac{\partial}{\partial t}(\rho E) + \frac{\partial}{\partial x_i}(u_i(\rho E + p)) = \frac{\partial}{\partial x_i}\left(k_{eff}\frac{\partial T}{\partial x_i} - hJ + u_j(\tau_{ij})_{eff}\right) \tag{3}$$

$$p = \rho T \frac{R}{M} \tag{4}$$

where ρ is the density, τ is the viscous stress, p is the gas pressure, E is the summation of kinetic energy and internal energy, T is the gas temperature, J is the gas diffusion energy, h is the enthalpy, k_{eff} is the thermal conductivity. τ_{eff} is the viscous dissipation coefficient, R is the gas constant, M is the molecular weight.

The pressure-based solver was applied. In order to ensure the solution convergence, the residual of the energy equation was set to 10^{-6}. The residual of other equations was set to 10^{-5}. The simulation time step was 10^{-5}, and the number of iterations of each time step was set to 20.

2.5. Grid Analysis and Validation

The purpose of grid independence verification was to avoid numerical simulation errors caused by grid quality. According to the method in reference [41,42], four sets of grids were prepared to compute with the boundary condition of inlet velocity of 19.55 m/s. The cell number of grids was 231,551, 328,018, 636,614, and 1,123,518, respectively. In order to ensure the convergence of numerical calculation, the time steps of 231,551 and 328,018 were $2 \cdot 10^{-5}$ s, and those of 636,614 and 1,123,518 were 10^{-5} s. All the cased were simulated to 0.126 s, at which the pressure fluctuation amplitude was steady (Figure 5a). The normalized pressure p^* was calculated as Equation (5), where v was the mean flow velocity in the main pipeline, p was the pressure amplitude, ρ was the gas density. As the grid increased from 636,614 to 1,123,518, the pressure amplitude did not change much. In Figure 5a, compared with the pressure pulsation amplitude of grid number 1,123,518, the amplitude errors of grid number 636,614, 328,018 and 231,551 were 1.7%, 30.1% and 46.5%, respectively. Therefore, the result of the 636,614 grid was considered as a grid-independent solution.

$$p* = \frac{p}{(1/2)\rho v^2} \tag{5}$$

The flow conditions of reference [27] were simulated with the grid of 636,614 in case 1. The experimental data of Ziada et al. [27] and the current simulation results were basically consistent (Figure 5b), which proves the validity of the DDES model. At most flow velocities, the pressure fluctuation amplitude of the simulation results was larger than that of the experiment because the wall was rigid, and there was no radiation penetration of sound energy in the simulation.

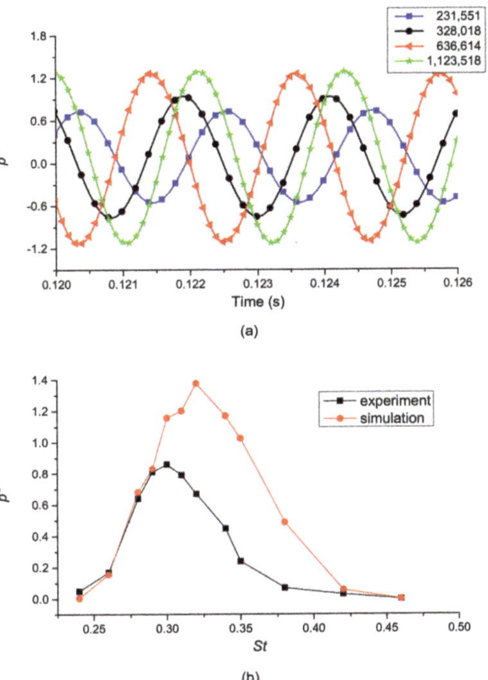

Figure 5. Verification for grid-independent solutions (**a**); validation between the experimental data of Ziada et al. [27] and the current simulation results (**b**).

In experiments, sound energy radiated through the pipe wall, and the vibration of the pipe wall attracted a certain amount of the sound source energy of the flow. In the simulation, the wall was only be set as a non-penetrating and rigid wall boundary; therefore, the pressure pulsation amplitudes obtained by the simulation were bigger than that obtained by the experiment. However, it was difficult to quantitatively determine the influence of wall boundary conditions.

3. Influences of Pipe Cross-Section Shape

In this section, the acoustic resonance characteristics of circular cross-section pipe and equivalent square cross-section pipe are compared to study the influences of pipe cross-section shape on the acoustic resonance. First, the time histories of static pressure at the monitoring point $p1$ and their frequency domain signals were compared, including the onset, offset and the maximum amplitude point of acoustic resonance in the middle of St. Second, the vortex modes of the exciting acoustic source at the branch mouth were compared and analyzed to determine the difference of the mechanisms between these two cross-section structures. Finally, the influence of the intersecting line shape on the z-direction oscillation of the acoustic vortex at the branch mouth was analyzed.

3.1. Onset and Offset of Acoustic Resonance

Figures 6 and 7 show the amplitude and frequency characteristics of the monitoring point $p1$ pressure of circular and square pipes for St = 0.26, 0.30, and 0.42, correspondingly. The normalized pressure p^* was calculated as Equation (5), where v was the mean flow velocity in the main pipeline, p was the pressure amplitude, ρ is the gas density. In Figure 6, the spectrum pressure (right) after Fast Fourier Transform (FFT) of the time history pressure (left) was also normalized according to Equation (5). The frequency f (left) in the FFT analysis was normalized by dividing the first-order acoustic resonance frequency f_1. The acoustic resonance in the circular pipe and the square pipe exhibit similar characteristics with the *St*. At a certain *St*, a stable acoustic resonance was formed in both circular and square branches. At low *St* (high velocity) and high *St* (low velocity), the amplitude of pressure fluctuation decreased finally. At high *St* (low velocity), the energy generated by the initial acoustic vortex was insufficient to make up for the dissipation of acoustic resistance and radiation and could not maintain sustained pressure pulsation; at low *St* (high velocity), the energy generated by the initial acoustic vortex was also large, but the flow in the main pipe took more energy from the branch inlet, so the pressure pulsation could not be sustained. It can be seen from the spectrogram that both the circular pipe and the square pipe could show pure tones when acoustic resonance occurred. When acoustic resonance did not occur, the noise of the circular pipe was more obvious than that of the square pipe. It shows that the cross-section shape of the branch mouth affects the frequency of the initial vortex shedding, but after the acoustic resonance was formed, the frequency of the pressure pulsation was dominated by the first-order acoustic resonance frequency of the branch.

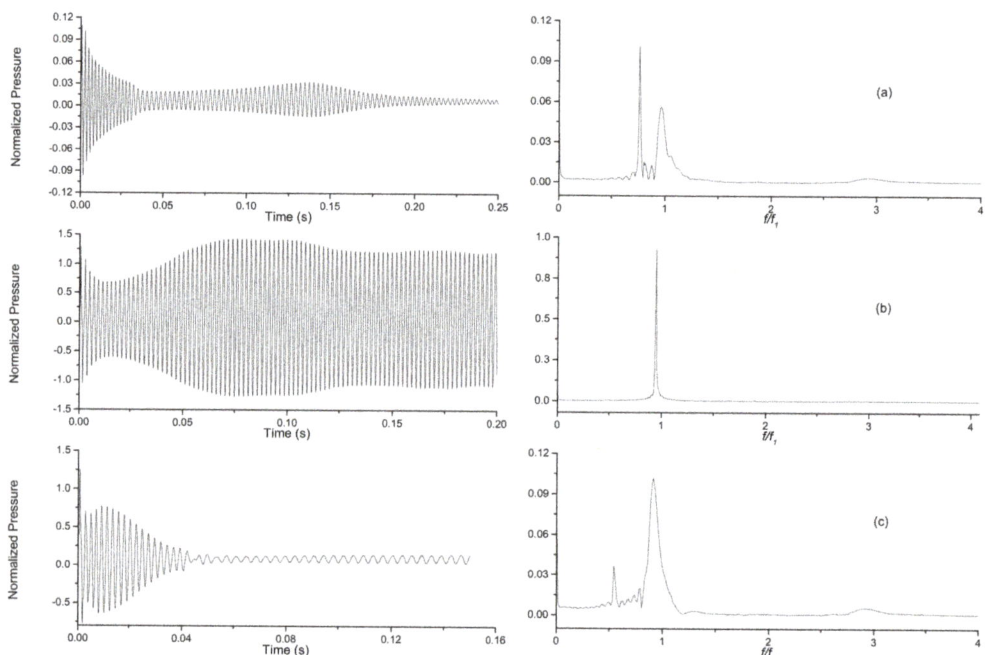

Figure 6. Time histories of static pressure at the monitoring point $p1$ of circular pipe (case 1), $St = 0.26$ (**a**), $St = 0.30$ (**b**), $St = 0.42$ (**c**).

It can be seen from Figure 8 that the modes of the circular section pipe and the equivalent square section pipe were basically the same. At the peak pressure value, the acoustic vortex beats the branch cavity from the main pipe, and at the nadir pressure value, the acoustic vortex was pushed to the main pipe by the gas in the branch. The difference was that the structure and strength of vortices in the equivalent square pipe were larger, which determines that the amplitude of pressure fluctuation was higher. Another subtle difference was that the vortices near the downstream branch wall of the circular cross-section pipe were more complex than that of the square cross-section pipe.

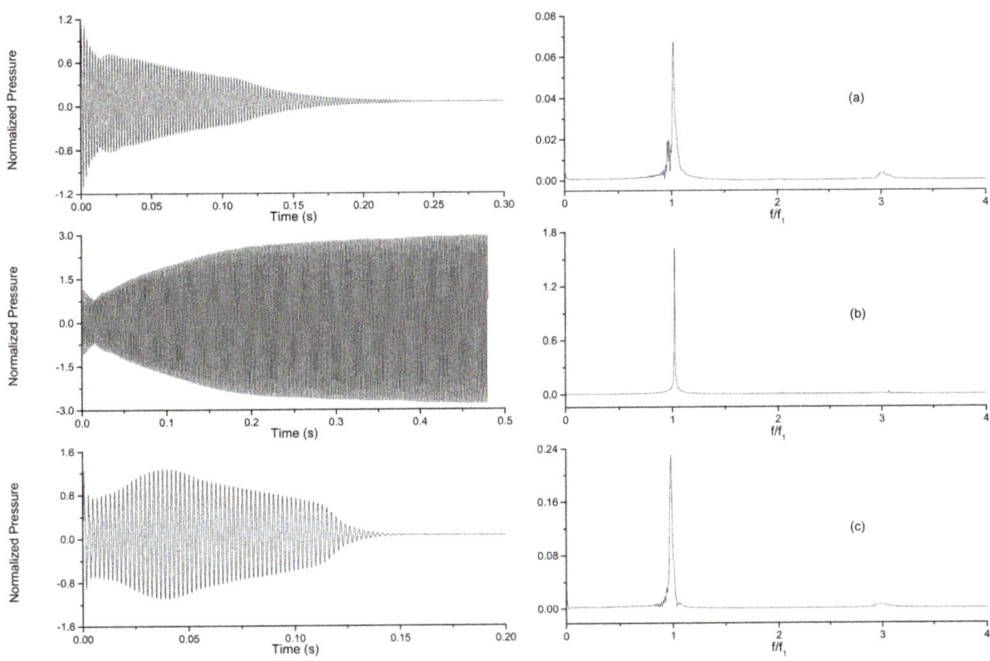

Figure 7. Time histories of static pressure at the monitoring point $p1$ of square pipe (case 2), $St = 0.26$ (**a**), $St = 0.30$ (**b**), $St = 0.42$ (**c**).

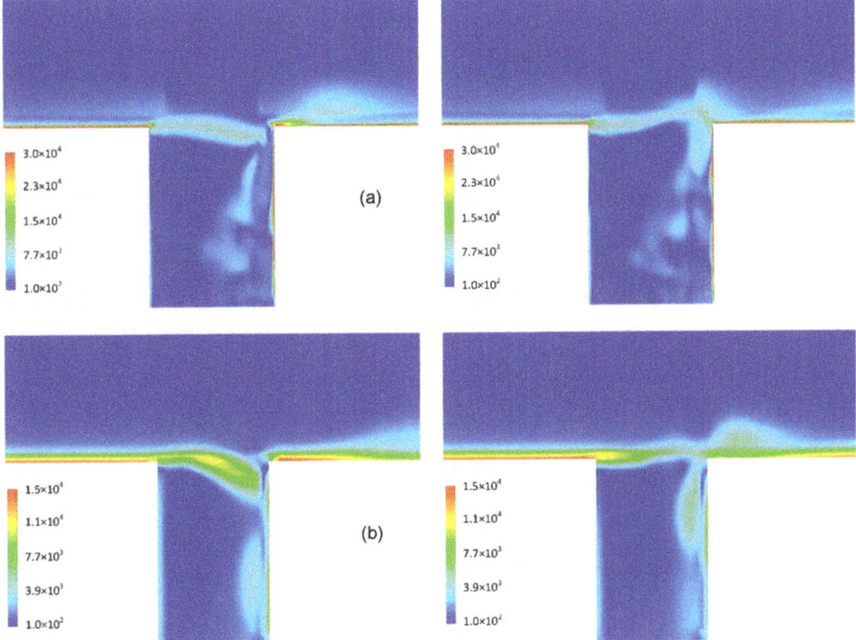

Figure 8. The vorticity (unit: s^{-1}) at the peak pressure value (**left**) and the nadir pressure value (**right**) of the circular section pipe (**a**) and the equivalent square section pipe (**b**), $St = 0.3$.

3.2. Vorticity Around the Branch Mouth

The onset and offset of acoustic resonance with *St* in the circular pipe and the equivalent diameter square pipe were almost the same, but the acoustic resonance pressure amplitude in the equivalent square pipe was generally larger than that of the circular one (Figure 9). Therefore, although the results of equivalent diameter square pipe were relatively overestimated as predicting the pressure amplitude in the circular pipe, they were effective in predicting the onset and offset acoustic resonance. The increase in the pressure pulsation amplitude was beneficial to the efficiency improvement for the wind energy harvesting equipment or the thermo-acoustic engine [7,43–45]. Therefore, the design of a square cross-section branch in wind energy utilization equipment with can greatly improve the efficiency in a wide range of flow velocity.

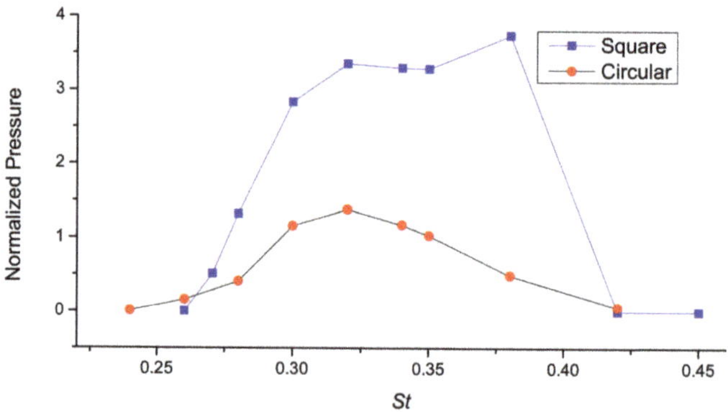

Figure 9. Variation of pressure fluctuation with *St*.

3.3. Influence of Intersection Line on Vortex Oscillation of z-Velocity

Figure 10 shows the z-velocity at the center of the branch entrance of cases 1–2. The z-velocity $v_z{}^*$ was normalized by $v_z{}^* = v_z/v$, where v_z was the actual z-velocity, v was the mean flow velocity in the main pipeline. Although in both cases, the low regions were symmetric with respect to the $z = 0$ planes, the z-velocity was not zero, indicating the asymmetry of flow. The vortex oscillated up and down along the y-direction at the branch mouth (Figure 8) but at a certain speed along the z-direction at the same time. When acoustic resonance was stable, the FFT analysis of the oscillation velocity along the z-direction shows that the oscillation amplitude was the largest in case 1 and the swing amplitude in case 2 was smaller. It was revealed that the circular shape had a greater influence on the z-direction oscillation, while the square shape had a smaller influence. In a circular pipe, more energy was consumed in the z-direction, which explains the higher pressure pulsation amplitude in the equivalent square pipe (Figure 9).

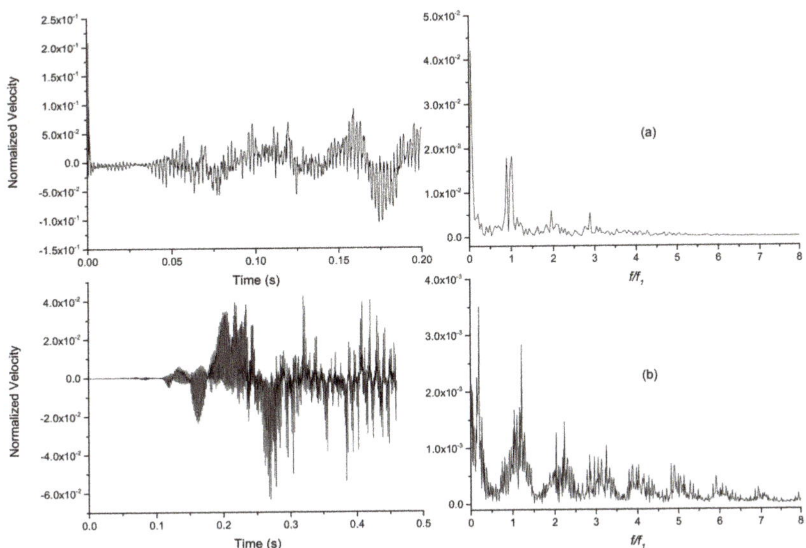

Figure 10. The z-velocity at the center of the branch entrance. (**a**) Case 1; (**b**) Case 2.

4. Effects of Branch Angle

It was necessary to avoid the occurrence of acoustic resonance in the pipeline design stage, so the range of St of acoustic resonance occurrence was focused on. According to the results in Section 3, the shape of the branch only affects the amplitude of the pressure fluctuation when the acoustic resonance occurred, and the onset and offset of the acoustic resonance of St was consistent in the equivalent pipe. Therefore, in order to obtain a high-quality grid and save the hardware resources and time cost of calculation, in this section, the configuration of equivalent square cross-section pipe was used to study the influence of branch angle on acoustic resonance. The abnormal beat phenomenon in acoustic resonance in different branch angles was analyzed. The influence of the branch angle on the pressure amplitude and frequency of the acoustic resonance and vortex modes at the branch mouth and the pressure contours are analyzed in this section.

4.1. Beats Phenomena

In the flow velocity of $v = 70$ m/s, the acoustic resonance in the perpendicular branch ($\alpha = 90°$) and the backward-swept branch ($\alpha = 60°$) exhibited beat vibration (Figure 11), which indicates the instability of pressure pulsation as the flow velocity increases. This shows that the appearance of beat vibration was not determined by the branch angle but by the flow velocity.

However, as the flow rate increased, the acoustic resonance was no longer stable, and when $St < 0.08$, the acoustic resonance ends (Figure 12). It can be seen from the spectrogram that the pressure pulsations of both were affected by the second-order acoustic resonance frequency of the branch. The difference was that the acoustic resonance in the backward-swept branch was the first-order dominant frequency, and in the perpendicular branch, the acoustic resonance was the third-order dominant frequency.

4.2. Acoustic Resonance Characteristics

The branch angle α had a significant effect on the acoustic resonance (Figure 12). In the cased of $\alpha > 90°$, acoustic resonance was suppressed. In the case of $\alpha = 120°$, acoustic resonance was greatly suppressed at all flow velocities. In the case of $\alpha = 100°$, to a certain extent, the pressure pulsation amplitude of the first-order acoustic resonance was suppressed, while the pressure pulsation of the second-order acoustic resonance pressure pulsation was completely suppressed, and no acoustic resonance occurred. In the field where acoustic resonance needs to be suppressed, the design of branch angle $\alpha = 120°$ was simple and effective, showing great application potential. In the cased of $\alpha < 90°$, acoustic resonance shows a tendency to be enhanced. In the case of $\alpha = 60°$, the velocity range and pressure fluctuation amplitude of the acoustic resonance were greatly enhanced, compared with acoustic resonance in the case of $\alpha = 90°$, which could provide about 2 times of the velocity range and several times of the amplitude of pressure fluctuation of acoustic resonance. It should be pointed out that at $St = 0.07$–0.15, acoustic resonance occurred in the case of $\alpha = 90°$, but in the second and third-order acoustic resonance frequencies, respectively (Figure 12). Acoustic resonance in the case of $\alpha = 60°$ keeps the first-order frequency acoustic resonance unchanged, which could simplify the design of wind energy harvesting equipment.

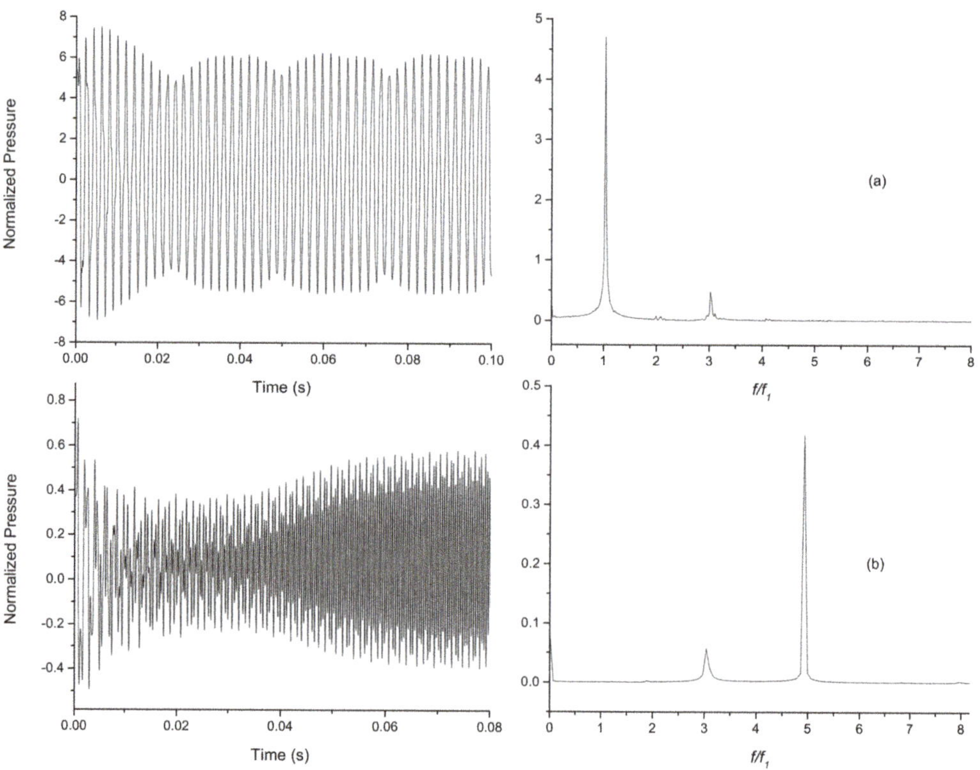

Figure 11. The beating phenomenon at the flow velocity of $v = 70$ m/s, (a) $\alpha = 60°$, (b) $\alpha = 90°$.

4.3. Vortex Modes and Pressure Contours

The vortex modes of branch mouths are important characterizations for studying the mechanism of acoustic resonance. Figure 13 shows the vortex modes of acoustic resonance in the different branch angles. Compared to in other cases, in the case of $\alpha = 60°$, the distance of the vortex pressed in (or pushed out) the branch distance was larger, and the vortex area was also larger (Figure 13a). In the case of $\alpha = 90°$, the sound vortex took on the form of whipping (Figure 13b), and the sound vortex exhibited the similar mode in the case of $\alpha = 80°$. In the cases of $\alpha > 90°$, the shear layer shed from the upstream main pipe wall propagated backward along the flow direction, and it was basically no swing

(Figure 13c). The results show that the acoustic vortices were more likely to fall into the branches and accumulate the sound energy more easily in the backward-swept branch structure ($\alpha < 90°$), which caused greater pressure fluctuation amplitude in a wider range of velocity, while in the cases of $\alpha > 90°$ the results were just the opposite. It was revealed that the acoustic vortex at the branch mouth was the exciting source of acoustic resonance.

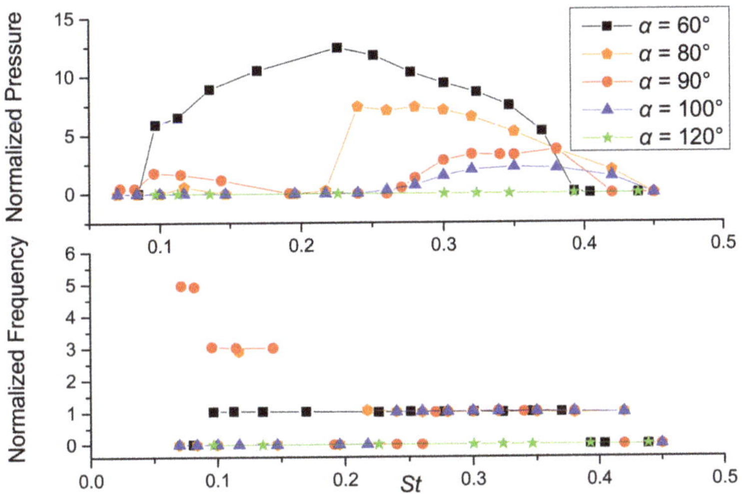

Figure 12. Variation of pressure fluctuation with *St*.

Figure 14 shows the pressure contours in the pipe of different branch angles in the flow velocity of $v = 50$ m/s. In the case of $\alpha = 60°$, a 1/4 wavelength standing wave of the first-order acoustic frequency was formed. In the case, $\alpha = 90°$, a 3/4 wavelength standing wave of the second-order frequency was formed, and similar phenomena occur in the case of $\alpha = 80°$. However, in the case of $\alpha = 120°$, the pressure in the branch was uniform, the standing wave did not exist, and only a continuous high-pressure point was formed at the vertex of the branch mouth downstream wall (Figure 14c), and similar phenomena occur in the case of $\alpha = 100°$.

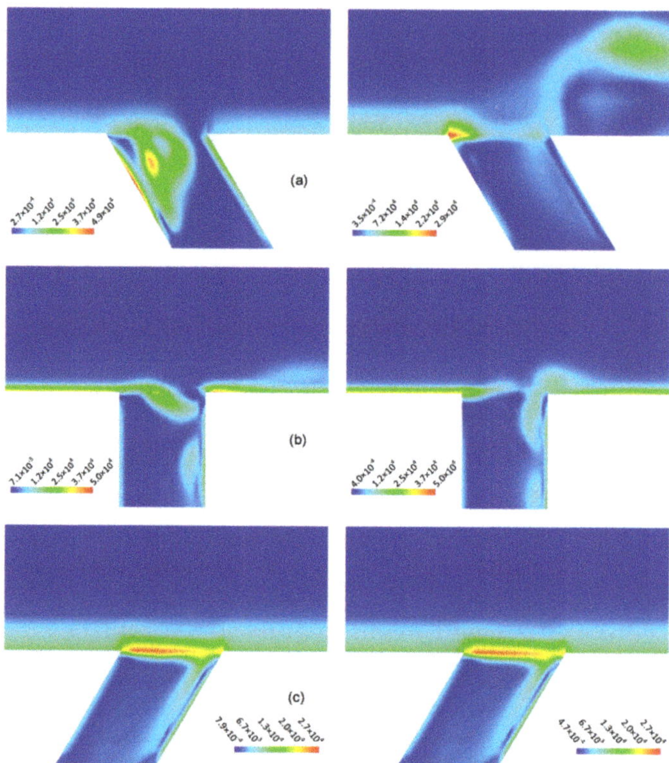

Figure 13. The vorticity (unit: s^{-1}) at the peak pressure value (**left**) and the nadir pressure value (**right**) of $\alpha = 60°$ (**a**), $\alpha = 90°$ (**b**) and $\alpha = 120°$(**c**), $v = 50$ m/s.

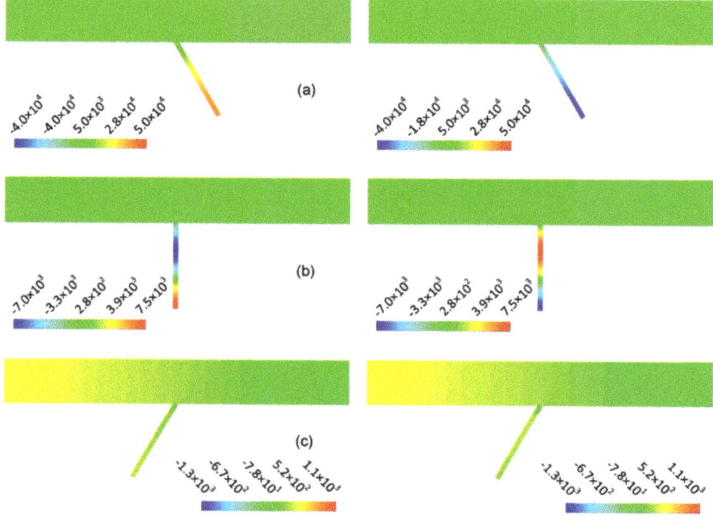

Figure 14. The pressure (unit: Pa) contour at the peak pressure value (**left**) and the nadir pressure value (**right**) of $\alpha = 60°$ (**a**), $\alpha = 90°$ (**b**) and $\alpha = 120°$ (**c**), $v = 50$ m/s.

5. Conclusions

The realizable k-ε DDES model was used to simulate the acoustic resonance in different shape closed side branches. The results show that the realizable k-ε DDES model could effectively predict the acoustic resonance, and it was found that the shape of the branch had a significant effect on the acoustic resonance. The conclusions are summarized as follows:

1. The Strouhal number range from onset to offset of acoustic resonance were consistent in the circular branch and the equivalent diameter square branch, which indicated that the results studied in these two shapes pipe were universal for predicting the occurrence of acoustic resonance. The vortex modes in the branch mouths were similar, which indicated the similarity of the mechanisms in these two shapes;
2. The intersection line of the circular pipe intensified the z-direction oscillation of the sound vortex at the branch mouth, which led to energy dissipation and reduced the pressure fluctuation amplitude of the acoustic resonance caused by the sound source excitation. Therefore, compared with the circular branch, acoustic resonance in the equivalent square branch provided higher pressure pulsation and was more efficient in wind energy harvesting. The design of a square wind energy harvester was better than that of a circular one;
3. At a certain flow rate, acoustic resonance show unstable beats phenomena. These phenomena were independent of the shape of the pipe and were related to whether the flow velocity in the main pipe was close to the frequency conversion point;
4. The branch angle α had a significant effect on the acoustic resonance. It was found that changing the branch angle could increase or suppress the acoustic resonance, which could improve the utilization efficiency of the acoustic resonance and provide a new method for suppressing the acoustic resonance. As $\alpha = 60°$, the acoustic resonance maintains the first-order frequency in a wide range of flow velocity, which could be more efficient in wind energy harvesting. As $\alpha = 120°$, the acoustic resonance was suppressed in a wide range of velocity, which could be more efficient and economical to suppress the acoustic resonance in the gas transmission pipeline system;
5. In the present simulation, all cases were initialized from the inlet with the corresponding mass-flow rate. Taking the quasi-steady acoustic resonance state of one flow velocity as the initial state of another flow velocity was expected to reduce the cost of the solution. However, the impact of this operation should be further studied. In addition, the effect of changing the branch angle shows great potential application value. Therefore, it is worthy of further study to verify this effect by experimental method.

Author Contributions: Conceptualization, L.J. and H.Z.; methodology, L.J., Q.D. and H.Z.; validation, L.J., H.Z., Q.D. and X.L.; data curation, L.J.; writing—original draft preparation, L.J.; writing—review and editing, H.Z.; supervision, H.Z.; project administration, H.Z.; funding acquisition, H.Z. and X.L. All authors have read and agreed to the published version of the manuscript.

Funding: This research was funded by China National Key Research and Development Plan (no. 2016YFC0802105).

Institutional Review Board Statement: Not applicable.

Informed Consent Statement: Not applicable.

Data Availability Statement: The data used to support the findings of this study are available from the corresponding author upon request.

Conflicts of Interest: The authors declare no conflict of interest.

References

1. Sun, D.; Xu, Y.; Chen, H.; Shen, Q.; Zhang, X.; Qiu, L. Acoustic characteristics of a mean flow acoustic engine capable ofwind energy harvesting: Effect of resonator tube length. *Energy* **2013**, *55*, 361–368. [CrossRef]
2. Sun, D.; Xu, Y.; Chen, H.; Wu, K.; Liu, K.; Yu, Y. A mean flow acoustic engine capable of wind energy harvesting. *Energy Conver. Manag.* **2012**, *63*, 101–105. [CrossRef]

3.   Yu, Y.S.W.; Sun, D.; Zhang, J.; Xu, Y.; Qi, Y. Study on a Pi-type mean flow acoustic engine capable of wind energy harvesting using a CFD model. *Appl. Energy* **2017**, *189*, 602–612. [CrossRef]

4.   Yu, Y.; Sun, D.; Wu, K.; XU, Y.; Chen, H.; Zhang, X.; Qiu, L. CFD study on mean flow engine for wind power exploitation. *Energy Convers. Manag.* **2011**, *52*, 2355–2359. [CrossRef]

5.   Slaton, W.V.; Zeegers, J.C.H. Acoustic power measurements of a damped aeroacoustically driven resonator. *J. Acoust. Soc. Am.* **2005**, *118*, 83–91. [CrossRef] [PubMed]

6.   Slaton, W.V.; Zeegers, J.C.H. An aeroacoustically driven thermoacoustic heat pump. *J. Acoust. Soc. Am.* **2005**, *117*, 3628. [CrossRef] [PubMed]

7.   Luo, E. Experimental Study on Thermoacoustic Refrigerators Driven by a Travelling-Wave Thermoacoustic Engine. In *AIP Conference Proceedings*; AIP Publishing: New York, NY, USA, 2003.

8.   Tonon, D.; Hirschberg, A.; Golliard, J.; Ziada, S. Aeroacoustics of pipe systems with closed branches. *Noise Notes* **2011**, *10*. [CrossRef]

9.   Kriesels, P.C.; Peters, M.C.A.M.; Hirschberg, A. High amplitude vortex-induced pulsations in a gas transport system. *J. Sound Vibrat.* **1995**, *184*, 343–368. [CrossRef]

10.   DeBoo, G.; Gesior, R.; Ramsden, K.; Strub, B. Identification of quad cities main steam line acoustic sources and vibration reduction. In *American Society of Mechanical Engineers, Pressure Vessels and Piping Division (Publication) PVP*; American Society of Mechanical Engineers: New York, NY, USA, 2008; Volume 4, pp. 485–491.

11.   Baldwin, R.M.; Simmons, H.R. Flow-induced vibration in safety relief valves. *J. Press. Vessel Technol.* **2009**, *120*, 267–272. [CrossRef]

12.   Okuyama, K.; Tamura, A.; Takahashi, S.; Ohtsuka, M.; Tsubaki, M. Flow-induced acoustic resonance at the mouth of one or two side branches. *Nuclear Eng. Des.* **2012**, *249*, 154–158. [CrossRef]

13.   Xiao, Y.; Gu, H.; Gao, X.; Zhang, H.; Zhao, W. Flow visualization study of flow-induced acoustic resonance in closed side branches. *Ann. Nuclear Energy* **2018**, *149*, 107783. [CrossRef]

14.   Xiao, Y.; Zhao, W.; Gu, H.; Gao, X. Effects of branch length and chamfer on flow-induced acoustic resonance in closed side branches. *Ann. Nuclear Energy* **2018**, *121*, 186–193. [CrossRef]

15.   Tamura, A.; Okuyama, K.; Takahashi, S.; Ohtsuka, M. Development of numerical analysis method of flow-acoustic resonance in stub pipes of safety relief valves. *J. Nuclear Sci. Technol.* **2012**, *49*, 793–803. [CrossRef]

16.   Bravo, R.; Ziada, S.; Dokainish, M. Aeroacoustic Response of an Annular Duct with Coaxial Closed Side Branches. In *Collection of Technical Papers, Proceedings of the 11th AIAA/CEAS Aeroacoustics Conference, Monterrey, CA, USA, 23–25 May 2005*; ARC: Reston, VA, USA, 2005.

17.   Arthurs, D.; Ziada, S.; Bravo, R. Flow induced acoustic resonances of an annular duct with co-axial side branches. In *American Society of Mechanical Engineers, Pressure Vessels and Piping Division (Publication) PVP*; American Society of Mechanical Engineers: New York, NY, USA, 2006.

18.   Arthurs, D.; Ziada, S. Flow-excited acoustic resonances of coaxial side-branches in an annular duct. *J. Fluids Struct.* **2009**, *25*, 42–59. [CrossRef]

19.   Ziada, S. Flow-excited acoustic resonance in industry. *J. Press. Vessel Technol. Trans. ASME* **2010**, *132*, 015001. [CrossRef]

20.   Ziada, S.; Lafon, P. Flow-excited acoustic resonance excitation mechanism, design guidelines, and counter measures. *Appl. Mech. Rev.* **2014**, *66*, 010802. [CrossRef]

21.   Dequand, S.; Hulshoff, S.J.; Hirschberg, A. Self-sustained oscillations in a closed side branch system. *J. Sound Vibrat.* **2003**, *265*, 359–386. [CrossRef]

22.   Li, Y.; Someya, S.; Okamoto, K.; Inagaki, T.; Nishi, Y. Study on flow-induced acoustic resonance in symmetrically located side-branches using dynamic PIV technique. *J. Fluid Sci. Technol.* **2014**, *9*. [CrossRef]

23.   Li, Y.; Someya, S.; Okamoto, K.; Inagaki, T.; Nishi, Y. Visualization study of flow-excited acoustic resonance in closed tandem side branches using high time-resolved particle image velocimetry. *J. Mech. Sci. Technol.* **2015**, *29*, 989–999. [CrossRef]

24.   Salt, E.; Mohamed, S.; Arthurs, D.; Ziada, S. Aeroacoustic sources generated by flow-sound interaction in a T-junction. *J. Fluids Struct.* **2014**, *51*, 116–131. [CrossRef]

25.   Salt, E.; Mohamed, S.; Arthurs, D.; Ziada, S. Identification of aeroacoustic sources in a T-junction. In *American Society of Mechanical Engineers, Pressure Vessels and Piping Division (Publication) PVP*; American Society of Mechanical Engineers: New York, NY, USA, 2014.

26.   Ziada, S. A flow visualization study of flow- acoustic coupling at the mouth of a resonant side-branch. *J. Fluids Struct.* **1994**, *8*, 391–416. [CrossRef]

27.   Ziada, S.; Shine, S. Strouhal numbers of flow-excited acoustic resonance of closed side branches. *J. Fluids Struct.* **1999**, *13*, 127–142. [CrossRef]

28.   Bruggeman, J.C.; Hirschberg, A.; van Dongen, M.E.H.; Wijnands, A.P.J.; Gorter, J. Self-sustained aero-acoustic pulsations in gas transport systems: Experimental study of the influence of closed side branches. *J. Sound Vib.* **1991**, *150*, 371–393. [CrossRef]

29.   Jiang, L.; Zhang, H.; Duan, Q.; Zhang, Y. Numerical study on acoustic resonance excitation in closed side branch pipeline conveying natural gas. *Shock Vibrat.* **2020**, *2019*, 8857838. [CrossRef]

30.   Jungbauer, D.E.; Eckhardt, L.L. Flow-induced noise and vibration in turbocompressors and piping systems. *Hydrocarb. Process.* **2000**, *79*, 7.

31. Yu, Y.; Sun, D.M.; Ma, J.F.; Xu, Y.Q.; Chen, H.; Wu, K.; Ao, W.; Yan, W.; Qiu, L. Experimental study on acoustic characteristics of mean flow acoustic engine. *J. Eng. Thermophys.* **2012**, *33*, 23–26.
32. Morita, R.; Takahashi, S.; Okuyama, K.; Inada, F.; Ogawa, Y.; Yoshikawa, K. Evaluation of acoustic- and flow-induced vibration of the BWR main steam lines and dryer. *J. Nuclear Sci. Technol.* **2011**, *48*, 759–776. [CrossRef]
33. Radavich, P.M.; Selamet, A.; Novak, J.M. A computational approach for flow–acoustic coupling in closed side branches. *J. Acoust. Soc. Am.* **2001**, *116*, 105–112. [CrossRef]
34. Spalart, P.R.; Deck, S.; Shur, M.L.; Squires, K.D.; Strelets, M.K.; Travin, A. A new version of detached-eddy simulation, resistant to ambiguous grid densities. *Theor. Comput. Fluid Dyn.* **2006**, *20*, 181–195. [CrossRef]
35. Tan, X.M.; Xie, P.P.; Yang, Z.G.; Gao, J.Y. Adaptability of turbulence models for pantograph aerodynamic noise simulation. *Shock Vibrat.* **2019**, *2019*, 6405809. [CrossRef]
36. Zhang, Y.; Zhang, J.; Li, T.; Zhang, L.; Zhang, W. Research on aerodynamic noise reduction for high-speed trains. *Shock Vibrat.* **2016**, *2016*. [CrossRef]
37. Jeong, S.M.; Choi, J.Y. Combined diagnostic analysis of dynamic combustion characteristics in a scramjet engine. *Energies* **2020**, *13*, 29. [CrossRef]
38. Sedano, C.A.; Berger, F.; Rahimi, H.; Lopez Mejia, O.D.; Kühn, M.; Stoevesandt, B. CFD validation of a model wind turbine by means of improved and delayed detached eddy simulation in openFoam. *Energies* **2019**, *12*, 1306. [CrossRef]
39. Li, J.; Hu, J.; Zhang, C. Investigation of vortical structures and turbulence characteristics in corner separation in an axial compressor stator using DDES. *Energies* **2020**, *13*, 2123. [CrossRef]
40. ANSYS. *ANSYS Fluent User's Guide*; ANSYS Inc.: Canonsburg, PA, USA, 2019.
41. Yuan, Q.; Yu, B.; Li, J.; Han, D.; Zhang, W. Study on the restart algorithm for a buried hot oil pipeline based on wavelet collocation method. *Int. J. Heat Mass Transf.* **2018**. [CrossRef]
42. Zhang, K.; Li, J.; Yu, B.; Han, D.; Chen, Y. Fast prediction of the replacement process of oil vapor in horizontal tank and its improved safety evaluation method. *Process Saf. Environ. Protect.* **2019**, *122*, 298–306. [CrossRef]
43. Shao, Z.; Zhou, T.; Zhu, H.; Zang, Z.; Zhao, W. Amplitude enhancement of flow-induced vibration for energy harnessing. *E3S Web Conf.* **2020**, *160*. [CrossRef]
44. Zhu, H.; Zhao, Y.; Zhou, T. CFD analysis of energy harvesting from flow induced vibration of a circular cylinder with an attached free-to-rotate pentagram impeller. *Appl. Energy* **2018**, *212*, 304–321. [CrossRef]
45. Zhu, H.; Zhao, Y.; Hu, J. Performance of a novel energy harvester for energy self-sufficiency as well as a vortex-induced vibration suppressor. *J. Fluids Struct.* **2019**, *91*, 102736. [CrossRef]

energies

Article

Strategies to Facilitate Photovoltaic Applications in Road Structures for Energy Harvesting

Yiqing Dai [1,2], Yan Yin [3,*] and Yundi Lu [4]

[1] College of Civil Engineering, Fuzhou University, Fuzhou 350108, China; yiqing.dai@monash.edu or yiqing.dai@fzu.edu.cn
[2] Department of Civil Engineering, Monash University, Clayton, VIC 3800, Australia
[3] Department of Publicity, Ding Huai Men Campus, Jiangsu Open University, Nanjing 210036, China
[4] School of Physics and Technology, Wuhan University, Wuhan 430072, China; luyundi2018@whu.edu.cn
* Correspondence: yiny@jsou.cn

Abstract: Photovoltaic (PV) facilities are sustainable and promising approaches for energy harvesting, but their applications usually require adequate spaces. Road structures account for a considerable proportion of urban and suburban areas and may be feasible for incorporation with photovoltaic facilities, and thereby have attracted research interests. One solution for such applications is to take advantage of the spare ground in road facilities without traffic load, where the solar panels are mounted as their conventional applications. Such practices have been applied in medians and slopes of roads and open spaces in interchanges. Applications in accessory buildings and facilities including noise/wind barriers, parking lots, and lightings have also been reported. More efforts in existing researches have been paid to PV applications in load-bearing pavement structures, possibly because the pavement structures cover the major area of road structures. Current strategies are encapsulating PV cells by transparent coverings to different substrates to prefabricate modular PV panels in factories for onsite installation. Test road sections with such modular solar panels have been reported, where inferior cost-effectiveness and difficulties in maintenance have been evidenced, suggesting more challenges exist than expected. In order to enhance the power output of the integrated PV facilities, experiences from building-integrated PVs may be helpful, including a selection of proper PV technologies, an optimized inclination of PV panels, and mitigating the operational temperature of PV cells. Novel integrations of amorphous silicon PV cells and glass fiber reinforced polymer profiles are proposed in this research for multi-scenario applications, and their mechanical robustness was evaluated by bending experiments.

Keywords: photovoltaics; solar panel; energy harvesting; highway; urban street; experimental investigation

Citation: Dai, Y.; Yin, Y.; Lu, Y. Strategies to Facilitate Photovoltaic Applications in Road Structures for Energy Harvesting. *Energies* **2021**, *14*, 7097. https://doi.org/10.3390/en14217097

Academic Editor: Dibin Zhu

Received: 2 October 2021
Accepted: 25 October 2021
Published: 30 October 2021

Publisher's Note: MDPI stays neutral with regard to jurisdictional claims in published maps and institutional affiliations.

1. Introduction

Photovoltaic (PV) facilities are able to generate electricity from solar energies without emission of pollutions or greenhouse gases, providing a sustainable solution to energy acquisition [1]. They have attracted global interest, especially in the background, as countries having set targets of reduction in greenhouse-gas emission in response to the Paris Agreement since 2016 [2]. PV facilities usually cover large areas in order to receive adequate solar energy and thereby are usually limited to remote open areas, e.g., deserts [3,4]. In order to avoid long-distance transmission of electricity, efforts have been made to allocate the PV facilities in spare spaces of urban and suburban areas, especially the exteriors of buildings [5]. They have been mounted by frames on rooftops similarly to those on the ground [6]. In some cases, PV panels may also function as building elements apart from generating electricity, for example, they may be used to shade light [7,8], and transparent or semi-transparent PV cells may be used as windows [9,10] and skylights [11].

In some recent practices, PV cells are integrated into conventional building elements and function as facades [12,13] and roof tiles [14,15]. There are four main strategies for such integration in existing applications and researches. First, a direct approach is to adhesively bond the entire commercial PV panels with building profiles [16,17]. Second, commercial PV panels may be installed by mechanical structures to facades of buildings [5]. Third, PV cells without face coverings are placed into transparent adhesives (e.g., Ethylene Vinyl Acetate, EVA) applied on building profiles, then a transparent covering (e.g., glass, composites materials [11] and concrete [18]) is applied on top of the cells for encapsulation [19]. Fourth, photovoltaic materials are directly deposited to surfaces of building elements (e.g., roof tiles) in the factory [14], and the prefabricated elements with PV skin can be installed conveniently on-site.

However, several challenges hindered more extensive applications of such building-integrated or building-applied photovoltaics [20,21]. For example, building designs are usually associated with aesthetic considerations while convention PV panels are flat and associated with limited colors [22]; second, different structures or functions of buildings may require customized products or installation arrangements, which may impair their affordability and cost effectiveness [23]. In addition, the available exterior area of a single building is usually limited especially considering the building orientation and shadings in the urban environment (e.g., adjacent buildings and landscapes) [24].

In the research process optimizing the building-applied photovoltaics, attention has also been paid to the highway and urban street structures which cover a considerable area of urban and suburb districts and may have potentials for integration with PV facilities. Some strategies used for developing building-applied photovoltaics have also been applied to road structures. Compared to buildings, highways and urban streets are associated with less esthetic considerations; due to almost uniform cross-sections of road structures, the design for applicable PV facilities may be easier and a successful practice may be conveniently promoted to similar projects. Second, highways and urban streets are usually maintained by professional teams, which is favorable for sustaining long-term performance of PV facilities. In addition, the increasing number of electrical vehicles brings a demand for access to electricity in remote highways [25], where the PV facilities may be even more cost effective than constructing grid electricity facilities. An attractive application of PV facilities on the pavement is to realize wireless charging for electrical vehicles during driving with inductance coil [26–28].

However, it should be noted that the pavement structures may be subjected to traffic loads, while conventional PV cells (e.g., mono-crystalline and multi-crystalline silicon PV cells) are brittle and may break at even a small mechanical strain [29]. In existing researches, efforts have been paid to three main strategies as follows.

- To utilize the spare spaces without traffic load, e.g., medians and side slopes of highways, open space of interchanges [30], and side faces of noise barriers [31].
- To insulate the PV cells from mechanical loads through robust and transparent upper structures [32], e.g., glass, composites materials [11], transparent concrete [18] and high molecular polymers, such as polycarbonates, polymethyl methacrylate, and resin [33].
- To introduce PV cells with adequate resistance to mechanical loads [34].

Despite the extensive researches on the applications of PV facilities in highway and urban street structures, successful practices are still limited. This is possibly because the existing researches are independent and focus on certain detailed aspects without an overall consideration. PV facilities are not among conventional materials for pavement construction and such applications require interdisciplinary experiences in pavement engineering and photovoltaics. An extensive review may help to understand the difficulties to apply PV facilities in load-bearing scenarios and share experiences from different existing trials. Several reviews about utilizing road structures for energy harvesting have been available, but PV facilities are merely listed as a single brief section alongside piezoelectric, thermoelectric, wind, and other energy-harvesting facilities [28,35,36].

The main objective of this review is to present the approaches, in use or recently proposed, for applying PV facilities in road structures for energy harvesting. It starts with the PV facilities applied in spare spaces without traffic loads, which are similar to the conventionally applicable scenarios of PV panels. After that, existing practices of PV facilities acting as structural components bearing traffic loads are extensively reviewed, with their constitution, electrical and mechanical performance, and current service status presented. Limitations of current studies and research opportunities are discussed. In addition, a novel integration of amorphous silicon (a-Si) PV cells and glass fiber reinforced polymer (GFRP) profiles are proposed and their mechanical performance is evaluated by four-point bending experiments.

2. Applications in Spare Spaces without Traffic Loads

2.1. Spare Ground

Mounting PV panels on the spare ground is a direct application in road structures, and the PV panels may be installed at the optimal orientation similar to their conventional applications in PV power plants [37]. Such spare ground includes medians of roads, slopes along expressways, open ground among interchanges, and disused highways [30]. The feasibility of such applications has been verified by a wide range of practices and research. For example, an economic analysis of using two open spaces among two interchanges in South Korea was conducted considering the land lease fee for the ground, and one of them was suggested to be economically feasible [30]. PV applications in the spare ground of roads have been widely implemented but the majority of them were not reported in academic literature since their technical innovations from conventional PV applications are limited. In order to provide an overview of such applications, two large-scale distributed PV systems on the spare ground of expressways in China are introduced.

In 2018, an upgraded program for expressways was conducted in Hubei Province of China, where PV panels were installed to 16 pairs of service areas, 34 toll stations, and 36 medians adjacent to tunnels (see Figure 1a). The total investment was CNY ¥80 million (about €10.7 million), and the estimated annual power output was 10 million kWh. Considering the industrial electricity price is CNY ¥1.0 per kWh, the cost payback time is thereby 8 years without considering the performance degradation of the PV cells, the maintenance cost, or feed-in-tariff policies [38,39].

Figure 1. Solar array mounted on spare spaces of (**a**) a median of an expressway in Hubei, China and (**b**) an interchange of Ji-Qing Expressway in Jinan, China (117.00° E, 36.56° N).

In 2017, distributed PV systems were installed to three expressways (i.e., the Ji-Qing Expressway, the City Expressway, and the Airport Expressway) in Jinan, China. The project included PV applications on open spaces of several interchanges (see Figure 1b), 8 pairs of service areas, and 7 toll stations. For the Ji-Qing Expressway alone, the investment was CNY ¥180 million (about €24 million), and the estimated annual power output was 36.5 million kWh, leading to an estimated payback time of 5 years. All PV panels used in this project were 265 Wp poly-crystalline Si PV panels, and the system was connected to the grid.

It may be noticed that in the two projects, PV panels were usually installed only adjacent to energy-consumption facilities although spare space may exist along the whole length of the expressways. This is possibly because PV facilities in such locations are more cost effective due to the on-site consumption of electricity. A photo for the PV panels mounted at a median of an expressway in the aforementioned project in Hubei is provided as Figure 1a, and a capture of the satellite image from Baidu Map for the PV panels installed in an interchange of the Ji-Qing Expressway in Jinan is provided as Figure 1b. The PV systems in tolls and service areas are usually installed on rooftops, which may be reasonably categorized into building-applied photovoltaics instead of road-applied ones. Slopes were not considered in the former two projects possibly because the slope orientation is constrained by road alignment, and the optimal orientation may not be available for the PV panels. An evaluation approach to estimate the power output of PV panels installed on highway slopes has been put forward based on digital elevation model (DEM) and road alignment information, and it was used to examine the potentials of ten highways in South Korea [40].

2.2. Other Accessory Facilities and Buildings

Other accessory road facilities without major mechanical loads have also been investigated, as demonstrated in Figure 2a. PV facilities have been extensively applied in shadings of parking lots (see Figure 1a) due to the increasing number of electric vehicles [41], and they are used in conjunction with the charging piles. The largest solar parking project was reported to be the North Park Project located in Dhahran, Saudi Arabia, at the headquarters of the oil company Saudi Aramco [42], which covers 200,000 m^2 and has 4450 parking spaces, and the CIS (copper indium diselenide) PV panels with a total power of 10 MW are used as shadings.

Lights powered by PV facilities have been used for lighting [43], traffic lights, and traffic studs [31] (see Figure 2b,d,e) and they are usually used on occasions without convenient access to grid electricity and are integrated with battery energy storage systems to manage energy consumption.

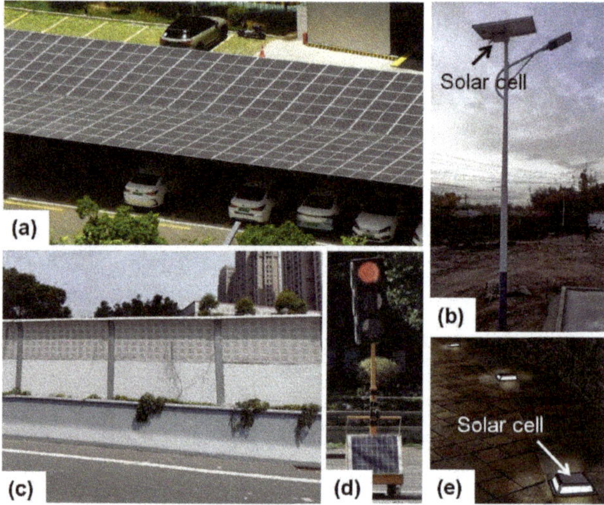

Figure 2. Applicable scenarios in road facilities: (**a**) parking lots, (**b**) lighting, (**c**) noise/wind barrier, (**d**) traffic lights and (**e**) traffic studs.

Installation of PV facilities on noise/wind barriers has a long history, and the first application in Europe was reported to be the one in the municipality of Domat/Ems, Switzerland in 1989 [31]. Shown in Figure 2c is a noise barrier without PV panels. It should be mentioned that, although noise/wind barriers are not subjected to vehicle loads, the barriers may deform at the effects of wind, and the mechanical strain of the barriers may be transferred to the integrated PV panels [44] therefore, the strain effects on the PV panels should be investigated before their applications. Some barriers may also extend to the top of roads and cover the whole top and side faces of roads, providing more space for PV application, for example, 4400 PV panels were installed on the rooftop of the Blackfriars Railway Bridge in London. Several existing researches have also investigated the feasibility to install a top roof for a whole length of a highway to allow installation of PV panels, but the cost effectiveness was not evaluated [45].

3. Applications in Pavement Infrastructures Bearing Traffic Loads

3.1. Associated PV Technologies

Although applications of PV facilities in scenarios without traffic loads have been developed, the applications to the pavement are still challenging since conventional PV cells may break at the effects of traffic loads. Existing researches have shown that crystalline silicon (c-Si) PV cells may break at 0.2 to 0.3% tensile strain [29,46], and the degradation in their electrical performance may happen at an earlier stage. In major associated applications and researches, c-Si PV cells are usually used, possibly due to the high conversion efficiency and easy availability from the market as presented in Table 1, where data for amorphous silicon (a-Si), CIGS (Copper Indium Gallium Diselenide), and CdTe (Cadmium Telluride) PV technologies is also provided. As presented in Table 1, c-Si PV cells are associated with the highest efficiency, and the PV modules and commercial panels made of c-Si PV cells also have higher efficiency than other technologies. The c-Si PV technology accounts for over 95% PV market and the mass production further makes their price competitive.

Table 1. Efficiency, temperature coefficient and market share of different PV technologies (unit: %).

PV Technologies	c-Si	a-Si	CIGS	CdTe
Cell efficiency [47]	26.7 ± 0.5	10.2 ± 0.3	23.35 ± 0.5	21.0 ± 0.4
Module efficiency [47]	24.4 ± 0.5	12.3 ± 0.3	19.2 ± 0.5	19.0 ± 0.9
Temperature coefficient [48]	-0.37 to -0.52	-0.10 to -0.30	-0.33 to -0.50	-0.18 to -0.36
Efficiency of commercial panel [49]	14.9 to 19.0	9.5	12.7 to 16.0	8.5 to 14.0
Market share [50]	95.70	0.05	1.13	3.12

The temperature coefficient is used to quantify the output reduction with elevated temperature, and the temperature coefficient of -0.37 to -0.52% for c-Si PV technology means the power output will decrease by 0.37 to 0.52% when the temperature surpasses 25 °C (i.e., the temperature in standard tests) by 1 °C. Therefore, the performance degradation at the effects of elevated temperature is more obvious compared to other PV technologies, while a-Si PV cells are favorable in this regard. During the operation of the PV cells, some of the solar energy absorbed by the PV cells will be converted into heat, while ventilation is almost prevented when they are encapsulated into pavement structures; therefore, the elevated temperature would be witnessed in PV cells during operation, leading to reduced power output than the rated efficiency.

3.2. Existing Applications

Several existing applications of PV facilities in pavement structures are listed in Table 2. The walkway in Ashburn was made by Onyx and installed on the Virginia Science and Technology Campus, George Washington University. The landscaped walkway was reported to be "the first installation of a walkable solar-paneled sidewalk in the world" [31], and it consisted of 27 PV panels with a slip-resistant surface. The PV panels are semi-transparent based on the a-Si PV technology, allowing 450 LED, powered by the solar panels, to backlight the walkway. It should be mentioned that the semi-transparent PV panels have also been installed on top of buildings as skylights.

The bicycle lane in Krommenie, Holland (see Table 2) was constructed by the installation of 57 prefabricated modular solar road elements, and each element was constructed by encapsulating poly-crystalline Si PV cells using a 1-cm thick glass layer to a concrete slab

(2.5 m × 3.5 m) [51]. The modular elements were inconvenient for installation due to the density of concrete and the large dimension. Delamination between the glass coverings and the PV cells was witnessed during the service of the lane [52]. Data for power-output observation and a prediction model are available from literature [53], and the actual power output for 2015 was 78 kWh/m^2, lower than the predicted 93 kWh/m^2; it was suggested that using mono-crystalline Si PV cells to replace the poly-crystalline cells would increase the output by 50% and a project site with more abundant solar radiation should be selected.

The countryside road in Tourouvre, France (see Table 2) was reported as the first large-scale solar road application [54]. It was constructed by adhesively bonding PV modules to road surfaces [55]. In the fabrication of each PV module, 28 mono-crystalline PV cells were encapsulated into a heave-duty resistant material, and a glass layer was then bonded by transparent resin to the upper face, while a glue layer was applied to the bottom, enabling the module to be conveniently bonded to existing road surfaces. Each module has a dimension of 1257 × 690 mm^2, a thickness of 6 mm, and a weight of 5.5 kg [54]. Unexpected noise caused by passing vehicles and sharp degradation in power output of the PV modules in service has been reported [56]. The PV modules are commercially available and have also been applied to other sites on smaller scales, allowing only pedestrians and bicycles [55].

The walkway in Sandpoint, US (see Table 2), was constructed by the installation of multifunctional hexagon modular panels. Prototypes of the modular panels were made earlier, and optimizations are still being made. The present version of the modular panels weighs about 70 pounds (~32 kg), covering an area of about 0.4 m^2 with a thickness of 3.6 cm. Textured heated glass was used as the upper face. Batteries, LED lights, and microprocessor boards were integrated into the modular panels to realize more functions, e.g., dynamic traffic signs. They were designed and tested to sustain vehicle loads, but currently applied only as walkways. In the development of the modular panels, a series of test procedures were introduced as described in detail in literature [57]; it includes freeze/thaw cycling, moisture conditioning, heavy vehicle loading, shearing testing, and the British Pendulum test, providing a reference for developing similar panels.

The expressway in Jinan, China (see Table 2) was the only reported application of solar pavement in an expressway, and it was also constructed by the installation of solar panels. Each panel had a transparent concrete covering with a dimension of 1.50 × 1.06 m^2 [51]. Despite the numerical and laboratory experiments before the application, several panels were broken after only several days of open service. The expressway section was closed four times for maintenance in the following year, and the solar panels have been totally replaced by conventional asphalt pavement.

The road in the Hexi Service Area of the Yongtai Highway in Fuzhou, China (see Table 2) was open to the public in 2020, allowing passing and parking of vehicles. The square modular panel (about 0.5 × 0.5 m^2) has a light weight and can be installed by a single person. Its surface was made of modified epoxy resin, and the four edges were reinforced with metal frames. A damping layer was applied beneath the panel to reduce vehicle vibration.

Table 2. Existing applications of solar pavement.

Year	Location	Scenario	Annual Output	Investment	LCOE
2013	Ashburn, US [31,52]	10-m^2 walkway	647 kWh (design)	-	-
2014	Krommenie, Holland [54]	70-m bicycle lane	11 MWh (actual)	€1.5 M (€10,714/m^2)	€12.7/kWh
2016	Tourouvre, France [54]	1000-m road	150 MWh (actual)	€5 M (€1786/m^2)	€3.1/kWh
2016	Sandpoint, US [57]	14-m^2 walkway	780 kWh (estimated)	~€52,000 (€3714/m^2)	€1.3/kWh
2017	Jinan, China [54]	1080-m expressway	1000 MWh (design)	~€5.3 M (€1402/m^2)	€0.5/kWh
2020	Fuzhou, China	100-m^2 service area road	36 MWh (design)	-	-

4. Discussion

4.1. Capacity to Sustain Loads

The highway in Tourouvre and the expressway in Jinan (see Table 2) are the only two large-scale solar road applications, and they allow a relatively high speed of vehicles. The highway in Tourouvre was reported to have almost totally lost its capacity for electricity generation, while the expressway in Jinan has been destroyed. Although vehicle-load tests have been conducted before their implementations, severe damages were witnessed shortly after their opening to traffic. This is possibly because, in the vehicle-load tests, the load is applied by a static or smooth-running vehicle, while the impact loads caused by vehicle vibration and accidents were not considered. Based on this experience and in order to mitigate the vehicle vibration, in the application in Fuzhou (see Table 2), a damping layer was applied, and the base was specially designed with high-viscosity asphalt, but the road has been open to traffic for only a short time, and its current status has not been reported.

In addition, the performance tests for the panels are applied individually, and in several researches, each specimen is only subject to a single type of load (e.g., vehicle load, thermal load, or moisture conditioning), while in practice such loads and conditions are applied together, leading to different results comparing to the laboratory tests. Therefore, the application of PV facilities in scenarios with vehicle load is still challenging and requires improvement in both construction and test procedures of the panel specimens.

4.2. Investment

According to the estimation on the applications listed in Table 2, the investment is at least €1402/m^2, and the unit cost of the bicycle lane in Holland is exceptionally high, reaching €10,714/m^2. If the payback time was estimated considering the industrial electricity prices is CNY ¥1.0 per kWh as conducted in Section 2, the cost payback time is thereby about 40 years for the most cost-effective one listed in Table 2. It suggests the cost would never be paid back since the usual service life for PV panels is 25 years [58]. It should be mentioned that the payback time is 5–8 years for the large-scale appellations in spaces without traffic loads in Hubei and Jinan, demonstrated in Section 2. Therefore, from the aspects of cost effectiveness, load-bearing pavement is not a favorable scenario for PV applications.

Levelized cost of energy (LCOE) has also been used in existing researches to reflect the average cost of a power generation system [59,60]. It is defined as the ratio between the total life cycle cost and the total lifetime energy production, and it is presented as a net present value. LCOE for the PV pavements listed in Table 2 has also been provided, assuming that the systems would operate for 20 years with an annual output degradation of 1%, an annual maintenance cost of 5% of the initial investment, and a discount rate of 4%. The solar road in Holland is estimated to be associated with the highest LCOE of €12.7/kWh, and the road in Jinan has the lowest one of €0.5/kWh, which is yet competitive to grid electricity in most countries. Also, it should be noted that the estimation for the road in Jinan is based on the designed capacity, which is usually higher than their actual performance. For comparison, the LCOE for the two PV systems allocated in the spare space of roads as described in Section 2 is estimated as €0.10/kWh for the expressway in Hubei and €0.06/kWh for the one in Ji-Qing Expressway.

4.3. Compassion between PV Facilities in Spare Space and Pavement

In order to understand the differences between PV facilities in spare space and pavement, a summary is listed in Table 3, where their scale, space availability, applicable scenarios, technical maturity, and cost effectiveness are compared. In terms of technical maturity and cost effectiveness, practices in spare space are obviously more practical. However, the applications of PV facilities on the pavement are still attractive since the pavement structure is associated with large and continuous space, which provides opportunities for large-scale PV plants. In addition, the cross-section of the pavement structure is simple and uniform, enabling successful cases to be promoted conveniently to other projects, which also reduces the challenges in installation and maintenance.

Apart from the aforementioned investigations, several other modular solar panels for pavement application are under laboratory investigations [34,52]. For example, a thin solar pavement module with a thickness of 20 mm and a dimension of 50 mm × 50 mm has been developed for walkable road and roof [52]; it was fabricated by encapsulating PV cells into tempered glass panels. Robustness and thermal responses of the panels at the effects of sunlight have been investigated by theoretical and experimental approaches. Several approaches have proposed to mitigate the elevated temperatures of the solar panels on pavement based on phase change materials [15,61] and soil regenerator [32]. A target of researches concerning solar pavement is to realize wireless charging for electric vehicles during driving using solar energy; theoretical verification and prototype experiments have been completed in several recent researches [26–28].

Table 3. Compassion between PV facilities in spare space and pavement.

	PVs on Pavement	**PVs in Spare Space**
Scale	The largest test section is about 1 km, but it is associated with potentials for larger scale than those in spare space.	Determined by demand and space, and large applications reached an output power of 10 MW.
Space availability	Large and continuous space	Small and distributed space
Scenarios	Pavement only	Spare ground: medians, slopes and interchanges; Other accessory facilities and buildings: shadings, lights and noise/wind barriers.
Technical maturity	Still under investigations. The main technical route is prefabricated modular for onsite installation.	Mature and commercialized. Solar panels are mounted by frame as conventional applications or integrated into construction materials.
Cost effectiveness	Low; LCOE: €0.5/kWh to €12.7/kWh; Cost may not be paid back.	High; LCOE: €0.06/kWh to €0.10/kWh; Cost payback time: 5 to 8 years.

5. Innovative Integrations for Multi-Scenario Applications

5.1. Conceptual Design

It may be noticed that the PV panels used in different scenarios are associated with different configurations and it may hinder wide applications. Based on existing experiences, this section thereby aims to develop an integration of PV cells and civil engineering materials feasible for applications in different scenarios of road structures and road facilities. According to the review on the current applications, the main strategy is to develop prefabricated modular PV panels with a substrate for convenient on-site installation. The "prefabrication and modular installation" strategy is effective to reduce the cost for construction and may also reduce the errors in wiring [62]. In order to reduce the total weight of the panels, the substrates made of civil engineering materials should have a high strength-to-weight ratio. Electric insulation is also an essential property for the substrate materials for safety considerations. Therefore, glass fiber reinforced polymer (GFRP) profiles are selected as the substrates due to their superior mechanical performance [63], cost effectiveness, resistance to aggressive environmental conditions [64], and electric insulation [65]. GFRP profiles have been used in multiple civil engineering scenarios, including retaining walls of road slopes [66], bridges [67], wind barriers [44], floor systems [68], and reinforcement for pavement [69]. In terms of the type of PV cells, a-Si PV cells were selected for the integration since existing applications suggested that the c-Si PV cells may not sustain long-term vehicle loads while flexible a-Si PV cells may operate normally even after deformation [17].

Strain effects on the a-Si PV cell samples have been investigated by our research team and the results have been available in literature [65,70,71]. Specimens of the a-Si PV cells were first subjected to tensile experiments as shown in Figure 3a, which was conducted in accordance with ASTM D882 [72]. Artificial sunlight with an intensity of 1000 W/m^2 was applied [73], and the load was applied at 0.2 mm/min via a testing machine (Instron 4204). The open-circuit voltage (V_{OC}) of the PV cells was recorded continuously to indicate the operational status of the cells. Strains of the PV cells were obtained based on the spacing change of the bonded reflectors monitored by the laser extensometer (MTS LX500). According to the experimental results, no visible change in V_{OC} of cells was observed until a tensile strain of 1.4% [71], which was suggested to be the ultimate tensile strain for normal operation. It should be mentioned that c-Si PV cells may break at a tensile strain of only 0.2 to 0.3% [46].

The a-Si PV cells were then bonded by an epoxy adhesive layer to GFRP plates and then subjected to similar tensile tests as shown in Figure 3b, and the integration failed due to breakage of the GFRP plates at about 1% tensile strain [65], while the PV cells operated normally until the GFRP breakage. The a-Si PV cell samples have also been adhesively bonded with GFRP square hollow sections and subjected to compressive loads as shown in Figure 3c, where the specimens were not loaded up to failure and the maximum compressive strains applied was 0.8%. Results suggested that if a proper adhesive layer was applied, the a-Si PV cells may operate normally at even the maximum compressive strain applied [70].

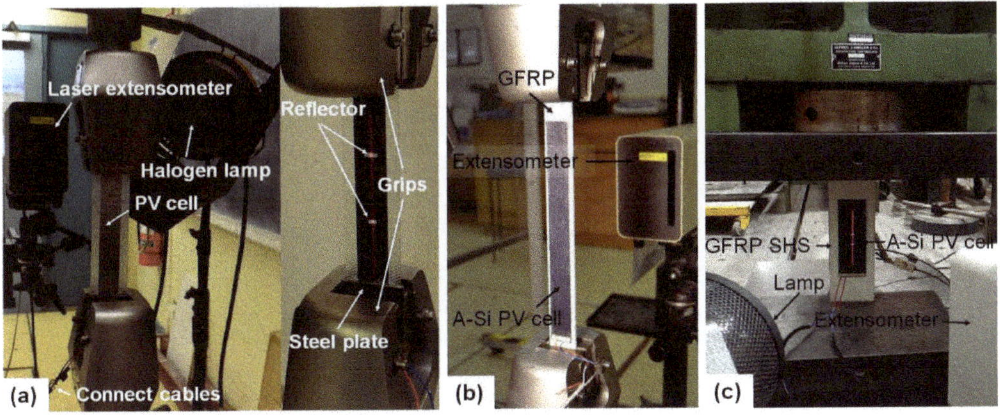

Figure 3. Experimental setup for (**a**) tensile test for a-Si PV cells, (**b**) tensile tests for integrations and (**c**) compressive tests for integrations.

5.2. Bending Experiment on Integration

The mechanical experiments on the a-Si PV cells and their integrations with GFRP profiles suggested that they may function normally for tensile strains within 1.4% and compressive strains within 0.8%. However, GFRP profiles are usually used as the sandwich form in order to provide robust support to structures. In addition, when such integrations are used as noise/wind barriers, walkable floors, and other facilities, bending loads may be transferred to the integrated PV cells, which may cause different effects from tensile and compressive loads. Therefore, an integration of GFRP sandwich and a-Si PV cells was fabricated as shown in Figure 4 and then subjected to four-point bending experiments as shown in Figure 5.

As shown in Figure 4a, the specimen was fabricated by two GFRP plates with a dimension of 550 mm (length) × 350 mm (width) × 6 mm (thickness) and three GFRP square tubes with a dimension of 550 mm (length) × 40 mm (width) × 40 mm (depth) × 4 mm (wall thickness) in between. The plates were bonded by an epoxy layer with a thickness of 1 mm. The a-Si PV cell with a dimension of 375 × 260 mm^2 was bonded to a surface of the GFRP sandwich, as shown in Figure 4b.

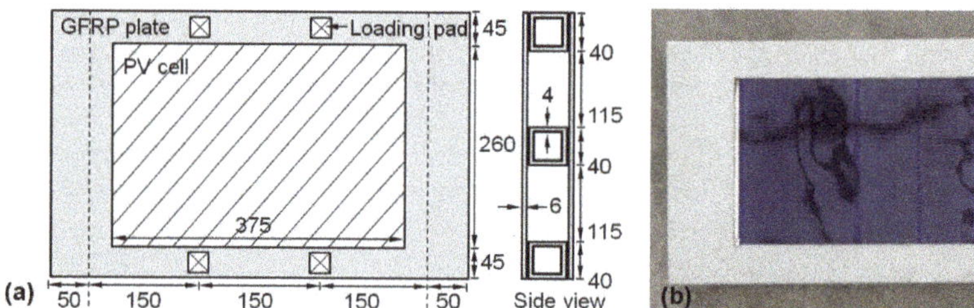

Figure 4. Integration of a-Si PV cell and GFRP sandwich: (**a**) dimension, in mm and (**b**) profile display.

During the bending experiments, the specimen was supported at 50 mm from both ends, and artificial sunlight was provided to the PV cell by a halogen lamp. The mechanical load was then applied through four GFRP loading pads to avoid direct contact with the PV cells, as shown in Figure 5a. The load was applied at a speed of 0.5 mm/min until the failure of the specimen. The voltage of the PV cell was continuously monitored during loading to show the operational status.

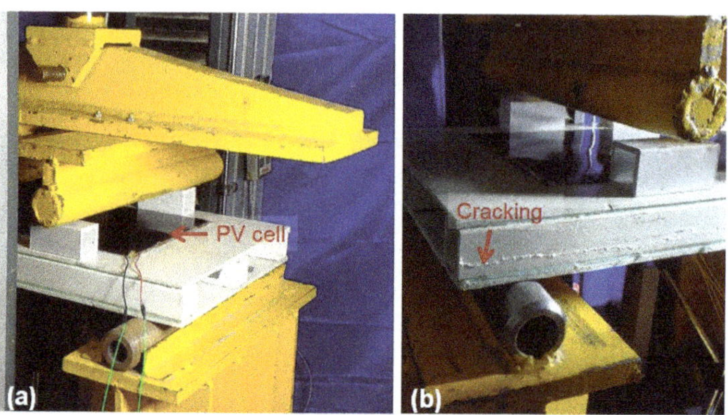

Figure 5. Four-point bending experiment on integration: (**a**) test setup and (**b**) specimen cracking after loading.

5.3. Experimental Results

During the loading process, longitudinal cracks along the side walls of the GFRP tubes were initiated from about 35 kN as shown in Figure 5b, and the cracks propagated with the increase of the load. The maximum load was 46 kN for the specimen, corresponding to a deflection of 4.2 mm at the bottom of the mid span. At the maximum load, the cracks developed into a throughout crack from one end of the GFRP tube to another at the end of the experiments as shown in Figure 5b, meanwhile, the degradation in voltage was 4.1% for the PV cell, and when the specimens were unloaded, the voltage was recovered to over 99% of their original voltage values before mechanical loading.

This result suggests that a load of 35 kN may not cause visible damage to the GFRP sandwich. The 46-kN load caused temporary voltage reduction to the PV cell, which may be recovered when the load was removed. It should be mentioned that the typical load of one tire of a vehicle is about 4 to 5 kN. This integration, therefore, shows potentials to sustain vehicle loads.

Due to wide applications of GFRP profiles in road engineering, such integrations of PV cells and GFRP profiles are promising elements. For example, integration of PV cells and GFRP plates may be used as noise/wind barriers; integrations with GFRP sandwiches may be used as bridge decks and walkable roofs, while the a-Si PV cells may also be integrated into GFRP retaining walls in road slopes. The mechanical experimental tests on the PV cells and the prototype integrations may provide guidance for these applications, but further laboratory and outdoor investigations are necessary before practice. Further research will be focused on the responses of PV cells to cyclic loadings, the thermal performance of the integrations, and the selection of transparent coverings.

6. Conclusions

In this research, existing applications of PV facilities in road structures and facilities are reviewed. It starts with the PV facilities applied in spare spaces without traffic loads, where two systematic large-scale applications in China were introduced. Potentials of other facilities without vehicle loads were also presented in brief. Then, existing practices of PV facilities in pavement structures with vehicle loads were presented and discussed. In addition, a novel integration of amorphous silicon (a-Si) PV cells and glass fiber reinforced polymer (GFRP) profiles is proposed. Existing mechanical experiments on the a-Si PV cells and their integrations with GFRP profiles were introduced, and their mechanical performance was evaluated in this study by four-point bending experiments. The following conclusions may be drawn.

1. In the presented two large-scale PV applications in expressways, PV facilities were applied to open spaces of several interchanges, service areas, toll stations, and medians adjacent to tunnels. Slopes were not considered in the two projects possibly because the slope orientation is constrained by road alignment, and the power output may be reduced. The cost payback time was five and eight years, respectively, for the two projects.
2. The main approach to applying PV facilities to load-bearing pavement structures is prefabricated modular PV panels with a substrate for convenient on-site installation. Further investigations are required before making out successful panels to sustain long-term vehicle loads. In addition, the current applications were not cost effective.
3. The integrations of amorphous silicon (a-Si) PV cells and glass fiber reinforced polymer (GFRP) are suggested to have potentials for multi-scenario applications in roads. According to experimental results, the PV cells may function normally at a tensile strain of 1.4% and a compressive strain of 0.8%. When a load of 46 kN was applied to the integration of a GFRP sandwich and the a-Si PV cell, a 4.1% reduction in voltage of the cell was witnessed, but it may be fully recovered when unloaded, suggesting its potentials to sustain vehicle loads.

Author Contributions: Conceptualization, Y.Y. and Y.D.; Methodology, Y.D.; Formal Analysis, Y.Y.; Investigation, Y.D. and Y.L.; Writing—Original Draft Preparation, Y.Y.; Writing—Review & Editing, Y.D. and Y.L.; Supervision, Y.D.; Project Administration, Y.D.; Funding Acquisition, Y.D. All authors have read and agreed to the published version of the manuscript.

Funding: This research was funded by the Qishan Scholar Project (511062) of Fuzhou University.

Institutional Review Board Statement: Not applicable.

Informed Consent Statement: Not applicable.

Data Availability Statement: The data presented in this study are available on request from the corresponding author. The data are not publicly available since this research is still ongoing and further analysis will be conducted.

Conflicts of Interest: The authors declare no conflict of interest. The funders had no role in the design of the study; in the collection, analyses, or interpretation of data; in the writing of the manuscript, and in the decision to publish the results.

References

1. Parida, B.; Iniyan, S.; Goic, R. A review of solar photovoltaic technologies. *Renew. Sustain. Energy Rev.* **2011**, *15*, 1625–1636. [CrossRef]
2. Goglio, P.; Williams, A.G.; Balta-Ozkan, N.; Harris, N.R.; Williamson, P.; Huisingh, D.; Zhang, Z.; Tavoni, M. Advances and challenges of life cycle assessment (LCA) of greenhouse gas removal technologies to fight climate changes. *J. Clean. Prod.* **2020**, *244*, 118896. [CrossRef]
3. Hanifi, H.; Pander, M.; Zeller, U.; Ilse, K.; Dassler, D.; Mirza, M.; Bahattab, M.A.; Jaeckel, B.; Hagendorf, C.; Ebert, M. Loss analysis and optimization of PV module components and design to achieve higher energy yield and longer service life in desert regions. *Appl. Energy* **2020**, *280*, 116028. [CrossRef]
4. Sher, H.A.; Murtaza, A.F.; Addoweesh, K.E.; Chiaberge, M. Pakistan's progress in solar PV based energy generation. *Renew. Sustain. Energy Rev.* **2015**, *47*, 213–217. [CrossRef]
5. Peng, C.; Huang, Y.; Wu, Z. Building-integrated photovoltaics (BIPV) in architectural design in China. *Energy Build.* **2011**, *43*, 3592–3598. [CrossRef]
6. Aelenei, D.; Lopes, R.A.; Aelenei, L.; Goncalves, H. Investigating the potential for energy flexibility in an office building with a vertical BIPV and a PV roof system. *Renew. Energy* **2019**, *137*, 189–197. [CrossRef]
7. Li, X.; Peng, J.; Li, N.; Wu, Y.; Fang, Y.; Li, T.; Wang, M.; Wang, C. Optimal design of photovoltaic shading systems for multi-story buildings. *J. Clean. Prod.* **2019**, *220*, 1024–1038. [CrossRef]
8. Yoo, S.-H. Optimization of a BIPV system to mitigate greenhouse gas and indoor environment. *Solar Energy* **2019**, *188*, 875–882. [CrossRef]
9. Chae, Y.T.; Kim, J.; Park, H.; Shin, B. Building energy performance evaluation of building integrated photovoltaic (BIPV) window with semi-transparent solar cells. *Appl. Energy* **2014**, *129*, 217–227. [CrossRef]
10. Leite Didoné, E.; Wagner, A. Semi-transparent PV windows: A study for office buildings in Brazil. *Energy Build.* **2013**, *67*, 136–142. [CrossRef]
11. Pascual, C.; de Castro, J.; Kostro, A.; Schueler, A.; Vassilopoulos, A.P.; Keller, T. Diffuse light transmittance of glass fiber-reinforced polymer laminates for multifunctional load-bearing structures. *J. Compos. Mater.* **2014**, *48*, 3621–3636. [CrossRef]
12. Alrashidi, H.; Issa, W.; Sellami, N.; Ghosh, A.; Mallick, T.K.; Sundaram, S. Performance assessment of cadmium telluride-based semi-transparent glazing for power saving in façade buildings. *Energy Build.* **2020**, *215*, 109585. [CrossRef]
13. Ascione, F.; Bianco, N.; Iovane, T.; Mastellone, M.; Mauro, G.M. The evolution of building energy retrofit via double-skin and responsive façades: A review. *Sol. Energy* **2021**, *224*, 703–717. [CrossRef]
14. Aguas, H.; Ram, S.K.; Araujo, A.; Gaspar, D.; Vicente, A.; Filonovich, S.A.; Fortunato, E.; Martins, R.; Ferreira, I. Silicon thin film solar cells on commercial tiles. *Energy Environ. Sci.* **2011**, *4*, 4620–4632. [CrossRef]
15. Alim, M.A.; Tao, Z.; Abden, M.J.; Rahman, A.; Samali, B. Improving performance of solar roof tiles by incorporating phase change material. *Sol. Energy* **2020**, *207*, 1308–1320. [CrossRef]
16. Chen, A.; Alateeq, A. Performance of solar cells integrated with rigid and flexible building substrates under compression. *J. Build. Eng.* **2020**, *34*, 101938. [CrossRef]
17. Chen, A.; Yossef, M.; Zhang, C. Strain effect on the performance of amorphous silicon and perovskite solar cells. *Sol. Energy* **2018**, *163*, 243–250. [CrossRef]
18. Li, Y.; Zhang, J.; Cao, Y.; Hu, Q.; Guo, X. Design and evaluation of light-transmitting concrete (LTC) using waste tempered glass: A novel concrete for future photovoltaic road. *Constr. Build. Mater.* **2021**, *280*, 122551. [CrossRef]
19. Keller, T.; Manshadi, A.P.V.; Behzad, D. Thermomechanical Behavior of Multifunctional GFRP Sandwich Structures with Encapsulated Photovoltaic Cells. *J. Compos. Constr.* **2010**, *14*, 470–478. [CrossRef]
20. Dai, Y.; Bai, Y. Performance Improvement for Building Integrated Photovoltaics in Practice: A Review. *Energies* **2021**, *14*, 178. [CrossRef]
21. Shukla, A.K.; Sudhakar, K.; Baredar, P.; Mamat, R. Solar PV and BIPV system: Barrier, challenges and policy recommendation in India. *Renew. Sustain. Energy Rev.* **2018**, *82*, 3314–3322. [CrossRef]
22. Ballif, C.; Perret-Aebi, L.-E.; Lufkin, S.; Rey, E. Integrated thinking for photovoltaics in buildings. *Nat. Energy* **2018**, *3*, 438–442. [CrossRef]
23. Petter Jelle, B.; Breivik, C.; Drolsum Røkenes, H. Building Integrated Photovoltaic Products: A State-of-the-Art Review and Future Research Opportunities. *Sol. Energy Mater. Sol. Cells* **2012**, *100*, 69–96. [CrossRef]
24. Goncalves, J.E.; Hooff, T.V.; Saelens, D. A physics-based high-resolution BIPV model for building performance simulations. *Sol. Energy* **2020**, *204*, 585–599. [CrossRef]
25. Zeb, M.Z.; Imran, K.; Khattak, A.; Janjua, A.K.; Pal, A.; Nadeem, M.; Zhang, J.; Khan, S. Optimal placement of electric vehicle charging stations in the active distribution network. *IEEE Access* **2020**, *8*, 68124–68134. [CrossRef]
26. Kandasamy, V.; Keerthika, K.; Mathankumar, M. Solar based wireless on road charging station for electric vehicles. *Mater. Today Proc.* **2021**, *45*, 8059–8063. [CrossRef]
27. Mohamed, N.; Aymen, F.; Issam, Z.; Bajaj, M.; Ghoneim, S.S.; Ahmed, M. The Impact of Coil Position and Number on Wireless System Performance for Electric Vehicle Recharging. *Sensors* **2021**, *21*, 4343. [CrossRef] [PubMed]

28. Venugopal, P.; Shekhar, A.; Visser, E.; Scheele, N.; Chandra Mouli, G.R.; Bauer, P.; Silvester, S. Roadway to self-healing highways with integrated wireless electric vehicle charging and sustainable energy harvesting technologies. *Appl. Energy* **2018**, *212*, 1226–1239. [CrossRef]

29. Kim, J.; Choi, I.; Kim, D.; Cheong, S. Development of Single Crystalline Silicon Solar Cells Lay-Down Process on Composites. In Proceedings of the 18th International Conference on Composite Materials, Jeju-do, Korea, 21–26 August 2011.

30. Kim, S.; Lee, Y.; Moon, H.-R. Siting criteria and feasibility analysis for PV power generation projects using road facilities. *Renew. Sustain. Energy Rev.* **2018**, *81*, 3061–3069. [CrossRef]

31. Subramanian, R. The current status of roadways solar power technology: A review. In Proceedings of the International Symposium on Systematic Approaches to Environmental Sustainability in Transportation, Fairbanks, AK, USA, 2–5 August 2015; pp. 177–187.

32. Xiang, B.; Cao, X.; Yuan, Y.; Hasanuzzaman, M.; Zeng, C.; Ji, Y.; Sun, L. A novel hybrid energy system combined with solar-road and soil-regenerator: Sensitivity analysis and optimization. *Renew. Energy* **2018**, *129*, 419–430. [CrossRef]

33. Northmore, A.B.; Tighe, S. Innovative pavement design: Are solar roads feasible? In Proceedings of the Conference & Exhibition of the Transportation Association of Canada (TAC) Transportation Association of Canada (2012), Fredericton, NB, Canada, 4–17 October 2012.

34. Dezfooli, A.S.; Nejad, F.M.; Zakeri, H.; Kazemifard, S. Solar pavement: A new emerging technology. *Sol. Energy* **2017**, *149*, 272–284. [CrossRef]

35. Zabihi, N.; Saafi, M. Recent developments in the energy harvesting systems from road infrastructures. *Sustainability* **2020**, *12*, 6738. [CrossRef]

36. Ahmad, S.; Abdul Mujeebu, M.; Farooqi, M.A. Energy harvesting from pavements and roadways: A comprehensive review of technologies, materials, and challenges. *Int. J. Energy Res.* **2019**, *43*, 1974–2015. [CrossRef]

37. Rehman, S.; Ahmed, M.A.; Mohamed, M.H.; Al-Sulaiman, F.A. Feasibility study of the grid connected 10 MW installed capacity PV power plants in Saudi Arabia. *Renew. Sustain. Energy Rev.* **2017**, *80*, 319–329. [CrossRef]

38. Martin, N.; Rice, J. The solar photovoltaic feed-in tariff scheme in New South Wales, Australia. *Energy Policy* **2013**, *61*, 697–706. [CrossRef]

39. Ye, L.-C.; Rodrigues, J.F.D.; Lin, H.X. Analysis of feed-in tariff policies for solar photovoltaic in China 2011–2016. *Appl. Energy* **2017**, *203*, 496–505. [CrossRef]

40. Jung, J.; Han, S.; Kim, B. Digital numerical map-oriented estimation of solar energy potential for site selection of photovoltaic solar panels on national highway slopes. *Appl. Energy* **2019**, *242*, 57–68. [CrossRef]

41. Chukwu, U.C.; Mahajan, S.M. V2G Parking Lot with PV Rooftop for Capacity Enhancement of a Distribution System. *IEEE Trans. Sustain. Energy* **2014**, *5*, 119–127. [CrossRef]

42. Almasoud, A.H.; Gandayh, H.M. Future of solar energy in Saudi Arabia. *J. King Saud Univ. Eng. Sci.* **2015**, *27*, 153–157. [CrossRef]

43. Qin, X.; Wei, Q.; Wang, L.; Shen, Y. Solar lighting technologies for highway green rest areas in China: Energy saving economic and environmental evaluation. *Int. J. Photoenergy* **2015**, *2015*, 1–10. [CrossRef]

44. Dai, Y.; Dai, X.; Bai, Y.; He, X. Aerodynamic Performance of an Adaptive GFRP Wind Barrier Structure for Railway Bridges. *Materials* **2020**, *13*, 4214. [CrossRef]

45. Sharma, P.; Harinarayana, T. Solar energy generation potential along national highways. *Int. J. Energy Environ. Eng.* **2013**, *4*, 16. [CrossRef]

46. Kang, J. *Structural Integration of Silicon Solar Cells and Lithium-ion Batteries Using Printed Electronics*; Hahn, H.T., Carman, G., Chiou, P.-Y., Pei, Q., Eds.; ProQuest Dissertations Publishing: Ann Arbor, MI, USA, 2012.

47. Green, M.A.; Dunlop, E.D.; Hohl-Ebinger, J.; Yoshita, M.; Kopidakis, N.; Hao, X. Solar cell efficiency tables (version 56). *Prog. Photovolt. Res. Appl.* **2020**, *28*, 629–638. [CrossRef]

48. Virtuani, A.; Pavanello, D.; Friesen, G. Overview of Temperature Coefficients of Different Thin Film Photovoltaic Technologies. In Proceedings of the 5th World Conference on Photovoltaic Energy Conversion, Valencia, Spain, 6–10 September 2010; pp. 4248–4252.

49. Lee, T.D.; Ebong, A.U. A review of thin film solar cell technologies and challenges. *Renew. Sustain. Energy Rev.* **2017**, *70*, 1286–1297. [CrossRef]

50. Feldman, D.; Margolis, R. *Solar Industry Update Q4 2018/Q1 2019*; National Renewable Energy Laboratory (NREL): Denver, CO, USA, 2019.

51. Wang, J.; Xiao, F.; Zhao, H. Thermoelectric, piezoelectric and photovoltaic harvesting technologies for pavement engineering. *Renew. Sustain. Energy Rev.* **2021**, *151*, 111522. [CrossRef]

52. Ma, T.; Yang, H.; Gu, W.; Li, Z.; Yan, S. Development of walkable photovoltaic floor tiles used for pavement. *Energy Convers. Manag.* **2019**, *183*, 764–771. [CrossRef]

53. Shekhar, A.; Kumaravel, V.K.; Klerks, S.; de Wit, S.; Venugopal, P.; Narayan, N.; Bauer, P.; Isabella, O.; Zeman, M. Harvesting roadway solar energy—performance of the installed infrastructure integrated PV bike path. *IEEE J. Photovolt.* **2018**, *8*, 1066–1073. [CrossRef]

54. Zhou, B.; Pei, J.; Xue, B.; Guo, F.; Wen, Y.; Zhang, J.; Li, R. Solar/road from 'forced coexistence' to 'harmonious symbiosis'. *Appl. Energy* **2019**, *255*, 113808. [CrossRef]

55. Wattway. Wattway Pack, an Autonomous Energy Solutionfor Roadside Equipment. Available online: https://www.wattwaybycolas.com/en/the-solar-road.html (accessed on 10 September 2021).
56. Benöhr, M.; Gebremedhin, A. Photovoltaic systems for road networks. *Int. J. Innov. Technol. Interdiscip. Sci.* **2021**, *4*, 672–684.
57. Coutu, A.R.; Newman, D.; Munna, M.; Tschida, J.H.; Brusaw, S. Engineering Tests to Evaluate the Feasibility of an Emerging Solar Pavement Technology for Public Roads and Highways. *Technologies* **2020**, *8*, 9. [CrossRef]
58. Ghosh, A. Potential of building integrated and attached/applied photovoltaic (BIPV/BAPV) for adaptive less energy-hungry building's skin: A comprehensive Review. *J. Prod.* **2020**, *276*, 123343.
59. Hu, H.; Vizzari, D.; Zha, X.; Roberts, R. Solar pavements: A critical review. *Renew. Sustain. Energy Rev.* **2021**, *152*, 111712. [CrossRef]
60. Branker, K.; Pathak, M.J.M.; Pearce, J.M. A review of solar photovoltaic levelized cost of electricity. *Renew. Sustain. Energy Rev.* **2011**, *15*, 4470–4482. [CrossRef]
61. Yang, M.; Zhang, X.; Zhou, X.; Liu, B.; Wang, X.; Lin, X. Research and Exploration of Phase Change Materials on Solar Pavement and Asphalt Pavement: A review. *J. Energy Storage* **2021**, *35*, 102246. [CrossRef]
62. Debbarma, M.; Sudhakar, K.; Baredar, P. Thermal modeling, exergy analysis, performance of BIPV and BIPVT: A review. *Renew. Sustain. Energy Rev.* **2017**, *73*, 1276–1288. [CrossRef]
63. Gu, X.; Dai, Y.; Jiang, J. Test and Evaluation for Bonding Property Between GFRP and Concrete. *J. Test. Eval.* **2016**, *44*, 878–884. [CrossRef]
64. Fang, H.; Bai, Y.; Liu, W.; Qi, Y.; Wang, J. Connections and structural applications of fibre reinforced polymer composites for civil infrastructure in aggressive environments. *Compos. Part B Eng.* **2019**, *164*, 129–143. [CrossRef]
65. Dai, Y.; Bai, Y.; Cai, Z. Thermal and mechanical evaluation on integration of GFRP and thin-film flexible PV cells for building applications. *J. Clean. Prod.* **2021**, *289*, 125809. [CrossRef]
66. Ferdous, W.; Almutairi, A.D.; Huang, Y.; Bai, Y. Short-term flexural behaviour of concrete filled pultruded GFRP cellular and tubular sections with pin-eye connections for modular retaining wall construction. *Compos. Struct.* **2018**, *206*, 1–10. [CrossRef]
67. Keller, T.; Schollmayer, M. Plate bending behavior of a pultruded GFRP bridge deck system. *Compos. Struct.* **2004**, *64*, 285–295. [CrossRef]
68. Zhang, L.; Dai, Y.; Bai, Y.; Chen, W.; Ye, J. Fire performance of loaded fibre reinforced polymer multicellular composite structures with fire-resistant panels. *Constr. Build. Mater.* **2021**, *296*, 123733. [CrossRef]
69. Benmokrane, B.; Sanni Bakouregui, A.; Mohamed, H.M.; Thébeau, D.; Abdelkarim, O.I. Design, Construction, and Performance of Continuously Reinforced Concrete Pavement Reinforced with GFRP Bars: Case Study. *J. Compos. Constr.* **2020**, *24*, 5020004. [CrossRef]
70. Dai, Y.; Bai, Y.; Keller, T. Stress mitigation for adhesively bonded photovoltaics with fibre reinforced polymer composites in load carrying applications. *Compos. Part B Eng.* **2019**, *177*, 107420. [CrossRef]
71. Dai, Y.; Huang, Y.; He, X.; Hui, D.; Bai, Y. Continuous performance assessment of thin-film flexible photovoltaic cells under mechanical loading for building integration. *Sol. Energy* **2019**, *183*, 96–104. [CrossRef]
72. ASTM. *D3090-17: Standard Test Method for Tensile Properties of Polymer Matrix Composite Materials*; ASTM International: West Conshohocken, PA, USA, 2017.
73. ASTM-G173-03. *Standard Tables for Reference Solar Spectral Irradiances: Direct Normal and Hemispherical on 37° Tilted Surface*; ASTM International: West Conshohocken, PA, USA, 2020.

energies

MDPI

Article

Silicon Particles/Black Paint Coating for Performance Enhancement of Solar Absorbers

Shwe Sin Han [1], **Usman Ghafoor** [2,*,†], **Tareq Saeed** [3,†], **Hassan Elahi** [4,†], **Usman Masud** [5,6], **Laveet Kumar** [1], **Jeyraj Selvaraj** [1] and **Muhammad Shakeel Ahmad** [1,*]

[1] Higher Institution Centre of Excellence (HICoE), UM Power Energy Dedicated Advanced Centre (UMPEDAC), Level 4, Wisma R & D, University of Malaya, Jalan Pantai Baharu, Kuala Lumpur 59990, Malaysia; shwesinhan2018@gmail.com (S.S.H.); laveet.kumar@gmail.com (L.K.); jeyraj@um.edu.my (J.S.)
[2] Department of Mechanical Engineering, Institute of Space Technology, Islamabad 44000, Pakistan
[3] Nonlinear Analysis and Applied Mathematics (NAAM)-Research Group, Department of Mathematics, Faculty of Science, King Abdulaziz University, P.O. Box 80203, Jeddah 21589, Saudi Arabia; tsalmalki@kau.edu.sa
[4] Department of Mechanical and Aerospace Engineering, Sapienza University of Rome, 00185 Rome, Italy; hassan.elahi@uniroma1.it
[5] Faculty of Electrical and Electronics Engineering, University of Engineering and Technology, Taxila 47050, Pakistan; usmanmasud123@hotmail.com
[6] Department of Electrical Communication Engineering, University of Kassel, 34127 Kassel, Germany
* Correspondence: usmanghafoor99@gmail.com (U.G.); shakeelalpha@gmail.com (M.S.A.)
† These authors have equally contributed as the second author in the manuscript.

Abstract: The availability of fresh drinkable water and water security is becoming a global challenge for sustainable development. In this regard, solar stills, due to their ease in operation, installation, and utilization of direct sunlight (as thermal energy), promise a better and sustainable future technology for water security in urban and remote areas. The major issue is its low distillate productivity, which limits its widespread commercialization. In this study, the effect of silicon (Si) particles is examined to improve the absorber surface temperature of the solar still absorber plate, which is the major component for increased distillate yield. Various weight percentages of Si particles were introduced in paint and coated on the aluminum absorber surface. Extensive indoor (using a self-made halogen light-based solar simulator) and outdoor testing were conducted to optimize the concentration. The coatings with 15 wt % Si in the paint exhibited the highest increase in temperature, namely, 98.5 °C under indoor controlled conditions at 1000 W/m^2 irradiation, which is 65.81% higher than a bare aluminum plate and 37.09% higher compared to a black paint-coated aluminum plate. On the other hand, coatings with 10 wt % Si reached up to 73.2 °C under uncontrolled outdoor conditions compared to 68.8 °C for the black paint-coated aluminum plate. A further increase in concentration did not improve the surface temperature, which was due to an excessive increase in thermal conductivity and high convective heat losses.

Keywords: water; solar still; absorber; silicon; temperature

Citation: Han, S.S.; Ghafoor, U.; Saeed, T.; Elahi, H.; Masud, U.; Kumar, L.; Selvaraj, J.; Ahmad, M.S. Silicon Particles/Black Paint Coating for Performance Enhancement of Solar Absorbers. *Energies* **2021**, *14*, 7140. https://doi.org/10.3390/ en14217140

Academic Editor: Dibin Zhu

Received: 29 July 2021
Accepted: 15 September 2021
Published: 1 November 2021

1. Introduction

Freshwater plays a crucial role in a sustainable environment. Therefore, the demand for drinkable water is increasing gradually every year due to population enlargement. Although 70% of the earth is covered with water, most of the accessible water cannot be utilized for drinking due to salinity and water contamination [1,2]. One of the potential alternatives to produce fresh water and decrease the cost of distillation plants is the utilization of solar energy directly. Various applications of solar energy use have been put forwarded, such as solar photovoltaics [3,4]. Solar distillation is one of the techniques to produce fresh water at a lower cost than the other possible seawater desalination processes [5,6]. The lower cost of manufacturing and installation is due to its simple design, use of less expensive and readily available raw material and tooling, and no moving components. More specifically, the solar distillation technique by using solar stills is the most feasible and environmentally friendly way to supply fresh water in arid and remote regions due to its ability to use thermal energy from the sun, little maintenance, simple design, little to no use of replacement components, and lower cost [7,8].

A solar still consists of a basin to contain water and a glass top cover that allows the water to be exposed directly to sunlight to stimulate the evaporation process, where the distillate is collected from the bottom side of the glass cover. A schematic illustration of a solar still is shown in Figure 1. It is worth mentioning here that the solar absorber in a conventional solar still consists of a metal (generally copper or aluminum) sheet coated with black paint to improve sunlight absorption. The production efficiency of a solar still is much lower compared to other desalination techniques. The freshwater production efficiency directly relates to the absorber plate temperature and temperature difference between the absorber plate and backside of the glass cover, which creates a temperature difference and improve the evaporation rate [9]. Therefore, various studies have been conducted to increase the productivity of solar stills.

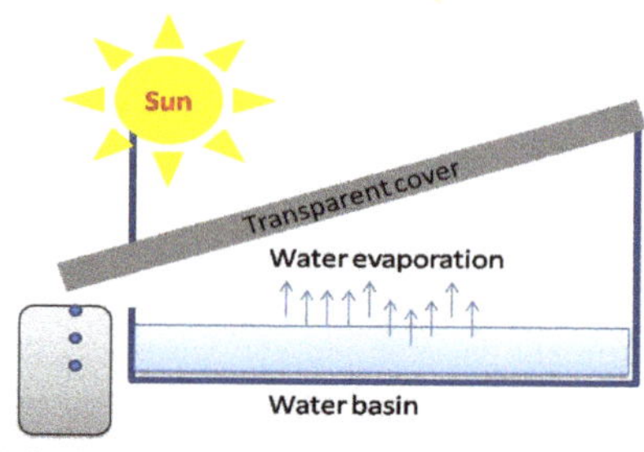

Figure 1. Schematic illustration of a solar still.

The solar still works on the evaporation and condensation principle: as the direct sunlight heats up the water and causes evaporation, the vapors condenses when coming into contact with the relatively colder surface of the glass cover, subsequently trickling down to a collection tank [10]. The depth of the basin water is one of the important factors in the evaporation rate and performance of a solar still; moreover, the salinity in water reduces the evaporation rate. This is because of the fact that impurities in water increase the boiling point, which, in turn, reduces the overall efficiency of the evaporation rate and performance. An experimental study investigated the effect of basin water depth on the performance of a single-slope (SS) basin-type solar still. The effect of water depth, varying from 2 cm to 10 cm, and the salinity of the water, which was 3000, 50,000, 10,000, 150,000, and 20,000 ppm on yield, were tested. The study showed that the thermal efficiency decreases with an increase in depth and salinity of the basin water. The maximum efficiency was obtained at the minimum water depth (2 cm) in the basin, producing the maximum volume of distilled water [11]. Another experiment was carried out on the performance of a single-slope (SS) solar still and double-slope (DS) solar still by changing the depth of the basin water at 1 cm, 2.5 cm, 5 cm, and 7.5 cm. From the experimental results, it was concluded that a decrease in water depth resulted in an increase in yield, and that the performance of the DS solar still outweighed the SS solar still at a similar water depth [12]. Another experiment was conducted between a single-basin double-slope (SBDS) solar still and double-basin double-slope (DBDS) solar still, with the use of different depths varying from 1 to 5 cm. It was observed that the DBDS insulated still had a 17.38% higher yield than the SBDS insulated still at a 1 cm depth of the basin water. The results prove that the DBDS solar still gave a higher performance than the SBDS solar still and the depth of the basin water has a considerable effect on the production rate of a solar still [13]. The study has experimentally analyzed the effect of water depth in a comparison of the performance of a pyramid solar still (PSS) and the single-slope solar still and concluded that a reduction in depth gives a higher productivity, and that the PSS has a higher performance than the SS solar still because of its greater condensation area [14].

One of the main components that contribute a lot to solar still production efficiency is the base plate, which acts as a solar radiation absorber and provides the required energy to water. Generally, aluminum or copper plates coated with black paints have been employed for maximum radiation absorption. In a recent study, TiO_2 nanoparticles have been mixed with black paint and coated on the absorbed plate in an attempt to increase the radiation absorption. A 1.5 °C increase in water temperature has been reported for TiO_2-modified black paint compared to bare black paint. This increase in temperature leads to a 6.1% increase in solar still distillate yield [12]. In another study, ZnO nano particulates have been investigated as reinforcement to the absorber plate. An absorber temperature of 70 °C has been reported when coated with nano tubular-shaped ZnO particles/black paint compared to a 60–65 °C temperature of the reference sample [13]. Various other nanomaterials, namely, carbon nanotubes (CNTs), carbon black, and graphene, have been employed to achieve the maximum temperature [15]. Similarly, CuO with various wt % from 10–40% has also been considered as a nanomaterial and mixed into paint to improve the distillate production efficiency. It has been claimed that the addition of CuO improves the efficiency by 16% and 25% as compared to the conventional solar still at 10% and 40% concentrations, respectively [16]. Another experimental study investigated the fumed SiO_2 nanoparticles in black paint at a varied concentration from 10–40% coated on the absorber plate of a stepped solar still for augmenting the freshwater yield. The average temperature of the water and absorber increased by 10.2–12.3% by adding the optimized concentration of nanoparticles of 20% with black paint. Results also concluded that the concentration increase from 10 to 40%, with 10% increment leading to a 27.2, 34.2, 18.3, and 18.4% increase in efficiency, respectively, compared to that of ordinary black paint [17]. Similarly, one of the research groups developed two similar solar stills with varying color of paints other than black. One still was painted white while the another was painted black for comparison. Results obtained with the white-painted solar still had a 6.8% improvement in efficiency compared to the ordinary black paint [18].

Furthermore, some studies suggested mixing various nanoparticles, i.e., CuO, Al_2O_2, TiO_2, etc., directly with water for simplicity of operation [19,20]. In one study, experimental comparisons were investigated between the solid nanoparticles of a CuO and Al_2O_3 nanofluid and concluded that the nanoparticles enhanced freshwater production by around 125.0% and 133.64%, respectively [21]. However, this approach (nanofluids) may contaminate the water due to its direct contact with water [13]. In another study, zinc microparticles were used to improve the surface absorption temperature [10]. Various other approaches for improved distillate yield include the use of wick materials, such as cotton, wool, nylon, wood pulp paper, styrene sponge, coral fleece, etc., and shape modification of the absorbed plate to improve the contact surface area, and have been employed with promising results.

In this study, metallic submicron-sized silicon particles are studied to improve the absorption property of absorber plates for solar still applications. Various concentrations of silicon were mixed with black paint and applied to the aluminum plate. Indoor and outdoor testing was conducted using an improvised apparatus for concentration optimization.

2. Experimental Investigation

2.1. Samples Preparation

The test samples were prepared using a simple chemical mixing method. Black paint was considered as the base matrix to paint onto an aluminum plate and Si nanoparticles in the various fractions 0.5, 1, 3, 5, 7, 10, 15, and 20 wt % were mixed in black paint. In brief, the nanoparticles were suspended in isopropyl alcohol (5 mL), sonicated for 1 h, and stirred for 5 h for proper suspension. The suspension was mixed with black paint using a mechanical mixer for 2 min and the now-formed formulation was applied to the aluminum plate using a hand brush. The samples were dried at room temperature for 24 h. The same procedure was repeated for each concentration. Figure 2 shows the prepared samples for testing.

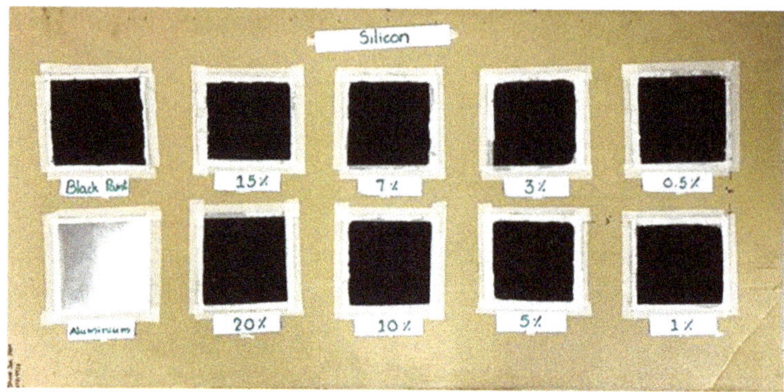

Figure 2. Samples under investigation.

2.2. Indoor Setup and Experimental Procedure

The indoor setup at the solar thermal laboratory, UMPEDAC, University of Malaya, was used. There were many components and instruments used in this experiment to fulfil the required indoor conditions. For providing the required irradiations, a solar simulator comprising 120 halogen bulbs (Brand: OSRAM, OSRAM GmbH Marcel-Breuer-Straße 680807 Munich, Germany) with power, voltage, and current capacity of 90 W, 12 V, and 7.5 A, respectively, was used. The solar simulator is controlled by three variable control AC-power-supply transformers, which can simulate the solar radiation from 100 W/m^2 to 1200 W/m^2. Further, a few other standard measuring instruments were used to record the data. Some key instrumentation and sensors used were a pyranometer (Brand: LI-COR, LI200R, LI-COR GmbH, Germany Siemensstraße 25A 61352 Bad Homburg Germany), data logger (Brand: Data taker (DT80), Thermo Fisher Scientific Australia Pty Ltd.), and thermocouples (K-type). The measuring and accuracy range of all the instruments and sensors are given in Table 1.

Table 1. Measurement ranges and accuracy of the instruments and sensors.

Instrument	Measuring Range	Accuracy
Pyranometer (Model: LI-COR, LI200R)	0–2000 W/m^2	±5%
Data Logger (Model: Data Taker DT80)		±2%
Thermocouple (K-Type)	−200–1000 °C	±1.50 °C

For the indoor experiment procedure, the samples of size 6 × 6 inches were placed on a cardboard in the solar simulator. The eight (08) samples with different concentrations of silicon (Si) nano particles-coated plates, one (01) black paint-coated plate, and one (01) bare aluminum plate were placed in two rows with a displacement of 3 inches apart on the cardboard. The K-type thermocouple was placed on each plate and a pyranometer in the center to record the change in temperature and varying radiations, respectively. Experiments were conducted in indoor working conditions with varying the radiation from 100 W/m^2 to 1000 W/m^2 using a solar simulator. It was observed that it took about 30 min in each irradiation level of the experiment to reach the stable conditions and then the solar simulator was switched off for about 15 min to record the decreasing temperature of all plates. The arrangement of the indoor experiment is systematically depicted in Figure 3.

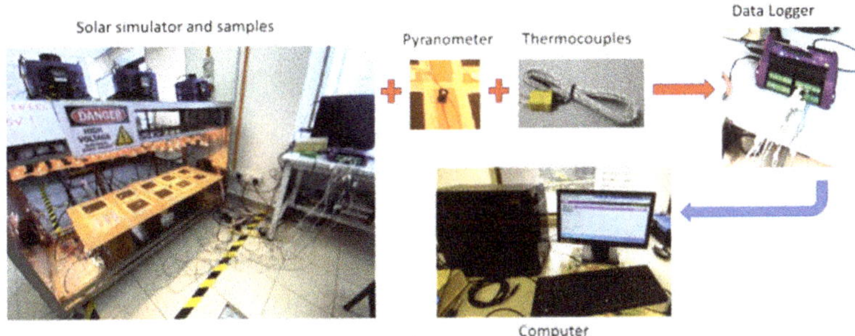

Figure 3. Indoor experimental setup.

2.3. Outdoor Setup and Experimental Procedure

The outdoor experimental setup was installed at the Solar Garden of Higher Institution Centre of Excellence, UM Power Energy Dedicated Advanced Centre (UMPEDAC), Kuala Lumpur, Malaysia. Experiments were carried out from 8.00 a.m. to 05:00 p.m. The same standard devices and instruments were used to record the data and the measuring and accuracy range of all the instruments and sensors has already been given in Table 1. Sensors were placed onto the systems at the necessary locations for recording the data. Other meteorological data, such as wind speed, etc., were measured using the weather station installed at the Solar Garden.

The same arrangement of samples, i.e., the silicon-coated nanoparticles, the plate coated with black paint, and the aluminum plate attached, were attached to a wooden sheet of a thickness of 1 mm and was mounted at an optimum inclined angle of 15 degrees. The layout of the outdoor experiment, including the position and connectivity of all the instruments and components, is systematically represented in Figure 4. The optimum angle was calculated using Cooper's Equation (1) to intercept the maximum radiation throughout the day [22]. Note that the experiment was performed for seven days from 8:00 a.m. to 5:00 p.m. and only the best data with the maximum sun hours are presented here.

$$\delta = 23.45 \ \sin \left[0.9863 \ (284 + n_1) \right] \tag{1}$$

where δ is the inclination angle and n_1 is the day of the year.

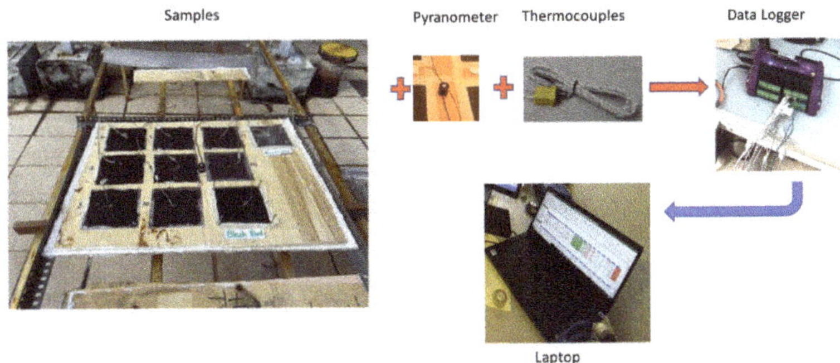

Figure 4. Setup of the outdoor experiment.

3. Results and Discussion

Figure 5a–c shows the scanning electron micrographs (SEM) of the as-received powder along with the X-ray diffraction (XRD) pattern and UV-vis spectra. The particles are of irregular shape with sizes ranging from micrometers to sub-micrometers; i.e., from 0.9–4 µm. The XRD pattern has been matched with the JCPDS#27-1402 card [23]. The peaks mentioned concerning the 2-theta axis corresponds to the silicon element. Figure 5c shows the light absorption properties of the as-received powder, which is of prime importance to this study; a sharp absorption in UV; and a visible range of the solar spectrum was identified, for which the levels in the near-infrared region and again a small absorption in the far-infrared region were also noticed. The corresponding light transmission curve is also presented in Figure 5c to better understand the optical properties.

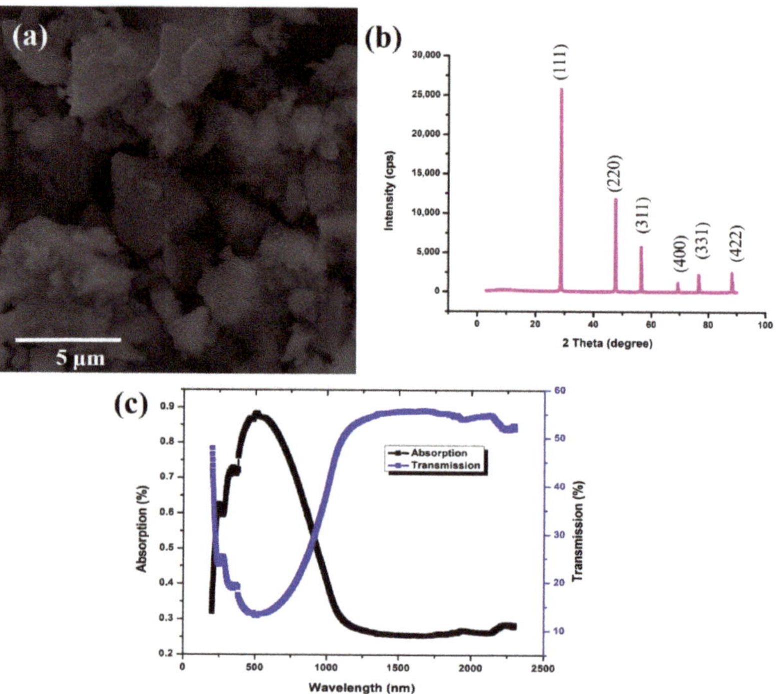

Figure 5. (**a**) SEM micrograph, (**b**) XRD pattern, and (**c**) UV-vis spectra of the as-received Si powder.

A notable outcome of the performance of the absorber plates is properly analyzed from the evidence and presented in detail with the results, followed by insightful discussion. In the following sections, the indoor and outdoor experiment results are presented, which demonstrate the complete statistical and graphical information about the beneficial effects of using various Silicon (Si) concentrations with black paint.

3.1. Indoor Analysis

In the present study, the indoor experiment was conducted by varying the irradiation from 100 to 1000 W/m^2; however, for the purpose of our analysis, the effect of temperature on the absorber plates under the increment of 200 W/m^2 and rate of drop in temperature is shown in Figure 6. The left-hand side of Figure 6 illustrates the rise in temperature at 200, 600, 800, and 1000 W/m^2, and the right-hand side illustrates the rate of a temperature drop on cooling. As illustrated in Figure 6a–d, the results showed a steady increase in surface temperature of the absorber plate with an increase in the concentration as well as solar irradiation. Furthermore, the cooling curves show an increase in the rate of cooling, as can be seen in Figure 6ai–di. This increased cooling rate corresponds to the increased thermal conductivity of specimens with an increasing concentration. The summary of the temperatures for various concentrations at maximum irradiation is presented in Table 2.

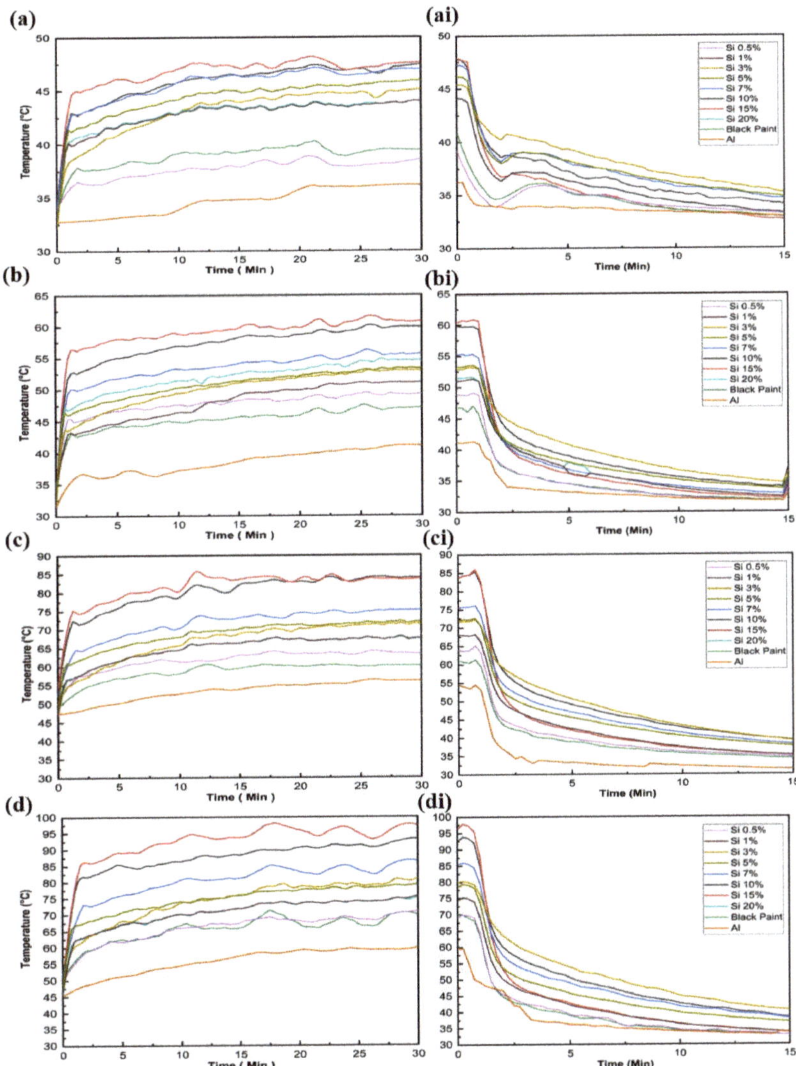

Figure 6. Effect on absorber temperature at an irradiation exposure of (**a**) 200 W/m^2, (**b**) 400 W/m^2, (**c**) 800 W/m^2, and (**d**) 1000 W/m^2, and the temperature-drop profile (**ai–di**).

Table 2. Summary of the maximum observed temperature on the absorber surface vs. concentration in indoor conditions.

Sr. No.	Matrix (Coated on Aluminum Sheet)	Silicon Concentration (wt %)	Maximum Temperature at 1000 W/m^2
1	Black paint	0.5%	72.8 ± 2%
2	Black paint	1%	75.7 ± 2%
3	Black paint	3%	79.4 ± 2%
4	Black paint	5%	81.8 ± 3%
5	Black paint	7%	87.8 ± 2%
6	Black paint	10%	93.4 ± 3%
7	Black paint	15%	98.5 ± 2%
8	Black paint	20%	79.3 ± 2%
9	Black paint	0%	71.9 ± 2%
10	Aluminum reference		59.4 ± 2%

The maximum temperature measured for specimens containing 0.5 wt % Si at 1000 W/m^2 irradiations was 72.8 °C, which is 22.58% higher compared to bare aluminum plate and 1.39% higher compared to the black paint-coated aluminum plate. Specimens containing 1 wt %, 3 wt %, 5 wt %, 7 wt %, and 10 wt % exhibited the maximum temperatures of approximately 75.6 °C, 79.3 °C, 81.7 °C, 87.8 °C, and 93.4 °C, respectively. The highest temperature was attained in the case of specimens containing 15 wt % Si at a 1000 W/m^2 irradiation level was 98.5 °C, which is 65.81% higher compared to the bare aluminum plate and 37.15% higher compared to the black paint-coated aluminum plate. A further increase in Si concentration in the black paint did not increase the surface temperature. For example, specimens containing 20 wt % Si in black paint exhibited 79.3 °C. This reduction in surface temperature is due to an excessive increase in thermal conductivity (cooling rate, Figure 6ai–di), which leads to an increase in thermal convective losses in the surrounding environment.

3.2. Outdoor Analysis

The weather of Kuala Lumpur experiences an equatorial climate illustrated by hot and humid weather throughout the whole year [24]. The Southwest monsoon takes place from April to September and the Northeast Monsoon occurs from October to March. The Southwest monsoon highlights a drier climate with less precipitation contrasted with the Northeast monsoon that brings more rainfall [25]. The outdoor experiments were conducted from 8:00 a.m. to 5:00 p.m. for about 7 days in September. The data for the best day are presented in Figure 7. It was noted that the tropical thunderstorms and rainfalls occurred usually in the afternoon although the weather conditions were generally good in the morning during the experiments. As shown in Figure 7a, it is observed that the solar intensity was lower in the morning and gradually increased, reaching its maximum at around noon and declined significantly because of cloudy weather conditions at this time. Then, it increased sharply and fluctuated before reaching the peak point at 1 p.m. and fell during the evening hours. It is also found that the wind speed steadily rose during the morning and hit to the maximum 7 m/s around 3:30 p.m. and was considerably reduced in the evening.

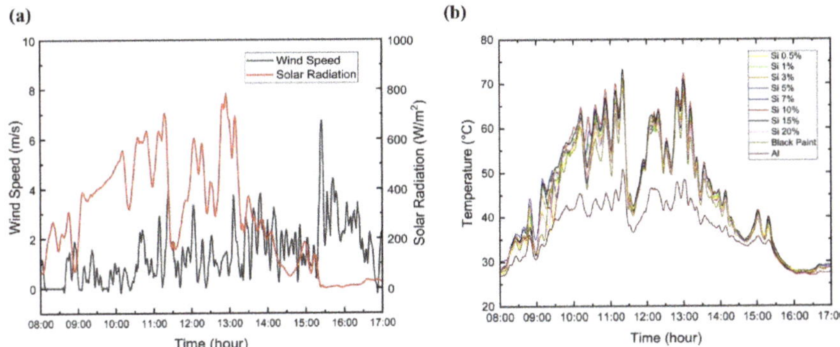

Figure 7. Outdoor experiment data at 27 September 2020: (**a**) weather profile, and (**b**) temperature of the absorber plates.

The outdoor temperature variations of the plates with silicon particles at different concentrations are demonstrated in Figure 7b, whereas the summary of the temperature is provided in Table 3. Outdoor results followed almost the same trend as indoor for the specimens containing 0.5 wt % Si, 1 wt % Si, 3 wt % Si, 5 wt % Si, and 7 wt % Si, exhibiting 68.8 °C, 69.4 °C, 70.4 °C, 70.8 °C, and 71.3 °C, respectively. Further, in outdoor conditions, specimens containing 10 wt % Si exhibited the maximum surface temperature of 73.2 °C, outperforming the specimen containing 15 wt % Si, which exhibited the highest temperature in indoor conditions. This behavior can be correlated with the cooling curves presented in Figure 6ai–di. With the increase in concentration, the thermal conductivity increased, which caused specimens to cool faster, giving heat energy to the surroundings at a faster rate. The cooling curve of the specimens containing 15 wt % Si shows a higher cooling rate compared to specimens containing 10 wt % Si. As the outdoor conditions are fluctuating from 200 W/m^2 to 800 W/m^2 on that particular day, the specimens with higher cooling rates lose heat energy as the convective loss at a faster rate compared to specimens having a less to optimum cooling rate. Further, an increase in concentration to 20 wt % exhibited 70.2 °C, which is in line with the indoor trend.

Table 3. Summary of the maximum observed temperature on the absorber surface vs. the concentration of silicon in black paint.

Sr. No.	Matrix (Coated on Aluminum Sheet)	Silicon Concentration (wt %)	Maximum Temperature (°C)	Average Temperature (°C)	Measuring Accuracy (°C)
1	Black paint	0.5%	68.9	44.0	±1.5
2	Black paint	1%	69.4	44.0	±1.5
3	Black paint	3%	70.5	44.5	±1.5
4	Black paint	5%	70.9	44.8	±1.5
5	Black paint	7%	71.3	45.6	±1.5
6	Black paint	10%	73.2	45.6	±1.5
7	Black paint	15%	72.7	45.6	±1.5
8	Black paint	20%	70.3	44.3	±1.5
9	Black paint	0%	68.8	43.4	±1.5
10	Aluminum reference		50.8	36.9	±1.5

In terms of average temperature throughout the particular day, the average temperature did not change much, and the temperature increase is within the accuracy range of thermocouples for specimens with only black paint and with varying concentrations. The maximum average temperature was exhibited by the specimen containing 10 wt % Si; i.e.,

45.6 °C, which is 23.8% higher compared to the bare aluminum plate and 5.13% high compared to only black paint-coated aluminum plate.

4. Conclusions

The technology of solar stills is improving at a great pace to avoid future scarcity fresh drinking water. In this study, silicon particles were investigated in an attempt increase the sunlight absorption of the solar still absorber plate. Various concentrations silicon particles were mixed with black paint and applied to the absorber plate. SEM a XRD were employed to examine the surface morphology and composition, respectiv UV-Vis spectroscopy was conducted to study the optical absorption properties of silic Indoor and outdoor tests were conducted using an improvised apparatus. With t introduction of Si particles into black paint, a temperature rise was observed, which can associated with the ability of silicon to absorb sunlight in UV and the visible range of t solar spectrum. The highest temperature of 98.5 °C was recorded for samples containi 15 wt % Si in indoor conditions at 1000 w/m^2 irradiation levels. In turn, the specimen w 10 wt % Si performed the best in outdoor conditions, achieving a maximum of 73.2 °C a 45.6 °C average temperature. This difference between the indoor and outdoor conditions due to the varying thermal conductions of the specimens due to the concentration of particles. A further increment in concentration did not effectively improve the absorpti characteristics due to excessive convective heat losses.

Author Contributions: Conceptualization, M.S.A. and H.E.; formal analysis, S.S.H. and L.K.; fundi acquisition, U.G.; investigation, U.M. and T.S.; methodology, L.K. and J.S.; project administrati J.S.; resources, T.S.; validation, U.M. and H.E; writing—original draft, S.S.H. and M.S.A.; writin review and editing, T.S. and U.G. All authors have read and agreed to the published version of t manuscript.

Funding: (1) This research and APC charges were funded by Japan International Cooperation Agen for AUN/SEED-Net on Collaboration Education Program UM CEP 1901, Japan ASEAN Collaborati Education Program (JACEP). (2) This research was partially funded by UM Power Energy Dedicat Advanced Centre (UMPEDAC) and the Higher Institution Centre of Excellence (HICoE) Progra Research Grant, UMPEDAC—2020 (MOHE HICOE-UMPEDAC), Ministry of Education Malays TOP100 UMPEDAC, RU003-2020, University of Malaya.

Informed Consent Statement: Not applicable.

Data Availability Statement: The prepared samples to support the findings of this study are availab from the corresponding authors upon reasonable request.

Acknowledgments: The authors would like to acknowledge the Japan International Cooperatic Agency for AUN/SEED-Net on Collaboration Education Program UM CEP 1901, Japan ASEA Collaborative Education Program (JACEP). The authors thank the technical and financial assistan of UM Power Energy Dedicated Advanced Centre (UMPEDAC) and the Higher Institution Cent of Excellence (HICoE) Program Research Grant, UMPEDAC—2020 (MOHE HICOE-UMPEDA(Ministry of Education Malaysia, TOP100 UMPEDAC, RU003-2020, University of Malaya.

Conflicts of Interest: The authors declare no conflict of interest.

References

1. Biswas, A.K.; Tortajada, C. *Water Crisis and Water Wars: Myths and Realities*; Taylor & Francis: Abingdon, UK, 2019.
2. Llácer-Iglesias, R.M.; López-Jiménez, P.; Pérez-Sánchez, M. Energy Self-Sufficiency Aiming for Sustainable Wastewater Systen Are All Options Being Explored? *Sustainability* **2021**, *13*, 5537. [CrossRef]
3. Fayaz, H.; Ahmad, M.S.; Pandey, A.K.; Abd Rahim, N.; Tyagi, V.V. A Novel Nanodiamond/Zinc Nanocomposite as Potenti Counter Electrode for Flexible Dye Sensitized Solar Cell. *Solar Energy* **2020**, *197*, 1–5. [CrossRef]
4. Pandey, A.; Ahmad, M.S.; Alizadeh, M.; Rahim, N.A. Improved electron density through hetero-junction binary sensitize TiO$_2$/CdTe/D719 system as photoanode for dye sensitized solar cell. *Phys. E Low-Dimens. Syst. Nanostruct.* **2018**, *101*, 139–14 [CrossRef]
5. Liu, S.; Wang, Z.; Han, M.; Wang, G.; Hayat, T.; Chen, G. Energy-water nexus in seawater desalination project: A typical wat production system in China. *J. Clean. Prod.* **2020**, *279*, 123412. [CrossRef]

Besha, A.T.; Tsehaye, M.T.; Tiruye, G.A.; Gebreyohannes, A.Y.; Awoke, A.; Tufa, R.A. Deployable Membrane-Based Energy Technologies: The Ethiopian Prospect. *Sustainability* **2020**, *12*, 8792. [CrossRef]

Liu, F.; Wang, L.; Bradley, R.; Zhao, B.; Wu, W. Highly efficient solar seawater desalination with environmentally friendly hierarchical porous carbons derived from halogen-containing polymers. *RSC Adv.* **2019**, *9*, 29414–29423. [CrossRef]

Alawi, O.; Kamar, H.; Mallah, A.; Mohammed, H.; Sabrudin, M.; Newaz, K.; Najafi, G.; Yaseen, Z. Experimental and Theoretical Analysis of Energy Efficiency in a Flat Plate Solar Collector Using Monolayer Graphene Nanofluids. *Sustainability* **2021**, *13*, 5416. [CrossRef]

Abujazar, M.S.S.; Fatihah, S.; Kabeel, A. Seawater desalination using inclined stepped solar still with copper trays in a wet tropical climate. *Desalination* **2017**, *423*, 141–148. [CrossRef]

). Ahmad, M.; Han, S.; Zafar, A.; Ghafoor, U.; Rahim, N.; Ali, M.; Rim, Y. Indoor and Outdoor Performance Study of Metallic Zinc Particles in Black Paint to Improve Solar Absorption for Solar Still Application. *Coatings* **2021**, *11*, 536. [CrossRef]

1. Srithar, K.; Rajaseenivasan, T. Recent fresh water augmentation techniques in solar still and HDH desalination—A review. *Renew. Sustain. Energy Rev.* **2018**, *82*, 629–644. [CrossRef]

2. Kabeel, A.; Sathyamurthy, R.; Sharshir, S.; Manokar, M.; Panchal, H.; Prakash, N.; Prasad, C.; Nandakumar, S.; El Kady, M. Effect of water depth on a novel absorber plate of pyramid solar still coated with TiO_2 nano black paint. *J. Clean. Prod.* **2019**, *213*, 185–191. [CrossRef]

3. Saleh, S.M.; Soliman, A.M.; Sharaf, M.A.; Kale, V.; Gadgil, B. Influence of solvent in the synthesis of nano-structured ZnO by hydrothermal method and their application in solar-still. *J. Environ. Chem. Eng.* **2017**, *5*, 1219–1226. [CrossRef]

4. Abdelal, N.; Taamneh, Y. Enhancement of pyramid solar still productivity using absorber plates made of carbon fiber/CNT-modified epoxy composites. *Desalination* **2017**, *419*, 117–124. [CrossRef]

5. Baticados, E.J.N.; Capareda, S.C.; Liu, S.; Akbulut, M. Advanced Solar Still Development: Improving Distilled Water Recovery and Purity via Graphene-Enhanced Surface Modifiers. *Front. Environ. Sci.* **2020**, *8*, 531049. [CrossRef]

6. Kabeel, A.; Omara, Z.; Essa, F.; Abdullah, A.; Arunkumar, T.; Sathyamurthy, R. Augmentation of a solar still distillate yield via absorber plate coated with black nanoparticles. *Alex. Eng. J.* **2017**, *56*, 433–438. [CrossRef]

7. Sathyamurthy, R.; Kabeel, A.E.; Balasubramanian, M.; Devarajan, M.; Sharshir, S.W.; Manokar, M. Experimental study on enhancing the yield from stepped solar still coated using fumed silica nanoparticle in black paint. *Mater. Lett.* **2020**, *272*, 127873. [CrossRef]

8. Tenthani, C.; Madhlopa, A.; Kimambo, C.Z.M. Improved Solar Still for Water Purification. *J. Sustain. Energy Environ.* **2012**, *3*, 111–113.

9. Modi, K.V.; Jani, H.K.; Gamit, I.D. Impact of orientation and water depth on productivity of single-basin dual-slope solar still with Al_2O_3 and CuO nanoparticles. *J. Therm. Anal. Calorim.* **2021**, *143*, 899–913. [CrossRef]

0. Arunkumar, T.; Murugesan, D.; Raj, K.; Denkenberger, D.; Viswanathan, C.; Rufuss, D.D.W.; Velraj, R. Effect of nano-coated CuO absorbers with PVA sponges in solar water desalting system. *Appl. Therm. Eng.* **2019**, *148*, 1416–1424. [CrossRef]

1. Kabeel, A.; Omara, Z.; Essa, F. Improving the performance of solar still by using nanofluids and providing vacuum. *Energy Convers. Manag.* **2014**, *86*, 268–274. [CrossRef]

2. Yadav, A.K.; Chandel, S. Tilt angle optimization to maximize incident solar radiation: A review. *Renew. Sustain. Energy Rev.* **2013**, *23*, 503–513. [CrossRef]

3. Wang, M.-S.; Fan, L.-Z. Silicon/carbon nanocomposite pyrolyzed from phenolic resin as anode materials for lithium-ion batteries. *J. Power Sources* **2013**, *244*, 570–574. [CrossRef]

4. Meteoblue. Available online: https://www.meteoblue.com/en/weather/historyclimate/weatherarchive/kuala-lumpur_malaysia_1735161?fcstlength=1y&year=2020&month=12 (accessed on 22 December 2020).

5. Tang, K.H.D. Climate change in Malaysia: Trends, contributors, impacts, mitigation and adaptations. *Sci. Total Environ.* **2019**, *650*, 1858–1871. [CrossRef]

MDPI
St. Alban-Anlage 66
4052 Basel
Switzerland
Tel. +41 61 683 77 34
Fax +41 61 302 89 18
www.mdpi.com

Energies Editorial Office
E-mail: energies@mdpi.com
www.mdpi.com/journal/energies